THE ARROW OF TIME

The "arrow of time," a concept first introduced by Sir Arthur Eddington, reflects the one-way flow of time and its association with various physical asymmetries in thermodynamics, cosmology, quantum mechanics, field theories, and beyond. Yet, the foundations of the arrow of time continue to challenge physicists and philosophers, having profound implications across multiple theories and disciplines. This volume compiles insights from the international colloquium "The Arrow of Time: From Local Systems to the Whole Universe" held in Buenos Aires in 2023. It explores diverse perspectives on the arrow of time in thermodynamics, quantum mechanics and cosmology, its relation to counterfactual reasoning, free will and the growing-block universe, the interplay between consciousness and time, and the implications of time-reversal invariance. Collectively, these contributions provide a rigorous and comprehensive analysis of the enduring enigma of time's unidirectional nature.

CRISTIAN LÓPEZ is Senior Postdoctoral Researcher with the Swiss National Science Foundation (FNS) at the University of Lausanne, Switzerland, and Assistant Researcher at CONICET, Argentina. He has a broad research background spanning philosophy of physics, metaphysics of science, and political philosophy. He has authored numerous articles on the arrow of time, symmetries, and quantum mechanics, and co-edited two books with Olimpia Lombardi for Cambridge University Press: *What Is Quantum Information?* (2017) and *Quantum Worlds* (2019).

OLIMPIA LOMBARDI is Superior Researcher at the University of Buenos Aires, Argentina, and Corresponding Member of the Academie Internationale de Philosophie des Sciences. She is also Honorary Fellow of the John Bell Institute for the Foundations of Physics, a member of the Foundational Questions Institute, and Associate Researcher at the Centre for Philosophy of Natural and Social Science at the London School of Economics and Political Science, UK. Lombardi has authored numerous articles in the philosophy of physics and chemistry, along with several books. Among her publications are two co-edited volumes with Cristian López for Cambridge University Press: *What Is Quantum Information?* (2017) and *Quantum Worlds* (2019).

THE ARROW OF TIME

From Local Systems to the Whole Universe

Edited by
CRISTIAN LÓPEZ
University of Lausanne and University of Buenos Aires

OLIMPIA LOMBARDI
University of Buenos Aires

CAMBRIDGE
UNIVERSITY PRESS

CAMBRIDGE
UNIVERSITY PRESS

Shaftesbury Road, Cambridge CB2 8EA, United Kingdom

One Liberty Plaza, 20th Floor, New York, NY 10006, USA

477 Williamstown Road, Port Melbourne, VIC 3207, Australia

314–321, 3rd Floor, Plot 3, Splendor Forum, Jasola District Centre,
New Delhi – 110025, India

103 Penang Road, #05–06/07, Visioncrest Commercial, Singapore 238467

Cambridge University Press is part of Cambridge University Press & Assessment,
a department of the University of Cambridge.

We share the University's mission to contribute to society through the pursuit of
education, learning and research at the highest international levels of excellence.

www.cambridge.org
Information on this title: www.cambridge.org/9781009581578

DOI: 10.1017/9781009581608

When citing this work, please include a reference to the DOI 10.1017/9781009581608

First published 2025

A catalogue record for this publication is available from the British Library

A Cataloging-in-Publication data record for this book is available from the Library of Congress

ISBN 978-1-009-58157-8 Hardback

Contents

Contributors

David Albert
Columbia University, New York, United States

Valia Allori
University of Bergamo, Italy

Craig Callender
University of California, San Diego, United States

Meir Hemmo
University of Haifa, Israel

Carl Hoefer
ICREA and University of Barcelona, Spain

Nick Huggett
University of Illinois Chicago, United States

Ruth E. Kastner
University of Maryland, College Park, United States

Barry Loewer
Rutgers University, New Brunswick, New Jersey, United States

Olimpia Lombardi
CONICET and University of Buenos Aires, Argentina

Cristian López
University of Lausanne, Switzerland, and University of Buenos Aires, Argentina

Elías Okón
National Autonomous University of Mexico, Mexico City, Mexico

Orly Shenker
The Hebrew University of Jerusalem, Israel

Daniel Sudarsky
National Autonomous University of Mexico, Mexico City, Mexico

Karim Thébault
University of Bristol, United Kingdom

Francesca Vidotto
Spanish National Research Council (IEM-CSIC), Madrid, Spain

David Wallace
University of Pittsburgh, United States

Preface

Time is as pervading as puzzling. The Argentinean writer, Jorge Luis Borges, wrote:

> Time is the substance I am made of. Time is a river which sweeps me along, but I am the river; it is a tiger which destroys me, but I am the tiger; it is a fire which consumes me, but I am the fire.

Prima facie, Borges is right. Time is everywhere and unrelenting. Yet, in a deeper reading, he is perhaps suggesting that *we* are time. Indeed, Borges believed that time was an illusion, that it was a contradictory notion. So did the English philosopher John McTaggart, and Parmenides before him. But it is not just the philosophers: cutting-edge physics also seems to suggest something along the same lines, depicting time (and space) as something nonfundamental that somehow emerges from a timeless reality. Notwithstanding all these respectable and lofty doctrines, most of us think, live, and feel as if time was real.

There is a point of tension here that has kept philosophers and scientists busy for centuries. On the one hand, time is so ordinary, so pervasive and part of our lives, that it *cannot* be unreal. On the other hand, when time is rigorously investigated conceptually and empirically, we face so many hurdles to grasp time (and all its complex features) that it dilutes on the way. A better approach to conduct this inquiry is perhaps to be less ambitious. Time is a complex concept, made up of several "ingredients," so to speak. Some of these ingredients have to do with our experience of time; some with its metaphysical, flow-like nature; and some with our view of ourselves as temporal, finite beings. So, time is perhaps not so colorful as it seems at first glance from ordinary experience, but some of its essential ingredients nonetheless persist. Among the plethora of characteristics associated with time, one of the most pervasive is its *directionality*, going from past to future but never the other way around.

But, why? Does the directionality of time come from ourselves? Or is the physical reality equipped with a directionality? If so, is this directionality a fundamental feature of the natural world, or is it rather emergent, nonfundamental? Physicists and philosophers have reflected on the relation between the evident directionality of time and the structure of the physical world since the late nineteenth century. Since Arthur Eddington, this class of questions in physics, foundations and philosophy of physics, aimed to delve into the physical basis of the directionality of time, came to be known as "the problem of the arrow of time." Is the arrow of time somehow "out there"?

The problem of the arrow of time is, however, not easy to formulate clearly. John Earman, in his celebrated 1974 paper ("An attempt to add a little direction to 'the problem of the direction of time'") already warned us that one of the problems with the direction of time is that it is not clear what the problem even is! This probably has many reasons, but one of them is that the problem is genuinely multifaceted. Not only does it involve many different physical theories, but also long-standing philosophical problems that are strongly related to physics. This explains why the problem of the arrow of time has branched off so wildly in the second half of the twentieth century, covering such different theories, disciplines, and issues as different as thermodynamics, quantum mechanics, philosophical theories of conditionals, philosophy of mind, and even our own free will. Any genuine representation of what goes by the label "the arrow of time" cannot be simply reduced to a single problem or formulation.

Readers will anticipate that, this being the case, every book and every volume on the arrow of time will necessarily be incomplete, partial, and insufficient. The aim of this volume is then not to exhaustively cover all the recent discussions around the arrow of time. Rather, it is to provide a snapshot of the state of play that shows not only the variety of issues that fall under the label "the arrow of time," but also how ubiquitous the arrow of time is, and how pervasive are the problems associated with it. This comprehends local systems as we find them in thermodynamics, philosophical issues involving the mind, our free will, symmetries, and our universe as a whole.

This volume gathers original work from salient and recognized physicists and philosophers of physics working in the field. These contributions show not only how rich and puzzling the arrow of time is, but also that progress can be made in bridging philosophy and physics. To unfold the variety of problems, proposals, and perspectives, the volume is structured in four parts.

Part I starts off with "Local Systems," the origin of the concerns about the arrow of time. David Albert opens this part with his article "The Direction of Time: From the Cosmos to Local Systems" (turning the subtitle of this book around!). In a very didactical and easy-to-follow way, Albert entertains himself with one of the most popular, and puzzling, explanations about the origin of many asymmetries (or arrows) of time: the Past Hypothesis. He defends and explains how this project relating today's observed temporal asymmetries with the macroconditions of the universe 15 billion years ago can be *seriously* entertained by anyone. In Chapter 2, David Wallace analyzes in depth the foundations of thermodynamics, the theory in which the first and main problems of the arrow of time are raised. In this multi-aim chapter, Wallace explores thermodynamics as a control theory, the microphysical origin of thermodynamics control theory, the connection between entropy and information, and the intrinsic limits that microphysics imposes on us regarding what can be done with thermodynamical systems. Part I is closed by Francesca Vidotto's contribution, "Thermodynamics without Time." In Chapter 3, she explores the foundations of relativistic thermodynamics, making an analogy with quantum theory and gravity. She argues that relativistic thermodynamics should dispense with an absolute notion of time and embrace a relational one.

Part II shifts the focus and delves into some of the many ramifications of the problem of the arrow of time in philosophy and philosophy of physics. In Chapter 4, Meir Hemmo and Orly Shenker tackle a philosophically crucial problem: if physics is time-reversal symmetric, and the mental is physical, how can the *psychological* arrow of time then be explained? They analyze three possible approaches, showing how the philosophy of mind plays a major role in one of the most puzzling problems in philosophy of physics. Barry Loewer, in Chapter 5, offers an in-depth analysis of the celebrated Consequence Argument that shows that if determinism is true, then we cannot affect the future. Loewer, in a first step, shows what is wrong in the Consequence Argument. In a second step, he develops in detail his view on counterfactuals based on the Mentaculus account of the foundations of statistical mechanics. His view, then, explains why we cannot control the past but can, to a certain degree, affect the future. In close relation to Loewer's paper, Carl Hoefer revisits his paper "Freedom from the Inside Out" (2002) in Chapter 6, "'Freedom from the Inside Out': Revisited." In his 2002 paper, Hoefer argued that the choices of free agents are the partial causes and explainers of the microstate of the world in both directions of time. This view heavily depends on adopting the so-called block universe. But what if this is not so? What if we live in a growing block universe? Here, Hoefer explores whether it is possible to recover, to some extent, his 2002 view in a growing block universe with deterministic laws. Finally, Ruth Kastner offers a critical view of how traditionally physics has treated the arrow of time. In Chapter 7, "The Arrow of Time Is Alive and Well but Forbidden under the Received View in Physics," Kastner analyzes sociologically and historically the received view in physics, arguing that "the Arrow of Time Problem" (or "Two Time Problem") is actually a myth.

Part III focuses on a concept that has occurred multiple times up to this point: time-reversal symmetry. Most of the problems around the arrow of time, in its different shapes and forms, were motivated by, or emerged from, having a time-reversal symmetric physics. In Chapter 8, "Time for Pancakes: Time Reversal and Ontology," Valia Allori contraposes two accounts of time reversal: the 'symmetry-first' account and the 'pancake' account. She then argues that this account is less problematic than the symmetry-first account and proposes an alternative view that combines the best of both worlds. In Chapter 9, Craig Callender digs into the problems of time reversal in quantum mechanics. In his chapter, he first shows how exactly the same problem that arises in quantum mechanics arises in classical mechanics as well. This, he argues, shows that the problem of time reversal in quantum mechanics is not in the theory per se, but in its representation. If quantum mechanics is provided with an ontology and laws as some quantum interpretations do, the puzzle disappears. In the last chapter of this part, "Saving the Appearances: The Direction of Time and Time-Reversal Invariance" (Chapter 10), Cristian López adopts phenomenal conservatism as an epistemic theory of justification and defends that the direction of time, as appearances suggest, is real and fundamental. He then argues that the main argument against this belief is based on time-reversal invariance. By rebuilding the structure of the argument, he offers three reasons that render the argument harmless, suggesting that it should not be taken too seriously when exploring the foundations of the arrow of time.

Parts I–III of this book deal with the arrow of time in subsystems of the universe. But what if the arrow of time depends on something much larger? What happens when we look at the biggest physical system? Part IV, "The Whole Universe," focuses on the arrow of time at the cosmological level. In Chapter 11, Olimpia Lombardi reviews the nonentropic and global approach to the direction of time. The chapter, titled "Following Earman's Time Direction Heresy: From the Global Arrow of Time to Local Irreversible Phenomena," argues that the arrow of time is a geometrical property of space-time, independent of entropic arguments and of any appeal to non–time-reversal invariance. The global arrow of time can be manifested at a local level as a tetradimensional flow of energy that breaks the symmetry of local theories. In addition, Lombardi considers some points that were not considered previously, making the approach stronger and more general. The next chapter, Chapter 12, deals with the entropic arrow of time in cosmology. Daniel Sudarsky and Elías Okón, in their chapter "The Seeds of Cosmic Structure, the Black Hole Information and the Entropic Arrow of Time," rely on their GRW-based theory of semiclassical gravity to explore the applications to the seeds of the cosmic structure, the black hole information puzzle, and the foundations of the entropic arrow of time at the cosmological level. The last chapter of the volume, by Nick Huggett and Karim Thébault, "Finding Time for Wheeler–DeWitt

Cosmology" (Chapter 13), analyzes the emergence of time in quantum cosmological models. As they argue, the emergentist model of Kiefer and Zeh can successfully account for the emergence of three temporal structures (chronordinal, chronometric, and chronodirected structures), avoiding some criticisms.

These 13 chapters illustrate the variety of puzzles that fall under the arrow of time as well as the diverse approaches and strategies for addressing it. Some readers might find solutions and clarity in these chapters; others, new grounds for puzzlement. After all, as Robert Nozick once said, philosophy is an epistemic exploratory venture in which many avenues, streets, and dead ends must all be traveled. We hope this volume helps readers in the venture that goes by the name of the arrow of time.

Acknowledgments

This volume is the fruit of the hard work that the Buenos Aires Group of the Philosophy of Special Sciences has been carrying out for many years. The group was born in the spirit of promoting philosophy of science, and philosophy of physics in particular, in a region where these areas have historically had less development. In these rather difficult conditions, academic funding is essential. We are therefore extremely grateful to the John Templeton Foundation for supporting our work over the past seven years. Their invaluable support has allowed us not only to develop our research at a high level and maintain a strong academic presence, but also to organize conferences and workshops in our country, and to edit books and volumes to promote the importance of philosophy of science and philosophy of physics in Argentina and in the region.

On February 20–23, 2023, and thanks to grant #61785 from the John Templeton Foundation, we organized in Buenos Aires (Argentina) the Workshop on The Arrow of Time: From Local Systems to the Whole Universe. We were very pleased to host some of the leading scholars in the philosophy of physics to discuss current issues (and solutions) to one of the oldest (and hardest) problems in the field: the foundations of the arrow of time. That was half of the project. The second half is the present volume, which collects all the contributions and discussions that took place during four days in Buenos Aires, featuring also original contributions from specialists who sadly could not attend. Both the workshop and this volume would not have been possible without them, so our first thanks go to David Albert, Valia Allori, Craig Callender, Elise Crull, Carl Hoefer, Nick Huggett, Ruth Kastner, Bryan Roberts, Elías Okón, Daniel Sudarsky, Karim Thébault, Francesca Vidotto, and David Wallace. For those interested in watching the workshop talks, you can find all the videos on YouTube.

Secondly, we would like to thank all the members of the Buenos Aires Group of the Philosophy of Special Sciences: Hernán Accorinti, Jesús Jaime Arriaga, Cecilia

Bejarano, Manuel Herrera, Juan Camilo Martínez, Matías Pasqualini, and Alfio Zambón. Their constant enthusiasm, commitment, and assistance made the organization of the workshop not only possible, but also enjoyable. We would also like to thank all those who attended the workshop. Some of them were researchers from Chile, Mexico, and Brazil who made every effort to attend the workshop. We greatly appreciate their efforts.

Last but not least, we also thank Cambridge University Press, in particular Nicholas Gibbons (Senior Commissioning Editor, Physics) and Aleksandra Serocka (Senior Editorial Assistant), for guiding us throughout the editorial process.

Part I

Local Systems

1

The Direction of Time: From the Cosmos to Local Systems

DAVID ALBERT

Our everyday macroscopic experience of being in the world is saturated with asymmetries – thermodynamic asymmetries, and radiative asymmetries, and epistemic asymmetries, and phenomenological asymmetries, and asymmetries of overdetermination, and asymmetries of influence, and what have you – between the past and the future.

And there is a long-cherished hope – something that has its origins in the work of Boltzmann and which has been pursued by any number of other investigators, through any number of fits and starts, and revelations, and wrong turns, ever since – that all of those asymmetries can ultimately be traced back to some relatively simple characteristic of the initial macrocondition of the universe. The thought (as people put it nowadays) is that all we need to do in order to account for these asymmetries is to add to the fundamental time-reversal-symmetric dynamical *laws*, and to the standard statistical-mechanical probability-measure over the space of possible fundamental physical *states*, a simple postulate – a so-called *Past Hypothesis* – to the effect that the world first came into being in whatever particular low-entropy macrocondition it is that the normal inferential procedures of cosmology are eventually going to present to us.

The business of working this thought out in detail is a large undertaking, which is still very much in its infancy, and which is still a topic of very lively debate – and I don't want to attempt anything along the lines of an overview of that project here. All I want to talk about today is a widespread, and fundamental, and perennial sort of puzzlement about how it is that such a project could even seriously be *entertained* – a puzzlement (that is) about how it is that the macrocondition of the universe 15 billion years ago – all by itself – could even *imaginably* be up to the job of explaining so much about the feel, today and on earth, of the passing of time.

This puzzlement takes a number of different forms and arises in a number of different contexts. On the most trivial level, there is a question of how the lowness of the entropy of the world 15 billion years ago can impose any genuinely profound and vivid constraints *whatever* on what the world is doing *now*. And all that needs to be said in order to make *that* sort of puzzlement go away is that although 15 billion years is a long time, the entropy of the universe at that time was very, very low – and that (in particular) 15 billion years is a great deal *shorter* than the expected *relaxation* time of the state in which our universe seems to have started out.

But there is a somewhat more *interesting* question, in the general vicinity of this first one, about how the lowness of the entropy of the world 15 billion years ago can have any genuinely profound and vivid effects or impose any genuinely profound and vivid constraints on what particular, localized, human-scale, quasi-isolated *subsystems* of the world are doing now. There is a worry (in particular) that runs like this: Let it be stipulated that the standard Boltzmannian arguments do indeed establish that the overall entropy of a universe which starts out in a low-entropy, Past Hypothesis sort of macro state is overwhelmingly likely to rise toward the future. People are fond of pointing out that that tells us *nothing at all*, in and of itself, and, considered as a purely logical matter, about the behaviors of the entropies of quasi-isolated *sub*systems of the world – just as the fact that dogs can run tells us nothing at all, in and of itself and, considered as a purely logical matter, about whether or not dogs' *heads* can run. And it is the behaviors of quasi-isolated *sub*systems of the world – after all – and not of the universe as a *whole*, that the science of thermodynamics is, primarily and in the first instance, *about*.

And all of this, considered as a purely logical matter, is surely true – but it seems to me to ignore, or to overlook, or to misunderstand how the Boltzmannian arguments actually *work*. The core of what the Boltzmannian tradition has given us is a general strategy for assigning specific numerical probabilities to *any propositions whatever* about the macroconditions of isolated thermodynamic systems at any time $t > t_0$, given the macrocondition of the system at t_0 – and the argument to the effect that thermodynamic systems in nonequilibrium macroconditions are overwhelmingly likely to *increase their entropies* as time goes forward is simply an *application* of that general strategy to the special case of *entropy*. The business of actually *calculating* those probabilities, in physically interesting cases, is usually going to be prohibitively complicated – but what Boltzmann and his various collaborators and inheritors are widely thought to have made plausible is that those probabilities are in good accord with the entirety of our everyday thermodynamic experience. If (for example) the isolated system in question consists of two gases, which are initially at different temperatures and are in thermal contact with one another, what the Boltzmannian arguments make plausible is not merely that the overall entropy of this system is overwhelmingly likely to go up toward the future, but (in addition) that the entropy of the hotter gas is overwhelmingly likely to go down toward the future. And if the isolated system in question consists of (say) 12 isolated gases, in 12 separate containers, each of which is initially far from its own individual equilibrium state, then what the Boltzmannian arguments are going to make plausible is not merely that the overall entropy of this system is overwhelmingly likely to go up toward the future, but (in addition) that the separate entropies of each of those 12 gases are all, individually, overwhelmingly likely to go up toward the future. Of course, the probability that the entropy of some particular one of those gases goes down toward the future will be much larger than the probability that the overall entropy of all 12 of them together goes down toward the future – but both of those probabilities are going to be fantastically small.

And there is also a third question in the general vicinity of these first two – which is the deepest and the most interesting, the most amorphous, and the most phenomenological form of the general puzzlement about the Boltzmannian project that I mentioned at the outset of this chapter – and it's that third question that I mainly want to talk about here.

Let me put it in four increasingly concrete, simple, and tractable ways:

(1) How can it seriously be imagined that my own sense of the passage of time, or (for example) that my own sense – right here and right now – of whether a specific baseball happens to be flying toward me or away from me, is somehow anchored in the lowness of the entropy of the universe 15 billion years ago?

(2) How it can *be*, how it can *work*, that the increase of the entropy of the world, or of myself, somehow constitutes the standard or the yardstick against which I judge the direction in which events are unfolding? How is it (that is) that the entropy gradient of anything ever comes into the picture? I am certainly not aware that I am checking on the entropy gradient of anything in the course of deciding whether a baseball is flying toward me or away from me. No comparison with anything else, so far as I am aware, is involved. I simply, directly, unmediatedly *see* that the baseball is flying either toward me or away from me.

(3) Consider the sense of the direction of time that is implicit in the operations of (for example) a simple mechanical realization of a Turing machine. Can anyone seriously believe that thermodynamical characteristics of the world somehow play a role in the way a machine like that distinguishes between what it has just done and what it is to do next? How so? How can that be? How would that work? Machines like that can apparently function perfectly well and apparently have no trouble at all distinguishing between what they have just done and what they are to do next, without the aid of special devices for measuring the entropy gradient of the world, or themselves, or anything else!

(4) Consider (finally) a simple mechanical device which has no business other than distinguishing between what it has just done and what it is to do next – the paradigmatic distinguisher, the distinguisher *par excellence*, between what it

has just done and what it is to do next. Think (that is) of a clock. And think (for the sake of concreteness, or of simplicity) of an old-fashioned, fully mechanical pendulum clock.

Good. Now we have our hands on something that we are in a position to analyze in detail. Note (to begin with) that in the course of the normal and intended operations of a clock of this kind there are going to be specific moments when the pendulum is precisely at the apogee of its swing – when each of its macroscopic moving parts is completely at rest, or (to put it slightly differently) that in the course of the normal and intended operations of a clock like that, there are going to be moments –the moments (again) when the pendulum is precisely at the apogee of its swing– when the macrocondition of the clock, in its entirety, is invariant under time reversal. And consider how it is, at such moments, that the clock manages to distinguish between what it has just done and what it is to do next.

The macrocondition of the clock, together with the microscopic dynamical equations of motion, together with the statistical postulate, is manifestly not going to do the trick. For if the present macrocondition of the clock together with the microscopic dynamical equations of motion and the statistical postulate makes it likely that the clock is going to read (say) 3:05 five minutes from now, and if the present macrocondition of the clock is invariant under time reversal, then the present macrocondition of the clock, together with the microscopic dynamical equations of motion and the statistical postulate – both of which are invariant under time reversal as well – is necessarily *also* going to make it likely, and to exactly the same degree, that the clock read 3:05 five minutes *ago*.

And all there is to break the symmetry, all there is that *stands in the way* of the clock's having read 3:05 five minutes ago, is the Past Hypothesis. The clock's ability to distinguish between what it did last and what it does next, and your ability to distinguish between a baseball's flying toward you and a baseball's flying away from you, is anchored in the entropy gradient of the universe. If we were to hold the present macrocondition of the world fixed and move the Past Hypothesis from the beginning of time to its end, *the clock would run backwards*.

Period. Case closed. End of story.

But the fact is that people sometimes find this hard to understand. Here (for example) is the reaction of an anonymous referee to an earlier version of this chapter:

It's uncharacteristic of Albert to pass over details. He could've described how a pendulum clock works (e.g., falling weight version) in a couple of minutes, but he didn't. The mechanism CANNOT run backwards. What drives it is the falling of the weight, pulling on the cord. If there's no pull on the cord the clock stops. If there's PUSH on the cord the clock stops. It'll ONLY work if there's tension on the cord and that will make the hands move clockwise because of the way the cord is wound around the drive axel. And that would be true even if the weight rose into the air and started pulling upwards. Putting the low entropy in the future of the universe can't make the clock run backwards, as he claims. The clock "knows" which way to go, if it's going to go at all, because the information is built into it in the way that the cord is wound around the axel.

And this will be worth pausing over and thinking about.

Let me begin by simply reiterating the very *general* point that I made a minute ago. *The referee must be mistaken*. No further details about the inner workings of the clock can make the slightest bit of difference. Consider (again) a moment when the macroscopic state of the clock is stationary. Consider (that is) a moment when the macroscopic state of the clock is invariant under a reversal of all the velocities of its microscopic particulate constituents. In the absence of a Past Hypothesis, the predictions of statistical mechanics about the evolution of the macrocondition of this clock away from that moment toward the future are going to be *identical* to its predictions about the evolution of its macroscopic condition away from that moment toward the past, for the simple reason that *there is nothing whatever in the situation to distinguish between them*. In the absence of a Past Hypothesis, the predictions of statistical mechanics are (in particular) that *as we proceed away from the present*, in *either* temporal direction, *the cord is always going to be unwinding, and the weight is always going to be going down, and the hands of the clock are always going to be turning in the clockwise direction*. And it is only because of the truth of the Past Hypothesis, it can only be because of the truth of the Past Hypothesis, that (as a matter of fact) those hands turn *counter*-clockwise, as we proceed away from the present, in the direction of the past. And that (again) is why a clock like this can be used – just as we might use a half-dispersed puff of smoke, or a half-melted block of ice – as an indicator of the temporal direction in which the entropy of the whole universe is increasing.

Good. But there is a puzzle here.

Consider a half-dispersed puff of brightly colored smoke in an otherwise empty, transparent, cubical container. The entropy of this puff of smoke is going to be a relatively simple monotonically increasing function of its volume. So, we can determine the temporal direction in which the entropy of this puff of smoke is increasing by simply noting the temporal direction in which the smoke's volume is increasing. And what that means – since the temporal direction in which the entropy of this puff of smoke is increasing is overwhelmingly likely to coincide with the temporal direction in which the entropy of the whole world is increasing – is that we can determine the temporal direction in which the entropy of the world as a whole is increasing by noting the temporal direction in which the volume of this puff of smoke is increasing. You might say that the puff of smoke wears its entropy – and, by extension, the entropy of the whole world – on its sleeve, or on its face.

And look at how utterly different everything is with the clock: We can determine the temporal direction in which the entropy of the world as a whole is increasing – or (at any rate) we are convinced that we can determine the temporal direction in which the entropy of the world as a whole is increasing – by noting the temporal direction in which the hands on the clock are moving clockwise. But how does that work? The entropy of the puff of smoke (remember) is a relatively simple and monotonically increasing function of its volume. But the entropy of the clock is not a function at all of the positions of its hands. And the reason that we can determine the temporal direction in which the entropy of the world is increasing by noting anything at all about the smoke is that the smoke is in thermodynamic disequilibrium – the reason that we can determine the temporal direction in which the entropy of the world is increasing by noting anything at all about the smoke is that the smoke is undergoing a process of dissipation. But clockmakers invariably go to a great deal of trouble precisely to eliminate any potential sources of dissipation – any rubbing or scratching or denting or heating or what have you – in the clocks they produce. And if I open up a pendulum clock and examine its inner workings, I find nothing that looks as if it has been designed to measure the volume or the temperature or the pressure of anything – I find nothing (that is) that looks as if it has been designed to track some process of thermodynamic equilibration. So, what is going on here? How does the increase in the entropy of the world, or of the clock, or anything at all, even get into the picture here? How does any of that ever become relevant to the direction in which the clock hands are turning? That's the puzzle – and that (I think) is what has the referee so confused – and it is, indeed, on the face of it, confusing.

Let's have a look (then) at the inner workings of the sort of pendulum clock that the referee has in mind – which are pictured in Figure 1.1.

Note that the cord to which the weight is attached is wound around the axle in the clockwise direction, so that the axle itself, and the clock hands that are attached to it, rotate in the clockwise direction as the weight descends.

The rest of the mechanism – the pendulum and the escapement – are apparently there simply to regulate the rate at which the axle rotates. If they weren't there, the axle, and the clock hands that are attached to it, would rotate more and more quickly as the weight accelerated, under the influence of gravitation, toward the floor. What the pendulum and the escapement do is to set a limit – call it θ – on the angle through which the axle is able to turn over the course of any single swing of the pendulum.

What happens – in a little more detail – is this: We begin in the macro state depicted in Figure 1.1. The pendulum is at the apogee of its rightward swing. As it begins to swing toward the left, the little metal tab on the left is lifted out of the way of gear tooth #1 and the weight exerts an uncanceled torque on the axle, pulling it in the clockwise direction. As the pendulum swings toward the left, the little metal tab

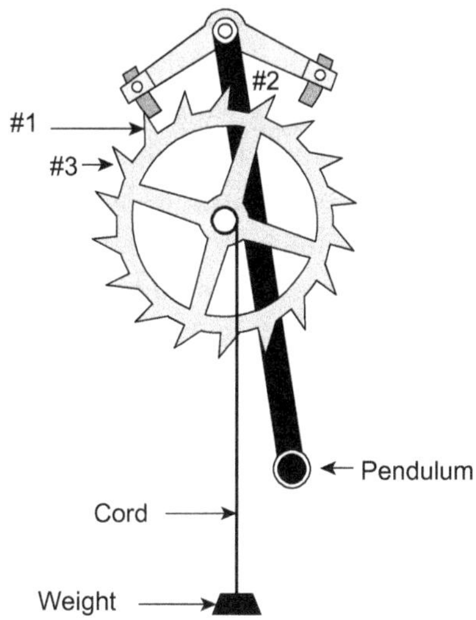

Figure 1.1 The inner workings of a pendulum clock.

on the right is lowered into the path of gear tooth #2, and the ensuing collision stops the rotation of the axle dead in its tracks, where it is held fast by the pressure of the weight, until the pendulum begins to swing back toward the right. As the pendulum swings toward the right, the little metal tab on the left is lowered into the path of gear tooth #3, and the ensuing collision stops the rotation of the axle dead in its tracks, where it is held fast by the pressure of the weight, and the process begins again. And what the referee seems to have overlooked, and this is the heart of the matter, is that all of this critically depends on the occurrence, at every swing of the pendulum, of phenomena like denting and scraping and heating in the operations of the escapement.

For suppose that no such things – no denting or scraping or heating – ever did occur. Suppose that the collisions between the gear teeth and the little metal things were perfectly elastic. We begin (again) in the macro state depicted in the picture. The pendulum is at the apogee of its rightward swing. As it begins to swing toward the left, the little metal tab on the left is lifted out of the way of gear tooth #1 and the weight exerts an uncanceled torque on the axle, pulling it in the clockwise direction. As the pendulum swings toward the left, the little metal tab on the right is lowered into the path of gear tooth #2, and in the ensuing elastic collision the gear tooth will bounce off the metal tab, and the axle will completely retrace its previous clockwise rotation in the *counter*clockwise direction, at the end of which the little tab on the left will be smoothly reunited, in exactly the configuration shown in the picture, with gear tooth #1! And then, of course, the whole process will simply begin again, and the axle will turn clockwise by θ, and then counterclockwise by θ, and then clockwise by θ, and then counterclockwise by θ, and so on, forever – and the clock will be rendered utterly incapable of distinguishing between the past and the future.

Good. Let's take one more look then – with all that in mind – at the words of the anonymous referee. The referee says "Putting the low entropy in the future of the universe can't make the clock run backwards, as he [that is: me] claims. The clock 'knows' which way to go, if it's going to go at all, because the information is built into it in the way that the cord is wound around the axel."

Well, let's see. Something is plainly right (to begin with) about the claim that "The clock 'knows' which way to go, if it's going to go at all, because the information is built into it in the way that the cord is wound around the axel." If the cord were wound around the axle in the opposite direction (after all) the axle would plainly turn in the opposite direction. But turn with respect to what? With respect to the temporal direction that we have arbitrarily, and as a matter of pure convention, labeled with a "+" sign? How could that be? With respect to the temporal direction that is picked out by some fundamental metaphysical "arrow of time" which makes no appearance whatever in the laws of physics? How could that be? What must be the case, and what we ought to have seen all along, and what we have just now managed to spell out in full mechanical detail is that the information that is built into the way that the cord is wound around the axle is information about which way to turn with respect to the entropy gradient of the escapement. What must be the case, and what we ought to have seen all along, and what we have just now managed to spell out in full mechanical detail, is that the information that is built into the way that the cord is wound around the axle is information about which way to turn with respect to the temporal direction *away from the Big Bang.*

And we can now see, in particular, that in the absence of dissipation – and notwithstanding the specific way that the cord happens to be wound around the axle – a clock like this is going to have no clue "which way to turn"; or (to put it slightly differently) the macroscopic evolution of a clock like this (not merely the *laws* of that evolution, mind you, but the evolution *itself*) is going to be completely invariant under time reversal! Suppose (for example) that we are shown a movie of the operations of a clock like the one we have just now been talking about – turning clockwise by θ and then counterclockwise by θ and then clockwise by θ and so on, ad infinitum. Nothing about the sequence of the images with which we are presented – and nothing (in particular) about the way that the cord happens to be wound around the axle – is going to tell us *anything at all* about whether the movie is being run forwards or in reverse.

So, it is perfectly true that clockmakers take great pains to eliminate every imaginable source of dissipation from the mechanisms they produce – and they are perfectly right to do so. But it is *also* true, and it is a beautiful and ironical and surprising feature of the way the world actually turns out to work, that if they were ever to perfectly *succeed*, their clocks would be perfectly *useless*.

And (to make a long story short) I suspect that something very much along these lines must explain how a typical mechanical realization of a Turing machine knows the difference between what it did last and what it does next. And it seems to me that once all this has been taken on board, the atmosphere of mystery around my *own* ability to sense the difference between a baseball flying toward me and a baseball flying away from me – to sense it (mind you) *immediately*, and without any conscious act of *inference*, and without my being aware of ever checking on the *entropy* of anything – entirely melts away.

2

Thermodynamics with and without Irreversibility

DAVID WALLACE

2.1 Introduction

Dynamical theories are concerned with what physical systems spontaneously do: they provide dynamical equations and dynamical laws relating a state of a system at one time to a state at another, all independent of outside intervention. Thermodynamics, the misleading name notwithstanding, is *not* a dynamical theory in this sense: if it has a dynamical law, it is simply that systems left to themselves go to equilibrium and stay there. Most of its content involves transitions between equilibrium states, and since equilibrium states by definition do not carry out any transitions at all if left to themselves, those transitions must be understood as externally induced. Thus understood, thermodynamics is a *control theory*, concerned not with what transitions spontaneously happen but with what externally induced transitions are and are not possible, or put another way, what *control operations* can or cannot be performed.[1]

Viewed as a control theory, thermodynamics is predominantly concerned with a somewhat more specific problem. Given some additive, conserved quantity – usually energy, but it could be angular momentum, or charge, or any other such quantity – how much of that quantity can be *extracted* from a given system by a given class of control operations?

An immediate answer might be: none of it. These are *conserved* quantities, after all: the quantity might flow from one part of the system to another, but it cannot increase or decrease. But, of course, among our control operations might be couplings of the collection to other systems – batteries, flywheels, lifted weights, or other storage devices – so that the collection's share of the quantity decreases even while the amount of it in the larger world remains fixed. Indeed, that is just

[1] The control-theory way of understanding thermodynamics is defended by Wallace (2014) and Myrvold (2020); it has been widely discussed in the recent physics literature under the title of "resource theory" (for a review, see, e.g., Gour et al. 2015).

what "extract" means for an overall-conserved quantity: the extractor has acquired some of it, so less of it remains.

That suggests a second answer: all of it. If the only constraint on extraction is compliance with the conservation laws, then the most we can extract is the difference between the *actual* amount of the quantity, and the *minimum* amount – for energy, for instance, the difference between the actual energy of the system and the energy of its ground state.

Phenomenological thermodynamics tells us that at least in some circumstances, the constraints are more stringent. If our system consists of a box of gas at uniform temperature, and in our control operations we are obliged to return the gas to the same volume at which it begins, then *none* of the gas's energy can be extracted. If the system instead consists of two such gases at different temperatures, we may run a heat engine between the two gases and extract some of the energy that thus flows – but not all of it, and only until the temperatures equalize. Indeed, the lesson of thermodynamics is that all such constraints can be systematized into two principles: that all control processes conserve energy (the first law), and that no control process decreases thermodynamic entropy (the second law). And the irreversibility of equilibrium thermodynamics is often said to lie in the fact that the second law's inequality need not be saturated: that although there are no control processes that decrease entropy, there are some which increase it, and so there are some operations which, once done, cannot be undone.

The first law sits on a solid microphysical foundation: thermodynamic energy is identified with dynamical energy (that is, the conserved quantity associated with time translation symmetry) and the principle that it is conserved is inviolate, no matter how sophisticated our control processes might be: any increase in energy in some external storage device we operate must be compensated, exactly, by a decrease of energy elsewhere, either in another external device or in the system on which we operate. (And the same applies mutatis mutandis to the extraction of other conserved quantities, albeit this is seldom a focus of equilibrium thermodynamics.)

The second law, on the other hand, is often attributed to the macroscopic nature of equilibrium thermodynamics, and/or to the crude limitations of the control operations that clumsy creatures like us have access to. The point was made vividly by Maxwell (1867, 1871: ch. 12) at the dawn of statistical mechanics: his infamous "demon" is a microscopic being, capable of tracking the movements of individual molecules in a gas and steering them, ever so delicately, so as to reduce the gas's entropy and allow work to be extracted. And ever since Maxwell's proposal the possibility has been raised and reraised that a subtler control theory, appropriate to a subtler agent, might transcend its limitations. This conception of the second law is closely tied to the idea that thermodynamics is in some way a higher-level, emergent, approximate theory, useful for large-scale systems no doubt but only statistically true even then, breaking down entirely on sufficiently small scales.

This way of thinking is in tension with another tradition, just as old, of seeing thermodynamics as exact and universal. Einstein (1949: 33) famously described thermodynamics as "the only physical theory of universal content concerning which I am convinced that, within the framework of applicability of its basic concepts, it will never be overthrown"; more recently, the physics consensus[2] about Maxwell's demon is that reasons of principle and not just practice make it impossible, and the defenders of that consensus consider microscopic systems in states far from equilibrium; meanwhile, recent advances in statistical mechanics[3] have seen physicists apply thermodynamics to extremely small systems, not just to macroscopically large ones, and to obtain theoretical results, which are then confirmed empirically.

This chapter has several goals. Firstly, I aim to disentangle these two aspects of thermodynamics: the emergent, high-level, large-system, approximate features, from the exact, microscopically derivable features. I do so by introducing the idea of a coarse-graining (a formal transformation of a system's state which washes out certain fine details of its microstructure) and of a control theory defined by that coarse-graining, not in the sense that any control operation actually performs the coarse-graining but that control operations are indifferent to details that the coarse-graining washes out. Each coarse-graining then specifies a class of control theories, and the finer the coarse-graining, the more powerful the control theory. The coarsest grain, which defines *equilibrium thermodynamics*, just replaces every system with its equilibrium state; the finest, which defines *reversible thermodynamics*, does not coarse-grain at all. I offer this framework as a general method both to derive phenomenological thermodynamics from microphysics, and to generalize to finer-grained control theories where agents have control options not present in phenomenological thermodynamics.

Secondly, I explore the microphysical origins of irreversibility in thermodynamic control theory (to be distinguished from the irreversibility of nonequilibrium statistical mechanics, which I do not consider here). This (I will argue) is closely tied to the limitations of our control of a system, and hence to our choice of coarse-graining.

Thirdly, I attempt to provide some clarity on the vexed question of how information connects to thermodynamic entropy. In some corners of physics, it is a commonplace that information and entropy are one and the same, but the idea has been met with skepticism in other corners and even more so in philosophy. In

[2] See Section 2.11 for details and references.

[3] See, e.g., Jarzynski (1997), Crooks (1998), Collin et al. (2005), Bustamante, Liphardt, and Ritort (2005); for a conceptual discussion, see Wallace (2020).

my account, gaining information about a system decreases its entropy, and allows us more control over it, only when that information tells us about the coarse-grained features of a system: coarse-grained, that is, with respect to the coarse-graining that characterizes an agent's particular control theory.

Finally, I argue that there are hard limits on what can be done to thermodynamic systems even by an agent with total, microscopic control of the system, limitations that apply not because of our limited ability to influence systems at the microscopic level but because of microscopic physics itself. Specifically, the most important phenomenological versions of the second law continue to apply, even in the case of a fully reversible thermodynamics.

The structure of the chapter is as follows. In Sections 2.2 and 2.3, I review first the structure of phenomenological thermodynamics (with an emphasis on the central role of the equation of state) and then the "canonical recipe" by which the equation of state is derived from a system's microphysics (which at this stage I simply state without defense). In Section 2.4, I provide a general framework for discussing thermodynamic control theories and, in Section 2.5, I introduce the idea of coarse-graining in general and the equilibration coarse-graining that justifies the canonical recipe in particular. In Sections 2.6–2.8, I further develop the microphysical basis for phenomenological thermodynamics, in the process introducing the important concepts of "free energy of equilibration" (Section 2.7) and "heat baths" (Section 2.8) which apply in a much more general context. In Sections 2.9–2.10, I consider some more powerful thermodynamic control theories, firstly one inspired by Boltzmannian statistical mechanics and secondly a fully reversible thermodynamics in which the controlling agent can apply arbitrary unitary (or, in a classical context, Hamiltonian) transformations. In Section 2.11, I consider the circumstances in which collecting information about a system can be thermodynamically useful; in Sections 2.12–2.13, I return to the second law in the context of the very general notions of thermodynamics I am considering and then distinguish ways in which Maxwell demons are or are not possible. Section 2.14 is the conclusion.

This chapter builds on previous work by myself and others. Wallace (2014) defends the idea of thermodynamics as control theory and develops the microscopic details in the specific context of equilibrium thermodynamics. Myrvold (2011, 2020) offers a similar defense, but introduces the important idea that there are different forms of thermodynamics corresponding to different levels of control an agent might have (or rather, reintroduces it, since he provides strong historical evidence that the idea was well understood by many of the physicists who developed thermodynamics and statistical mechanics); he too works in the context of equilibrium thermodynamics. Maroney (2007) develops a fully reversible form of thermodynamics, separated entirely from considerations of equilibrium, and my development here of "reversible thermodynamics" is indebted to his version. The physics I discuss is very standard and I do not attempt to provide original references; for textbook discussions see, for example, Tolman (1938), Landau and Lifshitz (1980), Kittel and Kroemer (1980), Blundell and Blundell (2010), and Throne and Blandford (2017).

Four technical notes. Firstly, I work in natural units in which Boltzmann's constant k_B is set to 1. Secondly, I develop my account for the most part neutrally as to whether the underlying dynamical theory is classical or quantum, though for certain technical proofs I specialize to quantum mechanics. (In most cases there is an analogous classical proof but, in any case, classical mechanics is valid only insofar as it successfully approximates quantum mechanics.) Thirdly, when discussing quantum mechanics, I assume (i) that the normal quantum formalism is complete; (ii) that closed-system dynamics are always unitary; (iii) that processes of observation and measurement can in principle be modeled mechanically within the quantum formalism. In my view (Wallace 2012) this more or less amounts to assuming the Everett (many-worlds) interpretation, but I make no use of Everett-specific ideas and most of what I say here can probably be taken over to other approaches to quantum mechanics with little change. Finally, by "phenomenological thermodynamics" I mean the macroscopically developed theory of thermodynamics, understood in isolation from its microphysical roots; I will later construct a microscopic control theory, (partial) equilibrium thermodynamics, which provides a microphysical justification of phenomenological thermodynamics.

2.2 Phenomenological Thermodynamics

Phenomenological thermodynamics is concerned with those transitions which can be induced by an external agent (put another way, those control operations that move a physical system or collection of systems from one equilibrium state to another), where "equilibrium" is understood operationally as the state to which an undisturbed system settles down to after some time. Different possible states of a thermodynamic system are labeled by the system's total energy U, by some (possibly empty) set of parameters describing externally controllable features of the system, and by some (possibly empty) set of conserved quantities other than energy. For instance:

- The thermodynamic state of a piece of rock of fixed size and shape is characterized by its energy alone.
- The thermodynamic state of a fixed quantity of gas in a box is characterized by the energy of the gas and the volume of the box. If the amount of gas in the box is also adjustable, the thermodynamic state is in addition characterized by the total number of particles (or, equivalently, moles) of the gas.

- The thermodynamic state of black-body radiation in a box is characterized by its energy and the box volume (but not by the number of photons, since that is not a conserved quantity).
- The thermodynamic state of a ferromagnet in an external magnetic field is characterized by its energy and the field strength.
- The thermodynamic state of a black hole, described in its rest frame, is characterized by its mass (= energy), charge, and angular momentum.

For simplicity I will usually write as if there was one parameter, V, and one conserved quantity, N; everything that follows generalizes straightforwardly to the case of none or several parameters and/or conserved quantities.

The quantitative features of a thermodynamic system are determined – at least as far as phenomenological thermodynamics is concerned – entirely by one function, the *state function* (or *equation of state*), which is a real-valued function $S(U, V, N)$ of U, V, and N whose value for a particular state is called the *thermodynamic entropy* of the state. It is normally assumed that (for fixed V, N) S is a strictly increasing function of U, so that this may be inverted to write $U = U(S, V, N)$ ("state function" is often used to describe this function also). Given a system that can be decomposed into non- or weakly interacting subsystems, each has its own equation of state, and the entropy of the total system is the sum of the entropies of its subsystem; energy and the other conserved quantities are likewise assumed to be additive. (The approximate additivity of the energy is more or less constitutive of two systems being weakly interacting.)

The operational significance of this machinery is then given by the two *laws of phenomenological thermodynamics* and by a secondary *accessibility principle*:

The First Law: For any transition $(U, V, N) \rightarrow (U', V', N')$ induced on an otherwise isolated system by an agent, the net energy required by the agent to perform the transition equals $(U' - U)$, the change in energy of the system. That is, the sum of system energy and controller energy is conserved. (As is standard in phenomenological thermodynamics, this assumes that we have an antecedent understanding of the energy of a mechanical system.)

The Second Law: Any transition $(U, V, N) \rightarrow (U', V', N')$ induced on an otherwise isolated system satisfies $S(U', V', N') \geq S(U, V, N)$.

The Accessibility Principle: If two states have entropies S, S' with $S' > S$, there is a transition from the first to the second.

Other thermodynamic quantities are definable from the equation of state. We can define the temperature $T(S, V, N)$, the generalized pressure $P(S, V, N)$, and the generalized potential $\mu(S, V, N)$, via the differential expression

$$dU(S,V,N) = T(S,V,N)dS - P(S,V,N)dV + \mu(S,V,N)dN \tag{2.1}$$

So that (for instance)

$$T = \left(\frac{\partial U}{\partial S}\right)_{V,N} \tag{2.2}$$

We define the *work done* to a thermodynamic system in a control process as the increase in the system's energy caused by the process. If the work done is negative, we speak of *work extracted*. If we think of a control process as ultimately physically realized in a larger system in which total energy is conserved, the work done by a control operation is the energy that the controller must provide in order to perform it, and the work extracted, conversely, is the energy gain to the controller in carrying out the control operation (which is often operationalized via some energy-storage device such as a raised weight, flywheel, or battery).

If, as is often the case, a thermodynamic system can be broken into some number of (for simplicity, let's say two) approximately isolated subsystems, it is generally possible to consider the total work W done as the sum of the work W_1 done on system 1 and W_2 done on system 2. If the energy changes in the two systems are $\Delta U_1, \Delta U_2$, then we define the *heat flow* Q_i into the ith system as

$$Q_i = \Delta U_i - W_i \tag{2.3}$$

We have $Q_1 = -Q_2$, so that heat can be thought of as flowing from one system to the other; definitionally, the total heat flow into our system is zero. (It is possible to construct versions of thermodynamics involving heat baths in which net heat flow can be nonzero – and I discuss these later in Section 2.8 – but for our purposes we can think of these as effective theories, reducible to phenomenological thermodynamics by including the heat bath as a subsystem of the thermodynamic system.)

It is worth noting that (except in the special case of an overall quasistatic, reversible process, which I will not need to consider here) the machinery of thermodynamics does not itself have the resources to cleanly define work and heat for subsystems, since it does not seek to characterize control processes directly, but only to constrain them via the second law.

As an illustration of how this framework has phenomenological consequences, let us define a transition as *cyclic* if it leaves parameters and conserved quantities (other than energy) unchanged. Then a cyclic transition operating on a single system, which extracts a small amount of net work from that system, satisfies

$$T\delta S = \delta U \leq 0 \tag{2.4}$$

and so, if T is positive, no such transition is possible. A small cyclic transition operating instead on a pair of systems (1 and 2) in which zero net work is done will have as its only effect the transfer of a small amount of heat between the systems; it satisfies

$$0 = \delta U = \delta U_1 + \delta U_2 \tag{2.5}$$

so that $T_1\delta S_1 + T_2\delta S_2 = 0$. The second law requires that $\delta S_1 + \delta S_2 \geq 0$, so that if $\delta S_1 < 0,\ \ T_1 \geq T_2$. That is: no cyclic process can induce heat flow from a cold body to a hotter body without some expenditure of external energy or other change in the systems. These two results can be recognized as, respectively, the Kelvin and Clausius statements of the second law. (I discuss these further in Section 2.12.)

If we instead consider allowing heat $đQ$ to flow from a higher-T to a lower-T body, while siphoning some quantity $-đW$ of it out altogether as extracted work, we have (assuming $T_2 > 0$)

$$\delta S_1 = -đQ/T_1 \tag{2.6}$$

and

$$\delta S_2 = +(đQ + đW)/T_2 \tag{2.7}$$

so that

$$0 \leq T_2(\delta S_1 + \delta S_2) = đQ\left(1 - \frac{T_1}{T_2}\right) + đW \tag{2.8}$$

or

$$-đW \leq đQ\left(1 - \frac{T_1}{T_2}\right), \tag{2.9}$$

which is the well-known Carnot limit on the efficiency of a heat engine. The accessibility principle tells us that this limit can be approached to arbitrary accuracy; we can define *reversible* processes (aka *Carnot* processes) as idealized limits of real processes which generate less and less entropy increase, and which in that limit obtain the Carnot efficiency.

2.3 Deriving the Equation of State

While it is contested to what extent equilibrium statistical mechanics offers a *conceptual* underpinning of phenomenological thermodynamics, there is no room to doubt that it provides methods to *quantitatively calculate* the thermodynamic description of a system from its microphysics. The recipe for doing so – let's call it the *canonical recipe*[4] (it would be premature to call it a "derivation") is

[4] There is also an microcanonical recipe, which I will not discuss here; in practice the canonical recipe is almost always the one used.

well known and can be used in both quantum and classical mechanics: it will be helpful to adopt as far as possible a notation neutral between the two. For these purposes, let's define a *dynamical state space* as either a phase space, that is, a symplectic manifold (classical) or a separable Hilbert space (quantum), and a *quantity* for a dynamical state space as either a smooth real function on the space (classical) or a self-adjoint operator on the space (quantum). Any such quantity induces a time-indexed flow on the dynamical state space, under either Hamilton's equations (classical) or the Schrödinger equation (quantum), in either case using the quantity as a Hamiltonian. And a *distribution* on a dynamical state space is either a probability distribution on the space (classical) or a density operator on the space (quantum). For any quantity X, the symbol

$$\int X$$

represents either the integral of X over phase space (classical) or the trace of X (quantum), and by definition the expectation value of a quantity X with respect to a distribution ρ is

$$\langle X \rangle_\rho = \int X\rho \tag{2.10}$$

The canonical recipe can now be summarized as follows:

1. An isolated thermodynamic system is represented by a dynamical state space and a parameterized family $H(V)$ of Hamiltonians for the space: each Hamiltonian is a quantity, and $V \rightarrow H(V)$ is a smooth map from the space of parameters into the space of quantities. Any other conserved quantity is represented by another such parameterized family $N(V)$ of quantities, which are conserved in the usual sense: for each $V, N(V)$ is invariant under the dynamical flow induced by $H(V)$. (Which is to say that the Poisson bracket or commutator, as appropriate, vanishes for each V between $H(V)$ and $N(V)$.)

   For future purposes, I take this as the *definition* within statistical mechanics of a thermodynamic system.
2. For given values U, V, N of the energy, parameter, and conserved quantity, the *canonical distribution* $\rho_c(U, V, N)$ is the distribution

$$\rho_c(U, V, N) = \frac{1}{Z(T, \mu)} \exp\left(-\frac{H - \mu N}{T}\right) \tag{2.11}$$

where the *partition function*

$$Z(T,\mu) = \int \exp\left(-\frac{H-\mu N}{T}\right) \tag{2.12}$$

is a normalizing factor, and T and μ are functions of U, V, N defined implicitly by the requirements that

$$U = \langle H(V) \rangle_\rho \tag{2.13}$$

and

$$N = \langle N(V) \rangle_\rho \tag{2.14}$$

(The generalization to several or no conserved quantities is trivial; it is common to call this expression the *grand* canonical distribution where conserved quantities other than energy are present, but I eschew this as adding unneeded complexity.)

3. The equation of state for a thermodynamic system is given by

$$S(U,V,N) = S_G[\rho_c(U,V,N)], \tag{2.15}$$

where $S_G[\rho]$ is the Gibbs entropy,

$$S_G[\rho] = -\int \rho \ln \rho \tag{2.16}$$

(The notation continues to be systematically ambiguous between classical and quantum: in the quantum case this is the von Neumann entropy.)

4. Explicit calculation from (2.16) gives

$$S = U/T + \ln Z \tag{2.17}$$

and hence

$$\left(\frac{\partial S}{\partial T}\right)_{V,N} = \frac{1}{T}\left(\frac{\partial U}{\partial T}\right)_{V,N}. \tag{2.18}$$

By comparison with (1), T is indeed the thermodynamic temperature, as my notation suggests; a similar calculation shows that μ is likewise the generalized potential.

Why the canonical recipe works is contested. *That* it works is amply demonstrated by the last century of work in equilibrium statistical mechanics.

2.4 Characterizing Control Theory

To explain why the canonical recipe works, we need to consider the control theory of a thermodynamic system in microphysical terms. That theory ought to consist of a set of allowable transformations, the control operations, on microphysically characterized states of the thermodynamic system; the hope would be to find some independently justified way of specifying the allowable control operations such that the resultant control theory explains the recipe's success.

What *are* the 'microphysically characterized states' of a thermodynamic system? A natural guess (working for the moment specifically in classical mechanics) would be that they are the points of phase space, each of which describes exactly the properties of the system (say, the positions and momenta of all its composite particles). A slightly more careful guess would be that the states are ordered pairs (x, V), where x is a phase space point and V is a parameter value, since we ought to include the parameter values as part of the specification of the state.[5]

This is not in fact the normal practice in (at least contemporary) statistical mechanics. Here the standard move is instead to treat the state as a probability distribution, giving the probability density for the system to be represented by any given phase-space point (again, in our framework the more careful move would be to take a state as an ordered pair (ρ, V) where ρ is a distribution and V a parameter value). Often, especially in textbooks and foundational presentations, this distribution is given an epistemic gloss: it represents the controlling agent's partial knowledge of the system's true state.

This move to a probabilistic characterization of the state has been heavily criticized by philosophers. David Albert, for instance,[6] writes:

> Can anyone seriously think that our merely being ignorant of the exact microconditions of thermodynamics plays some part in bringing it about, in making it the case, that (say) milk [mixes into[7]] coffee? *(Albert 2000, p. 64)*

However:

1. The epistemic interpretation of the probabilities is for the most part[8] optional. There are numerous ways to interpret them as objective – as long-run relative

[5] In the case of volume in particular, it may seem strange to think of the volume as a parameter in the Hamiltonian, rather than just part of the specification of the phase space. But ultimately, the particles are confined to the box not by a priori kinematics but by their dynamical interaction with the walls, and we can model this conveniently by treating the box as a potential, zero within the box and then increasing extremely rapidly to a value much larger than any relevant particle kinetic energy outside the box. Any particle position is then kinematically possible, but positions outside the box will be dynamically prohibited. Thanks to Katie Robertson for pressing me on this point.

[6] For other examples, see, e.g., Goldstein (2001); Callender (2001).

[7] Albert writes "dissolves in" here, but milk is not water-soluble.

[8] In statistical mechanics (as distinct from thermodynamics) it is entirely optional. In thermodynamics the existence of an agent who intervenes on a system on the basis of their information about the system means that some notion of the state as partly representing an agent's information about the system cannot entirely be dispensed with.

frequencies (Tolman 1938); as the result of a Hume–Lewis best-systems analysis (Lewis 1980, Loewer 2002, Ismael 2009); as the consequence of primordial chance events (Demerast 2016); as the classical limit of quantum indeterminacy (Albert 2000, Wallace 2016).

2. In any case, Albert's criticism applies to statistical-mechanical explanations of what systems do *by themselves*. In the present chapter, I am concerned with the control theory of mechanical systems, not their spontaneous behavior – and there is nothing inherently paradoxical about the fact that my having more information about a system gives me better ability to control it. (Of course, whether a piece of information about a system's state is useful to me depends on whether I have access to control operations that can exploit that particular piece of information; this is a theme which will recur several times in my analysis.)
3. Use of distributions in at least the formal sense is unavoidable once we start doing quantum rather than classical mechanics. The classical limit of a quantum state (even a pure state) is a classical probability distribution, not a classical microstate; and even if a system's state is pure, its subsystems' states are in general mixed; even if a composite system begins in a product state, dynamical evolution will generically entangle those states. (For more on this point, in the context of nonequilibrium statistical mechanics, see Wallace 2016.)

For these reasons (and also given the urgent goal of providing a justification for methods actually used in physics) I continue to follow physics orthodoxy: by definition a *state* of a thermodynamic system is an ordered pair (ρ, V) of a distribution and a parameter value (where we have returned to our systematic ambiguity between classical and quantum mechanics: in the quantum case, states are ordered pairs of density operators and parameter values).

We can now give a formal definition of a control theory for a thermodynamic system (or a *thermodynamic control theory*):

- A *control operation* for the system is a triple $C = \langle \Pi_C, V, V^C \rangle$ where Π_C, the *control map* of C, is a map from distributions to distributions and V, V^C are parameter values. It defines a partial map on states: C maps the state (ρ, V) to $C(\Pi_C, V, V^C) = (\Pi_C \rho, V^C)$. (I adopt the systematic notation that Π_C and V^C are respectively the distribution map and final parameter value associated with C.)

  Given two control operations $C_1 = \langle \Pi_{C_1}, V_1, V^{C_1} \rangle$ and $C_2 = \langle \Pi_{C_2}, V_2, V^{C_2} \rangle$ with $V^{C_1} = V_2$, their composition is

$$C = C_2 \cdot C_1 = \langle \Pi_{C_2}, \Pi_{C_1}, V_1, V^{C_2} \rangle \tag{2.19}$$

and the compositions of three or more operations can be defined inductively in the obvious way: $C_3 \cdot C_2 \cdot C_1 = C_3 \cdot (C_2 \cdot C_1)$, etc.

- A *control theory* for the system consists of
  - (i) A set of *allowable control operations* for that theory, containing at least the identity operations $\langle \mathrm{id}, V, V \rangle$ for all V.
  - (ii) A set of *allowable initial states* for that theory.

    (Note that I do not assume the allowable control operations are closed under composition; indeed, this will generally prove impossible.)

In phenomenological terms, the control operations for thermodynamics include (but are not limited to) operations like "vary the parameters over some period of time" and "place two isolated subsystems in dynamical contact and allow them to come partially to equilibrium, then decouple them."[9] Microphysically, we can characterize the control map of the former operation by dynamical flow under the time-dependent Hamiltonian $H(V(t))$, where $V(t)$ is a path through parameter space, and of the latter by dynamical flow under a time-dependent Hamiltonian that begins and ends at the total Hamiltonian of the isolated subsystems but passes through a region in quantity space representing Hamiltonians, which dynamically couple the subsystems. More generally, any time-dependent path $h(t)$ through the space of quantities satisfying $h(T) = H(V), h(t') = H(V')$ for some $t' > t$ determines a control operation in which the state flows under the time-dependent Hamiltonian $h(t)$, applied between times T and T'. Standard results of classical and quantum dynamics tell us that a control operation like this – where the Hamiltonian is varied along some smooth path from $H(V)$ to $H(V^C)$ and the system evolves under the resultant time-dependent dynamics – determines a dynamical map which is the push-forward of a symplectomorphism (i.e., a canonical transformation) in the classical case, and the adjoint action of a unitary operator in the quantum case; I call maps of this kind *canonical* (continuing with our systematically ambiguous notation). Conversely, any canonical map that is topologically connected to the identity can be generated (nonuniquely) by some choice of time-dependent Hamiltonian. We call a control operation *Hamiltonian* if its control map is canonical, and a control theory Hamiltonian if all its control operations are Hamiltonian.

[9] In some treatments (e.g., Myrvold 2020) these are taken to exhaust the range of available operations, but normally operations like vigorously stirring a fluid are included in the range of operations available even in phenomenological thermodynamics.

Are Hamiltonian control theories the most general that we can consider? To some extent it is a matter of definition. There are certainly things we could do to the systems that are not Hamiltonian control operations: for instance, we can bring in a new system and let it dynamically interact with the old ones so that not only does it induce a change in the Hamiltonian of the latter, but it allows new and old systems' microstates to become correlated. Or we can measure some properties of the systems and choose our control operation accordingly. (These are actually quite closely related, as we will see in Section 2.11.) But in each case, we can absorb the operation in our overall framework by zooming out. In the first case, we can simply choose to include the "new" system as a subsystem of a larger thermodynamic system and consider the control theory of that system. In the second, we can automate the process of measuring and applying the measurement-dependent operation and include the machine that so automates it as one more system in the collection; the externally applied control operation just consists of turning the machine on, which can be modeled just fine in the Hamiltonian class of control operations. So, for the most part we will avail ourselves of these tricks to restrict attention to the Hamiltonian operations; I develop the details in Sections 2.8 and 2.11. (It is, however, sometimes useful to consider certain formal extensions of the class of Hamiltonian control theories, such as idealized limits of Hamiltonian control operations, which are not Hamiltonian.)

A key feature of a Hamiltonian control map Π_C is that it conserves the Gibbs entropy: $S_G[\Pi_C\rho] = S_G[\rho]$. The converse is not true; we will call a control operation *volume-preserving* if it conserves Gibbs entropy, whether or not it is Hamiltonian (since the underlying mathematical reason that canonical transformations leave S_G invariant is that they conserve phase-space volume or its analog, Hilbert space dimension), and a control theory as volume-preserving if and only if its control operations are all volume-preserving.

2.5 Justifying the Canonical Recipe

So much for the *general* framework for thermodynamic control theory. What *specific* assumptions about the theory are needed to recover phenomenological thermodynamics? We can get insight by asking why the canonical distribution plays the central role in the recipe for calculating the equation of state of a thermodynamic system. It has two crucial properties, related but distinct, that explain this. Firstly, $\rho_C(U, V, N)$ is the unique maximum Gibbs entropy distribution for given U, V, N: any other distribution defined for the same parameter value V and with the same expectation values of energy and other conserved quantities has a strictly lower Gibbs entropy. (This is a standard result; for completeness I prove it in the Appendix in the quantum case, under mild simplifying assumptions.) Secondly, near-universal practice in statistical mechanics treats $\rho_C(U, V, N)$ as the unique equilibrium distribution which an isolated system with those expectation values and parameters *spontaneously* approaches – approaches in the sense that it is this distribution that is used to calculate all observable features of a system at equilibrium, including not only thermodynamic features but statistical properties like multitime correlation functions, fluctuations, and the expected fraction of particles with a given energy.

This suggests a rather straightforward (indeed, as we will see, naïve) control theory for phenomenological thermodynamics. Suppose that the controlling agent can manipulate the bulk features of a system but cannot prevent that system from spontaneously approaching equilibrium. And suppose the approach to equilibrium is *literally* modeled by the evolution of a state with given U, V, N to the *canonical state* $(\rho_C(U, V, N), V)$. Then, (i) the allowable initial states for the control theory should comprise the canonical states, and (ii) the control operations should consist of some Hamiltonian operation applied to an initial state, followed by equilibration, so that the state after a control operation is also a canonical state.

At this point, we can derive phenomenological thermodynamics fairly directly. Firstly, if the actual equilibrium state at (U, V, N) is a canonical distribution, then it is no wonder that the expectation values of that canonical distribution correspond to their phenomenological values (at least, assuming a system large enough that fluctuations are small), and the first law follows straightforwardly from energy conservation. Secondly, since the canonical distribution maximizes Gibbs entropy for given expectation values, the equilibration process is Gibbs entropy-nondecreasing. And since Hamiltonian flow is Gibbs entropy-conserving, the overall process of Hamiltonian flow plus equilibration is Gibbs entropy-nondecreasing. The second law then also follows via the definition (15). The Accessibility Principle is somewhat harder to establish (and depends rather more on the details of the control theory) but if we assume that we can vary the parameters of the Hamiltonian arbitrarily slowly, the very fact that Gibbs entropy is maximized by the canonical distribution, and so small perturbations from that distribution have lower entropy only at second and higher order in the perturbation, establishes that we can carry out transitions with arbitrarily low entropy increase (see Wallace 2014: section 4 for the details).

Alas, it's (probably) not that simple. Precisely because equilibration thus defined is Gibbs entropy-increasing, equilibration cannot be a Hamiltonian process. And if equilibration is just a playing out of the ordinary Hamiltonian (or Schrödinger) dynamics of a system in isolation, then equilibration is a Hamiltonian process. Only if equilibration requires fundamentally new dynamics and/or relies essentially on the interaction of a system with its environment, could our naïve control theory be correct. Serious cases have been made for both,[10] but the general consensus in foundations of statistical mechanics, with which I concur, is that it ought to be

[10] See, e.g., Sklar (1993: 246–254) and references therein; see also Albert (2000): ch. 7.

possible to understand equilibration without either. If so, this naïve control theory cannot be the true underpinning of phenomenological thermodynamics.

But it can suggest a better control theory. In orthodox accounts of the approach to equilibrium the idea is not that an equilibrated system *really and truly* is described by the canonical distribution, but that *for all intents and purposes* it is, at least in macroscopically large systems: any remotely realistically measurable dynamical quantity, on any remotely realistic timescale, will have the same expectation values on the canonical distribution as on the true distribution. Only if *per impossibile* we were to measure something like an N-particle correlation function for N of the order of the entire system size could we detect any difference between the two. If so, it seems reasonable to suppose that remotely realistic control operations likewise fail to distinguish between the true post-equilibration state and a canonical state with the same expectation values and parameters.

To express this precisely, it will be helpful to be somewhat more general. A *coarse-graining map* J is an assignment to each parameter value V of a projection J_V on the space of distributions (that is, a map of distributions to distributions satisfying $J_V^2 = J_V$) with the additional properties that

1. The expectation value of the Hamiltonian at parameter V, and any other conserved quantity N, is conserved under $J_V : \langle H(V) \rangle_{J_{V\rho}} = \langle H(V) \rangle_\rho$ and $\langle N(V) \rangle_{J_{V\rho}} = \langle H(V) \rangle_\rho$.
2. The Gibbs entropy satisfies $S_G(J_V\rho) \geq S_G(\rho)$, with equality only if $J_V\rho = \rho$; that is, if $J_V\rho \neq \rho$, $S_G(J_V\rho) \geq S_G(\rho)$.

The coarse-graining map induces a projection on the set of states $J(\rho, V) = (J_V\rho, V)$ and defines a function, the *thermodynamic entropy with respect to J* ("*J*-entropy" for short) on states:

$$S^j(\rho, V) = S_G[J_V\rho]. \tag{2.20}$$

The most important example of coarse-graining is *equilibration coarse-graining*, defined by

$$J_V^{eq}\rho = \rho_C(\langle H(V) \rangle_\rho, V, \langle N(V) \rangle_\rho), \tag{2.21}$$

but many others will be useful in due course.

A control theory is *forward compatible with J*, or *J*-compatible for short, if for any allowable control operations $C_1, C_2, \ldots C_N$ whose composition is defined and allowable, and for any allowable initial state (ρ, V),

$$JC_NJCC_{N-1} \cdots JC_1J(\rho, V) = J(C_N \cdots C_1)(\rho, V). \tag{2.22}$$

It is *forward complete with respect to* J, or J-complete for short, if

i. For any distribution ρ and parameter V, (J_ρ, V) is an allowable initial state.
ii. For any allowable initial state (ρ_1, V), and any state (ρ_2, V_2) with

$$S^j(\rho_2, V_2) > S^j(\rho_1, V_1)$$

there is an allowable control operation that takes (ρ_1, V_1) to some state (ρ_2', V_2) with $J_{V_2}\rho_2{}' = J_{V_2}\rho_2$.

It is *strictly J-complete* if, in addition,

iii. For any allowable initial state (ρ_1, V), and any state (ρ_1, V_2) with

$$S^j(\rho_2, V_2) = S^j(\rho_1, V_1)$$

there is an allowable control operation that takes (ρ_1, V_1) to some state (ρ_2', V_2) with $J_{V_2}\rho_2{}' = J_{V_2}\rho_2$.

(In other words, if (ii) holds when the entropy is nondecreasing and not just strictly increasing.)

This will require some unpacking. Forward compatibility with J means that control operations give the same result, up to coarse-graining, whether they are applied to the actual initial state or a coarse-graining of it, and furthermore that this continues to be the case if control operations are composed (provided that the composition remains allowed). This is to say that the control operation can be understood as acting autonomously on the coarse-grained features of the system, irrespective of the fine-grained details. A control theory forward compatible with equilibration coarse-graining, for instance, takes effectively canonical distributions to effectively canonical distributions in a fashion that does not depend on the in-practice-inaccessible ways in which those distributions differ from the exact canonical distribution; this (I suggest) accurately captures the intuitive idea, stated earlier, that "remotely realistic control operations . . . fail to distinguish between the true post-equilibration state and a canonical state with the same expectation values and parameters."

The J-entropy is nondecreasing, and in general strictly increasing, under volume-preserving (and in particular Hamiltonian) control operations:

$$\begin{aligned} S^J(C(\rho, V)) &= S_G[J_{V^C}\Pi_C\rho] \\ &= S_G[J_{V^C}\Pi_C J_V\rho] \\ &\geq S_G[\Pi_C J_V\rho] \\ &= S_G[J_V\rho] \\ &\equiv S^J(\rho, V). \end{aligned} \tag{2.23}$$

J-completeness expresses, up to coarse-graining, the idea that this limit is the *only* limit on the control theory: in a J-complete control theory, given any state there is another allowable initial state with the same coarse-grained description, and given any pair of states where the second has higher J-entropy than the first, there is a control operation that transforms the first into one identical to the second up to coarse-graining. (And strict J-completeness means that this continues to hold even when the two states have the same J-entropy.)

Let's now specialize to the specific case of the equilibration coarse-graining map J^{eq}, which replaces any state with the canonical state at the same expected values of conserved quantities. I define an *equilibrium thermodynamics* of a thermodynamic system as any control theory which is forward compatible and forward complete with respect to J^{eq}. (We have seen that there are good (if heuristic) reasons to think that any control theory that allows arbitrarily slow variation of the parameters of a system's Hamiltonian will define an equilibrium thermodynamics; we have also seen that any two control theories forward-compatible and forward-complete with respect to a given coarse-graining are essentially equivalent, so that it will rarely matter which equilibrium thermodynamics we have in mind.)

The thermodynamic entropy, with respect to J^{eq}, is determined only by the expected value of the conserved quantities and the external parameters:

$$S^{J^{eq}}(\rho, V) = S_G[\rho_c(\langle H(V)\rangle_\rho, V, \langle N(V)\rangle_\rho)]. \tag{2.24}$$

And, indeed, any state can be characterized completely by parameters and conserved-quantity expectations. At this point it should be apparent that equilibrium thermodynamics simply reproduce the canonical recipe.

2.6 Partial Equilibrium

To unpick this further, suppose that we have a thermodynamic system which can be decomposed into subsystems 1, 2 (with dynamical state spaces S_1, S_2), and that each subsystem i has a parameter V_i such that the total Hamiltonian $H(V_1, V_2)$ can be written at least to a good approximation as

$$H(V_1, V_2) \simeq H_1(V_1) \times \mathrm{id}_2 + \mathrm{id}_1 \times H_2(V_2) \tag{2.25}$$

that is, the two systems are very weakly interacting. (For the moment, suppose no conserved quantities are present other than energy.) Then the equilibration coarse-graining map is

$$J^{eq}_{V_1,V_2}\rho = \rho_C(\langle H(V_1, V_2)\rangle_\rho, V_1, V_2) \tag{2.26}$$

or more explicitly (and writing $A \otimes B$ as a neutral expression either for the function $A \otimes B(x, y) = A(x)B(y)$ (classical) or for the tensor product of A and B (quantum)),

$$J^{eq}_{V_1,V_1}\rho = \frac{e^{-(H_1(V_1)\times \mathrm{id}_2 + \mathrm{id}_1 \times H_2(V_2))/T}}{Z(T)} \simeq \frac{e^{-H_1(V_1)T}}{Z_1(T)} \otimes \frac{e^{-H_2(V_2)T}}{Z_2(T)}. \tag{2.27}$$

where $T = T(\langle H(V)\rangle_\rho, V_1, V_2)$ is the thermodynamic temperature given the expected energy and parameters of ρ and

$$Z_i = \int_{S_i} e^{-Hi/T} \tag{2.28}$$

and the approximation becomes exact in the limit where the total Hamiltonian factorizes exactly. That is, equilibration coarse-graining transforms a distribution to the product of two canonical distributions at the same thermodynamic temperature.

This coarse-graining map is appropriate for the control theory of an agent who is unable to prevent interactions between the two systems, and who acts on timescales slow compared to the equilibration timescale of the joint system: by the time they are able to intervene on the system, the two subsystems have (effectively) reached their joint equilibrium state. But we can also consider an agent who can intervene more rapidly and who can control whether or not interactions occur between the two systems (but who cannot intervene rapidly compared to the equilibration timescales of the two systems considered separately and lacks fine-grained control of those separate systems). This latter agent is better modeled via the *partial equilibration coarse-graining*, where the coarse-graining map takes each subsystem to its own equilibrium state, as determined by its own expected energy.

More precisely: for a given distribution ρ for the system, we define

$$\rho|_1 = \int_{S_2} \rho \tag{2.29}$$

as the marginal distribution of ρ over S_1: again, this is a neutral notation, denoting either the marginal probability distribution (classical) or the partial trace (quantum). Then partial equilibration coarse-graining is defined by

$$J^{pe}_{V_1,V_2}\rho = J^{eq,1}_{V_1}\rho|_1 \otimes J^{eq,2}_{V_2}\rho|_2, \tag{2.30}$$

where $J^{eq,i}_{V_I}$ is the usual equilibration coarse-graining map for the ith system. Explicitly, we have

$$J^{pe}_{V_1,V_2}\rho = \frac{e^{-H_1(V_1)T}}{Z_1(T_1)} \otimes \frac{e^{-H_2(V_2)T}}{Z_2(T_2)}, \tag{2.31}$$

where $T_i = T_i(\langle H_i(V_i)\rangle_{\rho|i}, V_i)$ is the thermodynamic temperature of system i. So partial equilibration coarse-graining again transforms a distribution into a product of canonical distributions, but now they may be at different temperatures.

Since the canonical distribution of the whole system is the lowest-energy state for given energy and parameters, it follows that if we have access to a control theory that is complete and compatible with respect to *partial* equilibration – to *partial-equilibrium thermodynamics*, that is – we can extract energy in this situation in a cyclic process, just by transforming (31) reversibly to the full (same-temperature) canonical distribution at the same parameter values. Indeed, the explicit process for doing so is familiar: we simply run a Carnot cycle between the two systems and extract the energy, continuing until they are at the same thermodynamic temperature.

This is our first example of how a more powerful control theory may allow us to extract more energy from a system than is possible in equilibrium thermodynamics. Note that it requires two things: a control theory (approximately) forward complete with respect to a finer-grained coarse-graining, but also an initial state whose coarse-graining under the new theory's coarse-graining map is not canonical. If the initial state of the system is the product of two canonical distributions at the same temperature, the additional power of partial-equilibrium thermodynamics is useless; conversely, if we have access only to (full-)equilibrium thermodynamics then it is useless to us to know that the two subsystems are initially at different temperatures.

To illustrate further, suppose for simplicity that the two subsystems are identical and have no adjustable parameters, and we define $\rho(U)$ as the canonical distribution for either of the subsystems at expected energy U. Let $U_C < U_H$ be two energies (for a "cold" and a "hot" system), and let the initial distribution be

$$\rho = \frac{1}{2}(\rho(U_C) \otimes \rho(U_H) + \rho(U_H) \otimes \rho(U_C)). \tag{2.32}$$

Equilibration coarse-graining this system gives

$$J^{eq}\rho = \rho((U_C + U_H)/2) \otimes \rho((U_C + U_H)/2). \tag{2.33}$$

Partial-equilibration coarse-graining gives the same result, since the marginal distribution of each system is $1/2(\rho(U_C) + \rho(U_H))$. The thermodynamic entropies of ρ are the same in either full-equilibrium or partial-equilibrium thermodynamics and in both cases are maximal for the given energy, reflecting the fact that neither control theory can extract energy from the system. In the full-equilibrium case, this is just because the controller cannot intervene on the separate systems, at any rate not quickly enough; in the latter case, if only the controller knew which system was hotter, they could run a heat engine between them, but they do not.

But now suppose that the controller learns (say, from a helpful third party who had already measured it) that in fact the system's state is

$$\rho' = \rho(U_C) \otimes \rho(U_H). \tag{2.34}$$

This makes no difference to the effect of equilibration coarse-graining, and hence no difference to the equilibrium-thermodynamics entropy: the system's total energy is still $U_C + U_H$, and that alone determines the global equilibrium distribution. But ρ' is invariant under *partial* equilibration coarse-graining, and so has a lower thermodynamic entropy, reflecting the fact that energy can be extracted from the system: now that the controller knows which is the hotter system, they can run a reversible heat engine between the two and extract the resultant energy.

This is a special case of a general principle which will recur. Gaining information about a system always lowers the fine-grained Gibbs entropy, but it lowers the *thermodynamic entropy* only with respect to a control theory able to exploit that information.

(There is, to be sure, a certain arbitrariness as to whether we regard a given control problem as *partial*-equilibrium thermodynamics, or as *full*-equilibrium thermodynamics with additional conserved quantities. We could have set up our problem *ab initio* as being defined by the same Hamiltonian and parameters but by an additional conserved quantity $H_1 \otimes \mathrm{id}_2$, at least in the idealized limit where the two systems are exactly isolated from one another. The full-equilibrium control theory of the latter system is identical to the partial-equilibrium control theory of the former, though the definition of cyclic process is narrower for the latter.)

As a second illustration,[11] suppose our system contains two boxes of gas, and that the gas is a mixture of two difficult-to-distinguish isotopes (for definiteness let the gas be chlorine, and the isotopes be ^{35}Cl and ^{37}Cl). The numbers of the two isotopes are separately conserved, and we assume that both boxes begin at the same thermodynamic temperature. Now consider two controllers, Alice and Bob, both of whom hope to extract energy through cyclic processes. Alice has access only to the full-equilibrium control theory for the system; Bob possesses semipermeable membranes that allow him, for example, to let ^{35}Cl but not ^{37}Cl isotopes flow between the two boxes.

If the initial state of the system has a 50/50 mixture[12] of both isotopes present in each box separately, Bob's control theory grants him no advantage over Alice's

[11] Readers familiar with the Gibbs paradox (see, e.g., Saunders 2018 and references therein) will recognize its close connection to this example; for reasons of space, I do not explore this topic here.

[12] Ordinary chlorine has a 75/25 mixture of the two isotopes; I use 50/50 for simplicity.

(and both agree on the thermodynamic entropy). Likewise, if the initial state is an equally weighted mixture of (a) all the ^{35}Cl in the left-hand box, all the ^{37}Cl in the right-hand box, and (b) vice versa, Bob and Alice agree on the coarse-grained state, on the thermodynamic entropy, and on the impossibility of cyclically extracting energy. But suppose they learn that in fact all the ^{35}Cl is on the left and all the ^{37}Cl is on the right. The information is useless to Alice – but it allows Bob, using his semipermeable membranes, to let the ^{35}Cl reversibly expand into both boxes and then do likewise with the ^{37}Cl, in each case extracting energy from the pressure exerted on the membrane as it is slowly moved. (As always, genuine reversibility occurs only in the unphysical – and useless – limit of infinitely slow movement but can be approximated arbitrarily well.)

If, before either agent can act, the partition between the two boxes is removed and the gases are allowed to mingle freely, Alice and Bob will disagree about its significance. To Alice this makes no thermodynamically relevant difference to the state: coarse-graining before or after has the same effect. To Bob, it is an irreversible process that increases the thermodynamic entropy and wastes energy that could otherwise have been extracted

2.7 Free Energy

The common theme of these partial-equilibrium processes is that they involve extracting energy from a system in a cyclic process by reversibly guiding it to the canonical state. In fact, this is a perfectly general feature of thermodynamic control theories, as we can see by returning to the general coarse-graining framework. Suppose we are operating with a control theory forward-compatible with a coarse-graining J; then for any state (ρ, V), the *free energy of equilibration* of that state relative to J is defined by

$$E^J(\rho, V) = \langle H(V)\rangle_\rho - U\big(S^J(\rho, V), V, \langle N(V)\rangle_\rho\big). \tag{2.35}$$

Since $U\big(S^J(\rho, V), V, \langle N(V)\rangle_\rho\big)$ is the minimum expected energy of any state with the same J-entropy and $\langle N(V)\rangle$ as (ρ, V), and since any J-compatible control operation is J-entropy-nondecreasing, the free energy of equilibration is the maximum work extractable from (ρ, V) by J-compatible cyclic control operations. That maximum can be approached arbitrarily closely in a J-complete control theory (and reached in a strictly J-complete control theory).

Similarly, suppose $(\rho, V) \rightarrow (\rho', V')$ is the result of a *non*cyclic (but J-compatible) control operation, and suppose the J-entropy and conserved-quantity expectation values before and after are respectively S, N and S', N'. Then the work extracted by that process is

$$\begin{aligned} W &= \langle H(V)\rangle_\rho - \langle H(V')\rangle_{\rho'} \\ &= (E^J(\rho, V) + U(S, V, N)) - (E^J(\rho', V') + U(S', V', N')) \\ &= (E^J(\rho, V) - E^J(\rho', V')) + (U(S, V, N) - U(S', V', N')). \end{aligned} \tag{2.36}$$

That is, extracted work equals decrease in free energy of equilibration, plus decrease in canonical energy. And the difference between any control theory and equilibrium control theory is just given by the fact that states may have free energy of equilibration according to the first control theory, whereas no state in equilibrium control theory has free energy of equilibration.

We can use the free-energy concept to analyze the partial-equilibrium examples in the previous section. Given two systems at different temperatures, and taking J to be partial equilibration, the free energy of equilibration is just the work that is extractable by letting the two systems reversibly equilibrate, extracting work as we do so. If we let the systems *irreversibly* equilibrate, not all the free energy will be extracted as work; some will simply be lost. In the limit when we just put the systems in thermal contact and leave them to equilibrate, all the free energy will be lost. Quantitatively, we have (writing S_1 and S_2 as the thermodynamic entropies for the two systems separately and $U_i(S_i)$ as the canonical energy of the ith system)

$$E(S_1, S_2) = U_1(S_1) + U_2(S_2) - U(S_1 + S_2) \tag{2.37}$$

($(U_1(S_1) + U_2(S_2))$ is the actual energy of the system at entropies S_1, S_2; $U(S_1 + S_2)$ is the minimum energy it can have for that total entropy.) Since U is an increasing function of entropy, we can see directly that the work extracted from the system (= the decrease in $(U_1 + U_2)$) is less than or equal to the decrease in free energy, with equality only when the process is reversible, that is, entropy-conserving.

If the heat flow between the two systems is reversible, so that overall entropy is conserved, the change in the free energy when a small amount of entropy δS flows from system 1 to system 2 is just minus the total change in energy, so we have

$$\delta E = -(T_1 - T_2)\delta S. \tag{2.38}$$

Similarly, for the case of the two boxes of chlorine gas, if the two systems have chemical potentials μ_1, μ_2 with respect to the number of ^{35}Cl atoms and are both at temperature T, then if some small quantity δN of atoms flows from box 1 to box 2 and the overall process causes entropy increases δS_1, δS_2 in the two boxes, the change in energy of the boxes will be

$$\delta U_1 + \delta U_2 = (\mu_2 - \mu_1)\delta N + T(\delta S_1 + \delta S + 2) \tag{2.39}$$

and since the last term here is just the change in the canonical energy of the combined system,

$$\delta E = -(\mu_1 - \mu_2)\delta N, \tag{2.40}$$

which is the familiar result that work can be extracted from combining two samples of a fluid if and only if their chemical potentials differ. (Note that, since N is conserved even in irreversible processes, this expression holds whether or not the process is reversible; if it is irreversible, though, it will heat up the gas and so the energy extracted will be less than the decrease in free energy.)

In the partial-equilibrium case, we can express the free-energy idea another way: the largest amount of work we can extract from a system through partial-equilibrium control processes is the difference between their total energy and the minimum value of that energy at constant entropy (and other conserved parameters). Since the minimum-energy state is also the overall equilibrium state, we can also use this to characterize overall equilibrium for a collection of subsystems: at equilibrium, energy is minimized with respect to entropy-conserving processes and any process that transfers conserved quantities from one system to another.

2.8 Free Energy and Heat Baths

(This section lies somewhat outside the main line of development of this chapter and may be skipped on a first reading; other than the definition of a heat bath, its primary purpose is to connect the use of "free energy" in this chapter with its other uses in the physics literature.)

Suppose we have a system consisting of two subsystems, one of which is a *heat bath*: a system with no adjustable parameters, no conserved quantities other than energy, and whose initial state is the canonical state,[13] and so large that energy flows into and out of it negligibly affect its temperature, so that we can in practice treat it as permanently at thermodynamic temperature T. And suppose that we are doing equilibrium thermodynamics with respect to this system, so that the primary (i.e., non–heat-bath) system begins and ends every control process at (effective) thermal equilibrium with the heat bath. (We could generalize this to partial-equilibrium thermodynamics, with respect to various subsystems of the primary system, but with the additional constraint that all the subsystems remain in *thermal* equilibrium

[13] For the purposes of this section, it suffices to require that the heat bath's state coarse-grains to the canonical state.

with one another and with the heat bath, even if they do not equilibrate with respect to other conserved quantities.) Then we can derive an effective control theory for the primary system alone, treated now not as isolated but as in thermal contact with the heat bath.

To do so, consider a control operation that takes the primary system from an initial state with entropy, parameter values, and conserved quantities (S, V, N) to another where these quantities are (S', V', N'). Since the primary system must begin and end at temperature T, there will in general be some energy flow between the heat bath and the primary system during the control operation. If this flow changes the heat-bath energy by ΔU_{HB}, we must have $\Delta U = T\Delta S_{HB}$, where ΔS_{HB} is the entropy change of the heat bath. The work W extracted by the control operation is then $-\Delta U_{HB}$ plus the decrease in energy of the system itself:

$$W = -U(S', V', N') + U(S, V, N) - T\Delta S_{HB}. \tag{2.41}$$

But since $\Delta S_{HB} + (S' - S)$ is the total change in the entropy, which must be ≥ 0, this can be rewritten as

$$W \leq -(U(S', V', N') - TS') + (U(S, V, N) - TS) \tag{2.42}$$

If we invert the equation of state so that S may be written as a function of T, V, N, we can define the quantity

$$F(T, V, N) = U(S(T, V, N), V, N) - TS(T, V, N) \tag{2.43}$$

which is called the *Helmholtz free energy*. Up to an additive constant, it equals the free energy of equilibration of the total system. For a system in contact with a heat bath, the Helmholtz free energy plays a similar role as the energy does for a thermally isolated system:

- The maximum work extractable from a system by any transformation in equilibrium thermodynamics equals the decrease in Helmholtz free energy.
- The free energy of equilibration of a collection of subsystems in a partial-equilibrium thermodynamics (but with all systems in thermal contact with the heat bath) is the sum of the Helmholtz free energies of the subsystems, minus the value of that sum at its lowest value (for given T, V, N for the combined system).
- The maximum work extractable from a system by any transformation in partial-equilibrium thermodynamics equals the decrease in the total Helmholtz free energies of the subsystems under the transformation.

- The equilibrium state of a collection of subsystems can be characterized by the fact that it minimizes Helmholtz free energy under any transformation that leaves parameters and nonenergy conserved quantities unchanged.

Other effective control theories can be defined similarly:

- By considering a primary system separated by an adiabatic barrier from a very large "pressure bath" system at generalized pressure P, so that the barrier can adjust to equalize pressure between the two systems but the systems cannot otherwise interact, we find that the role of energy is played by the *enthalpy*,

$$H(S,P,N) = U(S, V(S,P,N),N) + PV(S,P,N). \tag{2.44}$$

- By combining heat and pressure baths together, so that the external path constrains the primary system to have both fixed pressure and fixed temperature, we find that the role of energy is played by the *Gibbs free energy*,

$$G(T,P,N) = U(S(T,P,N), V(T,P,N),N) + PV(T,P,N) - TS(T,P,N). \tag{2.45}$$

2.9 Beyond Equilibration: Boltzmannian Thermo dynamics

In this section, I want to consider a class of control theories not built on the assumption of even partial equilibration: control theories based on Boltzmann's idea of a partition of a system's state space. (These are of limited practical use but are often conceptually helpful.) To describe that class, suppose, for simplicity,[14] that we are dealing with a classical system with phase space P and that its phase space has been partitioned into *macrostates*: subsets of the phase space, of positive measure, such that intuitively two phase-space points in the same macrostate are indistinguishable to an external observer. (For instance, in a dilute gas, the macrostates can be defined by coarse-graining the phase space of a single particle into cells of equal Liouville volume and then individuating the macrostates by the number of particles in each cell.) More precisely, the partition may depend on any external parameter V: write $\mathcal{M}(V)$ for the partition for parameter V.

[14] For the quantum version, replace the partition with a collection of mutually orthogonal finite-dimensional subspaces whose direct sum is the full Hilbert space.

In the thermodynamic context, "macroscopically indistinguishable" has an operational meaning: it implies that a controlling agent can move a system from macrostate to macrostate but has no control of a system on a grain finer than that defined by the partition. We can incorporate this by defining a *Boltzmannian coarse-graining*, which smears out these fine-grained details. To state it, first define for any macrostate M the projection operator P_M, which acts on distributions as

$$P_M\rho(x) = \begin{cases} \rho(x) & \text{if } x \in M \\ 0 & \text{if } x \notin M \end{cases} \tag{2.46}$$

The probability of a microstate given a distribution ρ, $Pr(M|\rho)$, is

$$Pr(M|\rho) = \int P_M\rho \tag{2.47}$$

and now we can define our coarse-graining map:

$$J_V^B\rho = \sum_{M \in \mathcal{M}(V)} \Pr(M|\rho)\frac{\chi_M}{\mathcal{V}(M)} \tag{2.48}$$

where χ_M is the characteristic distribution of M, given by

$$\chi_M(x) = \begin{cases} 1 & \text{if } x \in M \\ 0 & \text{if } x \notin M \end{cases} \tag{2.49}$$

and $\mathcal{V}(M)$ is the Liouville volume of M.[15] In other words, $J^B\rho$ is the distribution that assigns the same probability to each macrostate as ρ itself, but which is uniform across each macrostate.

The thermodynamic entropy defined by this coarse-graining is the *generalized Boltzmann entropy*. In the special case where ρ has support entirely on one macrostate M, it is a function of M alone and reduces to the *Boltzmann entropy*:

$$S_B(M) = \ln \mathcal{V}(M). \tag{2.50}$$

In the more general case, it is the sum of two terms: the expected value of the Boltzmann entropy with respect to the probability distribution across macrostates, and the Shannon entropy of that distribution:

[15] Quantum mechanically: take χ_M to be the projector onto macrostate M, define $P_M\rho = \chi_M\rho\chi_M$, and take $\mathcal{V}(M)$ to be the dimension of M (equivalently, $\mathcal{V}(M) = Tr\chi_M$).

$$S^{J_B}(\rho, V) = \left(\sum_{M \in \mathcal{M}(V)} \Pr(M|\rho) S_B(M) \right) + \left(- \sum_{M \in \mathcal{M}(V)} \Pr(M|\rho) \ln \Pr(M|\rho) \right) \quad (2.51)$$

In the special case where (i) ρ has support only on one macrostate M, and (ii) the control operation maps it to another distribution with support only on another macrostate M' (call such a control operation *macrodeterministic* for M), it is immediate from volume preservation that $V(M)' \geq V(M)$, that is, Boltzmann entropy is nondecreasing. But, of course, the generalized Boltzmann entropy is nondecreasing under any control map that is forward compatible with Boltzmannian coarse-graining, as a consequence of our general framework.

Boltzmannian thermodynamics offers an agent significantly more control over a system than equilibrium thermodynamics, and so significantly more opportunity to extract work, given an appropriate initial state. For instance, suppose that our system consists of a gas of N particles in a box of volume V, and that in fact the initial distribution has all the particles localized in a smaller region of the box, with volume $V' < V$. This additional information is useless to a controller who only has control operations forward commuting with equilibration, for whom the situation is not interestingly different from one where the gas is at full thermal equilibrium. But a Boltzmannian agent can, for example, slam a partition into the box so that the particles remain confined to the region of volume V', and then slowly allow the gas to expand, doing work against the partition, until it returns to volume V. The work extracted is the free energy of equilibration of the state relative to Boltzmannian coarse-graining. Once it is extracted, though, the greater control power of Boltzmannian thermodynamics is of no further use, and we effectively return to equilibrium thermodynamics.

This difference between the control theories also shows up in the different entropies they assign. The equilibrium-thermodynamics entropy is as usual set only by the volume of the box and by the total energy and number of particles; the Boltzmannian entropy is set by the size of the volume in which the particles actually are (and so is lower than the equilibrium-thermodynamics entropy by $N\ln(V/V')$). And so, being told that the particles are in fact on one side of the box makes no difference to equilibrium-thermodynamic entropy but decreases Boltzmannian entropy. There is no fact of the matter as to which entropy is *correct*, but the *appropriate* choice is set by the control theory being used.

2.10 Reversible Thermodynamics

We have seen that coarse-grained thermodynamics has a very general form. The coarse-graining, together with the initial state, determines the free energy of equilibration. In evolving the system to a state which coarse-grains to the canonical state, it may be possible to extract some of that free energy as work (it can all be extracted only in the ideal limit of a control theory strictly complete with respect to J). Once it is extracted, we can extract more energy only by varying the system's control parameters and/or the values of other conserved quantities, and the amount of energy thus extractable does not depend on the coarse-graining operation.

The limiting case of all of this is *reversible thermodynamics*, where the control operations include *every Hamiltonian operation*. Reversible thermodynamics is forward-compatible with only the trivial coarse-graining, defined by the identity map; its thermodynamic entropy is just the Gibbs entropy. The inequality that thermodynamic entropy increases under control operations continues to hold, but is always saturated, and simply reflects conservation of Gibbs entropy under dynamical flow. Reversible thermodynamics is reversible both in the literal sense that the time-reverse of any allowed control operation is also an allowed control operation, and in the formal sense that – unlike our various coarse-grained versions of thermodynamics – the entropy is not only nondecreasing but is strictly conserved.

Insofar as thermodynamics just reflects the macroscopic limitations of what we can do to a system, we would expect that in reversible thermodynamics – where there are no such limitations – there are no constraints on how much energy we can extract beyond those set by the conservation laws, so that it should be possible to extract all of a system's energy through a cyclic operation. But this is not the case. Assume for simplicity that the energy has been scaled so that the system's ground state energy is zero. The free energy of equilibration of a state (ρ, V) with respect to the trivial coarse-graining is just the difference between its actual expected energy and its canonical energy with respect to the trivial coarse-graining:

$$E(\rho, V) = \langle H(V) \rangle_\rho - U\Big(SG(\rho), V, \langle N(V) \rangle_\rho\Big). \tag{2.52}$$

Only if the latter equals zero will the free energy equal the total system energy. And the canonical energy can be zero, in general, only if the initial distribution is known *exactly.* More precisely: working in quantum mechanics, and assuming the nondegeneracy of the ground state, only if ρ is pure will the canonical energy equal the ground-state energy. In this case, a control operation can steer the system's state reversibly into the ground state and extract all the energy. But if the initial distribution is mixed, then no unitary operation can so steer it, and the lowest-energy state will still assign some probability to excited states. Similarly, in classical mechanics, and assuming that the gradient of the energy function nowhere vanishes, the subset of lowest-energy states will have Liouville measure zero; only if the initial state likewise has support on a measure-zero region will it be possible to steer it into the lowest-energy region by a Hamiltonian control operation.

(Can all the free energy actually be extracted in reversible thermodynamics? In general, no: extracting it requires us to steer the system's distribution into a *canonical* distribution, and in general that will not be possible. Specializing for simplicity to quantum mechanics, and following Pusz and Woronowicz (1978) (see also the very helpful discussion in section 4 of Maroney 2007) we can define a quantum distribution as *passive* if and only if no unitary control operation decreases its expected energy; it can be shown that a distribution is passive if and only if it is a mixture of energy eigenstates where the probability of an energy eigenstate is nondecreasing with energy, but there are passive distributions with strictly higher energy than the canonical distribution at the same Gibbs entropy. For small systems, it is an open question exactly how to characterize the extractable energy (the associated research project is sometimes called "single-shot thermodynamics"; see Yunger Halpern and Renes 2016 for an introduction). However, for sufficiently large systems the difference is generally expected to be negligible, albeit I am not aware of rigorous theorems to this effect: such systems can generally be expected to factorize into a large number of approximately isolated subsystems so that the overall density matrix is approximately a product of identical density matrices, and Pusz and Woronowicz prove that the canonical distribution is the only distribution whose N-fold product is passive for all N).

Reversible thermodynamics is the limiting case of the successively more fine-grained thermodynamic control theories we have been considering. In those theories, we have seen that the finer our control over a system, the more valuable it is to have fine-grained information about the system's state. In reversible thermodynamics, *any such information* can be used to extract work: in the limiting case, if we are told the exact microstate of a system (or, in quantum mechanics, if the system's state is pure and we know that state), we can extract all of its energy.[16] So to an agent with access to reversible thermodynamics, any information about the state is entropy-decreasing. But the most any such agent can do is extract the free energy from the system. Once that is done, any further work extraction has to be done by varying parameters and conserved quantities, and all the power of reversible thermodynamics avails us nothing beyond what equilibrium thermodynamics already permitted.

Free energy also provides a third way in which reversible thermodynamics is indeed reversible (and our various coarse-grained thermodynamic theories are irreversible). In coarse-grained thermodynamics, the change in free energy in

[16] In classical physics, if we literally know the exact microstate of a system, we can use it as a resource to extract all the energy from any system: its entropy is $-\infty$. Of course, this is an artifact of classical mechanics; it has no quantum equivalent.

a process is an upper limit on the energy extractable by that process, but it is possible to dissipate free energy without extracting any work from the system. In partial-equilibrium thermodynamics, for instance, if we just let two systems at different temperatures equilibrate without running a heat pump between them, the free energy is simply and irreversibly lost. This never happens in reversible thermodynamics, where the decrease in free energy always equals the energy extracted. (In reversible thermodynamics, we could simply run the equilibration process backwards and return the systems to their distinct temperatures.)

2.11 Collecting Information

We have seen that the more fine-grained an agent's control over a system, the more important are the fine-grained details of the system's initial state, and the more the entropy they assign can drop if they receive information about the system. On the face of it, an obvious strategy presents itself: instead of just carrying out a Hamiltonian control operation, the agent should first gain more information about the system (causing them to update the state) and only then apply a control operation, tailored to the information they collect.

It is easy to see that this cannot actually work, at least if the information-collection process conforms to the laws of physics. As I noted in Section 2.4, any information-collection process can be mechanized, and the system can be expanded to include both the original system and that mechanism; we can then include "turning on the mechanism" as just another Hamiltonian control operation. Or put another way, the agent themselves can be mechanized and included in the system. From that agent's point of view, they are collecting information and carrying out control operations conditional on the result, but from the point of view of the *ur*-agent outside the box in which the agent and their system sit, telling the inside agent to do their thing is a control operation that involves no information-gathering and no conditional control.

Still, it is instructive to ask what actually goes wrong with the information-collection strategy. The answer is well known[17]: any such process involves an ancilla system that records the result of the information collection; that ancilla is effectively a finite resource which can become exhausted, so that no further information collection is possible without resetting the ancilla; the resource cost of the reset process undoes the gain of collecting the information in the first place. For completeness, and to connect with this chapter's approach to thermodynamics, I outline it here.

[17] See, e.g., Bennett (2003) or the overview and articles in Leff and Rex (2002). The topic has been rather controversial in recent philosophy of physics (Earman and Norton (1999), Bub (2002), Norton (2005), Ladyman, Presnell, Short, and Groisman (2007), Norton (2011), Ladyman and Robertson (2013), Norton (2013a), Norton (2013b), Ladyman and Robertson (2014), Myrvold (2021)); I do not attempt to engage with these controversies here.

Suppose, specifically, that we have some system S with no adjustable parameters, and which begins in the canonical distribution for some given temperature T, so that (assuming quantum mechanics for definiteness) the specific form of the distribution is

$$\rho_S = \frac{1}{Z}\sum_n e^{-E_n/T}|n,S\rangle\langle n,S|. \tag{2.53}$$

where $|n,S\rangle$ is an eigenstate of S's Hamiltonian H_S with eigenvalue E_n. Since ρ_S minimizes expected energy for given Gibbs entropy, no unitary process can extract energy from the system.

The collect-information strategy would have us measure the system in the energy basis, and then if the result is that the system's state is $|n,S\rangle$, apply a unitary transformation U_n which implements $U_n|n,S\rangle = |0,S\rangle$, so that the system ends up determinately in the ground state. But, of course, no single unitary process acting on S alone can have this effect for all n, since it would have to map orthogonal states to the same state. The only way to implement the procedure is to have some ancilla system A whose state after the process records the result. Concretely, if the ancilla system starts in some fixed state $|0,A\rangle$, we can find a unitary process U with the effect

$$U|n;S\rangle\otimes|0;A\rangle = |0;S\rangle\otimes|\psi_N;A\rangle \tag{2.54}$$

with $\langle\psi_n,A||\psi_m,A\rangle = \delta_{m,n}$. (Often this is illustrated by supposing that (i) the state of the system is measured and recorded in the ancilla, and then subsequently (ii) an operation is performed on the system conditional on the state of the ancilla. But we do not need to make this separation: the crucial point is that the ancilla must end up in a state that records the original state of the system.) At the level of distributions, this unitary transformation enacts

$$U(\rho_S\otimes|0,A\rangle\langle 0,A|)U^\dagger = |0;S\rangle\langle 0;S|\otimes\rho_A, \tag{2.55}$$

where

$$\rho_A = \frac{1}{Z}\sum_n e^{-E_n/T}|n;A\rangle\langle n;A|. \tag{2.56}$$

Does this *conditional process* extract net energy (measured as always with respect to the self-Hamiltonian of the system before or after the control intervention, that is, in the case the sum of the Hamiltonians H_S and H_A of system and ancilla)? It depends on the details of the record basis $|n;A\rangle$ and the ancilla's Hamiltonian; certainly, there is no systematic reason that it will not. (If the ancilla has enough states that each $|\psi_n;A\rangle$ has negligible energy compared to the original energy of the system, energy can certainly be extracted.) But whether or not energy is extracted, the Gibbs entropy of the ancilla post-process now equals the Gibbs entropy of the system pre-process, say *S*.

If ρ_A is not the canonical distribution for the ancilla at entropy *S*, then there is some free energy available by reversibly steering it to that distribution; we may as well suppose this done, so that we extract that free energy together with the energy extracted by the conditional process. (Of course, we may not be able to extract *all* the free energy; but if so, that only reduces the total work extractable.) The end result is that we reversibly transformed a system whose state was initially a product of a canonical distribution at entropy *S* and a pure state, to another system whose state is a product of a pure state and a canonical distribution at entropy *S*.

But we could have done that anyway, without the need for anything so complicated. Suppose we had instead (i) extracted the ancilla's initial free energy by reversibly transforming it into its ground state, and then (ii) run a Carnot cycle between system and ancilla to reversibly cool the system into its ground state and heat up the ancilla. The result is exactly the same and so extracts exactly the same energy – and clearly is in general strictly less efficient than just running the Carnot cycle until system and ancilla reach the same temperature. Other than the possibility of extracting the ancilla's initial free energy, our information-collection and conditional-operation process gains us nothing that we could not gain just by letting the system and ancilla reversibly equilibrate.

(If instead we wanted to return the ancilla to its initial state so as to allow us to use it again to extract energy from another system, we would have to dump its Gibbs entropy into still another system. If, for instance, we have available a heat bath at temperature *T*, doing so would have energy cost *TS*.)

The example is simple, but the point is perfectly general. An ancilla system is a potential thermodynamic resource. Its free energy of equilibration can be extracted by reversibly transforming it to a canonical distribution; having done so, if the resultant distribution has a thermodynamic temperature different from other systems, then further energy can be extracted through a reversible heat engine. Using the ancilla to record the result of measurements on another system and then cool that system is just one, complicated, way of running that heat engine: it will extract net energy if the thermodynamic temperature of the ancilla (that is, the temperature of the canonical distribution to which it can be unitarily transformed) is lower than that of the system, and will actually cost energy if it is higher. In the classic example where the ancilla starts off in a known pure state, its thermodynamic temperature is zero, and systems at absolute zero are genuinely useful resources to extract energy from a finite temperature. But that extraction possibility has nothing really to do with the possibility of gathering information and works only until the ancilla has been brought up to thermal equilibrium with the system.

I have been assuming reversible thermodynamics throughout this section, but the basic points generalize to a coarse-grained thermodynamics. In the latter theory, some of what reversible thermodynamics treats as free energy may after all not be extractable – may not be free according to the coarse-grained theory – and it might be that a specific thermodynamics can extract some part of the free energy only through processes naturally described as "measure and then conditionally act." But it is still the case that any such process can be thought of as extracting some part of the free energy of equilibration of the ancilla, and/or the joint free energy of equilibration of system and ancilla, and that the limits of reversible thermodynamics are upper limits for any thermodynamics.

2.12 The Second Law Again

In my summary of phenomenological thermodynamics, I described the second law in fairly formal terms as the principle that no transition decreases thermodynamic entropy. And we have seen a fairly general derivation of that result in thermodynamic control theory, of which equilibrium thermodynamics is only a special case: if J is a coarse-graining, then the thermodynamic entropy with respect to J is nondecreasing under any J-compatible control process.

Historically and pedagogically, though, one often finds two more directly phenomenological versions of the second law: the Kelvin and Clausius statements, reviewed in Section 2.2. Both have fairly straightforward translations into control-theoretic terminology. Recall from Section 2.8 that a *heat bath* is a system with no adjustable parameters, no conserved quantities other than energy, whose initial state is the canonical state, and which is so large that energy flows into and out of it negligibly affect its temperature, so that we can in practice treat it as being permanently at thermodynamic temperature T. Heat baths are supposed to represent large systems which have achieved thermodynamic equilibrium in an uncontrolled way, so that an agent has no initial information about the system other than that it has equilibrated. (On some interpretations of quantum statistical mechanics, using an exact canonical state to describe the heat bath may reflect an agent's epistemic limitations; on others, it may reflect the fact that a heat bath is uncontrollably entangled with the environment and that its actual quantum state is canonical; it matters for my purposes only that we in fact represent heat baths that way.)

Kelvin Statement (precise control theory version): No control process can extract energy from a single heat bath without in doing so decreasing the free energy of some ancilla system or the joint free energy of an ancilla system and the heat bath.

Clausius Statement (precise control theory version): No control process with no overall energy cost can cause energy to be transferred from a heat bath to another heat bath at a higher thermodynamic temperature, without in doing so decreasing the free energy of some ancilla system or the joint free energy of an ancilla system and one of the heat baths.

Thus stated, both statements are direct consequences of our results so far. The most we can do in any thermodynamic control theory is to extract all the free energy. A heat bath in isolation has no free energy (even in reversible thermodynamics), and so, any process which decreases the energy of a heat bath must do so by either decreasing the free energy of some other system by itself (say, to extract the energy to operate a heat pump) or by decreasing the free energy of the joint system of ancilla-plus-heat bath (say, by running a heat engine between the bath and the ancilla). Similarly, any control process run on two heat baths without ancilla will extract negative energy if it decreases the energy of the lower-temperature system, so we can induce energy transfer from lower to higher temperature only by exploiting free energy provided by some ancilla, either in its own right or in combination with one or other heat bath.

And of course, any free energy provided by an ancilla is a finite resource: once extracted, it is gone. An alternative, and somewhat less formal, version of the two statements would be:

Kelvin Statement (informal control theory version): No indefinitely repeatable control process can extract energy from a single heat bath.

Clausius Statement (informal control theory version): No indefinitely repeatable control process with no overall energy cost can cause energy to be transferred from a heat bath to another heat bath at a higher thermodynamic temperature.

Importantly, neither the Kelvin nor the Clausius statement (unlike the statement that thermodynamic entropy is nondecreasing) makes any reference to the coarse-graining operation: they apply to any thermodynamic control operation, even fully reversible thermodynamics.

2.13 Two Kinds of Maxwell's Demon

To illustrate the difference between the nondecreasing-entropy form of the second law, and the Kelvin and Clausius statements, let's consider Maxwell's famous demon, which (recall) was originally conceived of as a "very observant and neat-fingered being"[18] that could respond to the microscopic state of a gas by opening and closing a partition, and thus violate the second law.

It is useful to distinguish two kinds of demon: a *Maxwell's demon of the first kind* can decrease a system's thermodynamic entropy, while a *Maxwell's demon of the second kind* can bring about a violation of the Kelvin or Clausius statement of the second law.[19] The difference between the two can be seen starkly by considering the practical uses to which they could be put:

- The main use of a demon of the first kind is to refute someone who claims that thermodynamic entropy is nondecreasing.
- The main use of a demon of the second kind is to run the electrical grid off the ambient air temperature, remove any real constraints human civilization faces from the scarcity of usable energy, and become (should you like that kind of thing) the richest person in the world.

Demons of the first kind are not so hard to come by. They exploit the fact that thermodynamic entropy is defined relative to a choice of coarse-graining and is guaranteed to be nondecreasing only under control operations forward compatible with that coarse-graining. So, given the thermodynamic entropy defined by a coarse-graining *J*, all you need to build a demon of the first kind is some *J*-incompatible control operation (and an initial state noninvariant under *J*). For instance:

1. Maxwell's original "neat-fingered being" is a demon of the first kind with respect to equilibrium thermodynamics, provided it has a sufficiently large memory capacity to record the various measurements it needs to make of the system's microstate. Eventually that memory will fill, but until then the demon can certainly lower the temperature of the gas just as Maxwell proposed. (But it could not do so if its memory subsystem was already in a canonical state at the same thermodynamic temperature as the gas.)
2. Given two boxes containing different isotopes of chlorine at the same temperature and pressure, any agent whose control operations can distinguish between isotopes is a demon of the first kind with respect to a control theory that cannot so distinguish them: to an agent using the latter control theory, another agent using the former theory will appear to have achieved a miraculous cooling of the gases. (But they could not achieve it if the two gases were already fully mixed.)
3. In the spin-echo experiment,[19] a system of coupled spins apparently relaxes to thermal equilibrium but then returns to its original state when the time reverse of the original operation is applied. This can be thought of as a demon of the first

[18] This distinction was first made in print by Myrvold (2020: section 8); as he acknowledges, it is drawn from an early version of the present chapter.

[19] Hahn (1953); Rhim, Pines, and Waugh (1971); there is a good conceptually focused discussion in Sklar (1993: 219–222).

kind relative to ordinary equilibrium thermodynamics: the time-reversed operation is not forward compatible with equilibration coarse-graining. (Some of the researchers on spin-echo (Rhim, Pines, and Waugh 1971) described it as a "Loschmidt demon", referring to Loschmidt's famous time-reversal objection to Boltzmann.)

As a slight variant on these, we can consider a demon that can reduce Boltzmann entropy (here it will be helpful, to avoid issues with the quantum measurement problem, to work in classical physics). The Boltzmann entropy has the useful feature that it is a function of the system's microstate and not of its full probability distribution (taking the Boltzmann entropy of a phase-space point to be the Boltzmann entropy – that is, the logarithm of the phase-space volume – of the unique macrostate in which that phase-space point is situated). We saw in Section 2.9 that a control operation that is macrodeterministic for a macrostate M will be Boltzmann-entropy-nondecreasing (since all points in M must be mapped into some fixed macrostate M', and so phase-space volume conservation means that M' has at least as high a volume as M). But even a control theory compatible with Boltzmannian coarse-graining can decrease Boltzmann entropy, provided that it is not macrodeterministic:

(i) A process that decreases Boltzmann entropy on average (that is, decreases the expected value of Boltzmann entropy) might still, through fluctuations, happen to decrease it in a specific instance.

(ii) A sufficiently macro-indeterministic process could even decrease average Boltzmann entropy. We saw in Section 2.9 that the generalized Boltzmann entropy – which is strictly nondecreasing under Boltzmann-compatible control operations – is the sum of two terms: the expected value of the Boltzmann entropy and the Shannon entropy of the probability distribution over macrostates. By spreading the initial distribution very widely across macrostates, a process could increase the latter term enough that the former term might decrease, perhaps significantly. Albert (2000: ch. 5) and Hemmo and Shenker (2012: ch.13) give detailed constructions of demons of this kind.

Of course, for such demons the availability of many macrostates is itself a resource, which can be depleted: at least for systems contained in a finite region, the Shannon entropy cannot increase indefinitely, and when it has reached a maximum any further Boltzmann-compatible control operations must increase expected Boltzmann entropy. (And note that if one's system is not confined to a finite region, it is usually possible to extract all its energy even without a demon: adiabatically expanding a cylinder of gas to an arbitrarily large volume will reduce its temperature to arbitrarily close to absolute zero.)

Demons of the first kind are curiosities. The really interesting possibility would be a demon of the second kind – but this is not a real possibility as long as the underlying dynamics are unitary (or, in the classical regime, Hamiltonian). We saw in Section 2.12 that the Kelvin and Clausius statements are absolute, not relativized to any notion of coarse-graining, and that they follow from unitarity alone. And so, demons of the second kind are in principle impossible. It is not the crude imprecisions of our macroscopic tools that prevent us building the kind of perpetual motion machine that the Kelvin and Clausius statements forbid: it is the fundamental laws of physics.

2.14 Conclusions

This chapter offers a rather general framework in which to discuss various forms of thermodynamics. The notion of a coarse-graining projector, and of a control theory forward compatible and (at least in idealization) forward complete with respect to that projector, provides a powerful method to characterize that control theory without needing to attend to its microphysical details. The framework is broad enough to encompass all the concrete versions of thermodynamic control theory I know, from the various versions of equilibrium thermodynamics to Boltzmannian thermodynamics, to fully reversible thermodynamics.

The framework allows us to characterize a J-compatible control theory in terms of the thermodynamic entropy and free energy with respect to J, and to establish that the most any control operation can do is extract as work the free energy plus any work extractable in equilibrium thermodynamics. It also gives a criterion for when knowing a system's state more precisely is useful: if and only if the new state has lower thermodynamic entropy with respect to J, which is to say: if and only if the coarse-graining of the new state has lower Gibbs entropy than that of the old state. Information is thus useful only where it concerns degrees of freedom over which we have control. At one extreme, in equilibrium thermodynamics the entropy is set only by parameters and conserved quantities, and fine-grained information is useless; at the other extreme, in reversible thermodynamics any information is exploitable.

For a J-compatible control theory, there is a J-dependent specification of the second law in its entropy-is-nondecreasing form: the thermodynamic entropy S^J with respect to J is nondecreasing under J-compatible control operations. Assuming that the control theory contains at least some S^J-increasing control operations, this version of the second law characterizes the way in which the theory is irreversible: entropy can go up but never down. But a more powerful control theory might contain J-incompatible control operations and might very well be able to decrease S^J. This creates a general recipe for building Maxwell's demons of the first kind, which can decrease entropy under a given definition: all they need to do is to be able to carry out control operations not adapted to that definition.

But there are other forms of the second law – the Kelvin and Clausius statements – that do not depend on J and which hold in any control theory, with or without irreversibility. It is these versions of the second law that best express the hard constraints Nature puts on our ability to get work out from systems in the wild, and no Maxwell's demon of the second kind, which would allow us to bypass them, can be built.

As a closing comment, note that the crucial feature of the dynamics that underwrite the microphysical justification of the second law (in any of its forms) is that they preserve Hilbert-space dimension or phase-space volume. These are the characteristic features of *reversibility*, the natural generalizations to their respective continuum systems of the discrete-system idea that dynamical maps are 1:1. Genuinely irreversible (and deterministic) operations – operations that map many Hilbert-space vectors to the same vector, or phase-space regions to smaller regions – would, at least if combined with a sufficiently large class of reversible operations, make it straightforward to extract all of a system's energy and leave it in its ground state. The second law is often held up as the example *par excellence* of irreversibility: it is ironic that it is reversibility that provides its microphysical foundation.

Acknowledgments

I am grateful for conversations on earlier versions of this material with too many people to name over a number of years. Special thanks are due to David Albert, Owen Maroney, and Wayne Myrvold. I am also grateful to John Norton for help in locating the authoritative source for Einstein's famous quote about thermodynamics, and to Wayne Myrvold and Katie Robertson for helpful comments on an earlier draft.

Appendix: Properties of the Canonical Distribution

Theorem: Suppose $\mathcal{H}$ is a separable Hilbert space of dimension>1 and H is a self-adjoint operator on $\mathcal{H}$ (the Hamiltonian) with discrete spectrum whose lowest eigenvalue is nondegenerate (and whose expectation value I write as $\langle H \rangle$, or $\langle H \rangle_\rho$ when the specific density operator ρ needs to be stated). For each $\beta > 0$, define the canonical density operator $\rho_C(\beta)$ by

$$\rho_C(\beta) = \frac{e^{-\beta H}}{Z(\beta)} \tag{2.57}$$

where

$$Z(\beta) = \mathrm{Tr} e^{-\beta H} \tag{2.58}$$

Let $S(\beta) = S_G[\rho_C(\beta)]$ be the Gibbs entropy of $\rho_C(\beta)$ and $U(\beta) = \langle H \rangle_{\rho_C(\beta)}$. Then:

i. Both $U(\beta)$ and $S(\beta)$ are decreasing functions of β.
ii. The range of $S(\beta)$ is the interval $(0, ln \dim H)$ (taking $ln \dim H = \infty$ if H is infinite-dimensional) and if $s \in (0, ln \dim H)$, there is a unique β such that $s = S(\beta)$.
iii. The range of $U(\beta)$ is some interval (E_0, E_{max}), where E_0 is the lowest eigenvalue of H, E_{max} may or may not be infinite, and for any u in this range, there is a unique β such that $U(\beta) = u$.
iv. For any $u \in (E_0, E_{max})$, $\rho_C(U^{-1}(u))$ is the unique global maximum–Gibbs entropy density operator with $\langle H \rangle = u$.
v. For any $s \in (0, \ln \dim \mathcal{H})$, $\rho_C(S^{-1}(s))$ is the unique global minimum-$\langle H \rangle$ density operator with Gibbs entropy s.

Proof:

(i) Differentiate U to obtain

$$U'(\beta) = -(\langle H^2 \rangle - \langle H \rangle^2), \tag{2.59}$$

which is strictly negative (unless H is a constant, which it cannot be, as an operator with nondegenerate lowest eigenvalue on a Hilbert space of dimension > 1). Then observe that

$$S(\beta) = \ln Z(\beta) + \beta U(\beta), \tag{2.60}$$

and differentiate to obtain $S'(\beta) = \beta U'(\beta)$.

(ii), (iii) Let the eigenstates of H, labeled in increasing order of eigenvalue, be $|0\rangle, |1\rangle \ldots$. Then

$$\lim_{\beta \to \infty} \rho_C(\beta) = |0\rangle\langle 0| \tag{2.61}$$

so that

$$\lim_{\beta \to \infty} S(\beta) = 0 \tag{2.62}$$

and

$$\lim_{\beta \to \infty} U(\beta) = E_0 \tag{2.63}$$

Taking limits in (2.60) gives

$$\lim_{\beta \to \infty} S(\beta) = \ln \dim \mathcal{H} \tag{2.64}$$

and we define

$$E_{max} = \lim_{\beta \to \infty} U(\beta). \tag{2.65}$$

Uniqueness follows in each case because S and U are strictly decreasing.
(iv) Let $\mathcal{D}$ be the set of density operators on $\mathcal{D}$, and let $\rho \in \mathcal{D}$. Consider an infinitesimal variation $\rho \to \rho + \delta\rho$. The variations of Gibbs entropy, $\langle H \rangle$, and trace, respectively, are, to second order,

$$\delta S_G = -\mathrm{Tr}(\{\ln\rho + \mathrm{id}\}\delta\rho) - \frac{1}{2}\mathrm{Tr}(\delta\rho^2\rho^{-1}) + o(\delta\rho^2), \tag{2.66}$$

$$\delta\langle H \rangle = \mathrm{Tr}(H\delta\rho), \tag{2.67}$$

and

$$\delta\mathrm{Tr}\rho = \mathrm{Tr}\partial\rho. \tag{2.68}$$

Firstly, we establish that the stationary points of the Gibbs entropy under variations that leave $\langle H \rangle$ constant are exactly the canonical states (though possibly with $\beta < 0$). For ρ is a stationary point under variations that conserve trace and $\langle H \rangle$ if and only if the first-order variation of S_G is a linear combination of the first-order variations of $\langle \mathrm{H} \rangle$ and $\mathrm{Tr}\rho$ (this is the method of Lagrange multipliers). This forces

$$\mathrm{Tr}(\{-\ln\rho - \mathrm{id} + \beta H + \alpha\mathrm{id}\}\delta\rho) = 0 \tag{2.69}$$

for some α, β and all $\delta\rho$, and we can solve to get $\rho = \rho_C(\beta)$ (where again β need not be positive).

Secondly, we establish that ρ_C is a local *maximum* of S^G under these variations. This follows from (66): the first-order variation vanishes under the constraint that trace and $\langle H \rangle$ are invariant, and $\mathrm{Tr}(\delta\rho^2\rho^{-1})$ is a positive-definite quadratic function of $\delta\rho$, as can be verified explicitly by expanding in a basis of energy eigenstates: if $\delta\rho = \sum_{n,m} \delta\rho_{nm}$, then (writing $H|n\rangle = E_n|n\rangle$)

$$\mathrm{Tr}\left(\delta\rho^2\rho_C^{-1}(\beta)\right) = \frac{1}{Z(\beta)}\sum_{m,n} \delta\rho_{mn}e^{\beta E_n}\partial\rho_{mn} = \frac{1}{Z(\beta)}\sum_{m,n} |\delta\rho_{mn}|^2 e^{\beta E_n} \tag{2.70}$$

Finally, we establish that $\rho_C(U^{-1}(u))$ is a global maximum of entropy for $\langle H \rangle = u$. For suppose otherwise; then there would be another density operator ρ_0 with $S(\rho_0 0) \geq \rho_C(U^{-1}(u))$, and then the function

$$\rho(x) = x\rho_C(U^{-1}(u)) + (1-x)\rho_0 \tag{2.71}$$

would satisfy

$$S^G[\rho(x)] > S^G[\rho_C(U^{-1}(u)] \tag{2.72}$$

for $0 < x < 1$, which contradicts the claim that $\rho_C(U^{-1}(u))$ is a local maximum.

(v) Fix $\beta = S^{-1}(s)$, and suppose for contradiction that there is a density operator ρ with $\langle H\rangle_\rho \leq \langle H\rangle_{\rho_{c(\beta)}}$ and $S^G[\rho] = S^G[\rho_C(\beta)]$. Note firstly that by (iv) we must have $\langle H\rangle\rho < \langle H\rangle_{\rho_C(\beta)}$, since otherwise $\rho_C(\beta)$ would not be the unique maximum-entropy density operator at $\langle H\rangle = \langle H\rangle\rho_c(\beta)$. But then, again by (iv), there must exist β' such that $S^G[\rho_C(\beta')] \geq S^G[\rho]$; by (i) $\beta' > \beta$. But then $S^G[\rho_C(\beta')] \geq S^G[\rho_C(\beta)]$, in contradiction with (i).

References

Albert, D. Z. (2000). *Time and Chance*. Cambridge, MA: Harvard University Press.

Bennett, C. H. (2003). "Notes on Landauer's principle, reversible computation, and Maxwell's Demon." *Studies in the History and Philosophy of Modern Physics 34*: 501–510.

Blundell, S. J., and K. M. Blundell (2010). *Concepts in Thermal Physics* (2nd ed.). Oxford: Oxford University Press.

Bub, J. (2002). "Maxwell's demon and the thermodynamics of computation." *Studies in the History and Philosophy of Modern Physics 32*: 569–579.

Bustamante, C., J. Liphardt, and F. Ritort (2005). "The nonequilibrium thermodynamics of small systems." https://arxiv.org/abs/cond-mat/0511629.

Callender, C. (2001). "Taking thermodynamics too seriously." *Studies in the History and Philosophy of Modern Physics 32*: 539–553.

Collin, D., F. Ritort, C. Jarzynski, S. B. Smith, I. Tinoco, and C. Bustamante (2005). "Verification of the Crooks fluctuation theorem and recovery of RNA folding free energies." *Nature 437*: 231–234.

Crooks, G. E. (1998). "Nonequilibrium measurements of free energy differences for microscopically reversible Markovian systems." *Journal of Statistical Physics 90*: 1481–1487.

Demerast, H. (2016). "The universe had one chance." *Philosophy of Science 83*: 248–264.

Earman, J., and J. Norton (1999). "EXORCIST XIV: The wrath of Maxwell's demon. Part II: From Szilard to Landauer and beyond." *Studies in the History and Philosophy of Modern Physics 30*: 1–40.

Einstein, A. (1949). "Autobiographical notes." In P. A. Schilpp (ed.), *Albert Einstein: Philosopher-Scientist* New York: MJF Books, pp. 1–95. English translation by Paul Arthur Schilpp.

Goldstein, S. (2001). "Boltzmann's approach to statistical mechanics." In J. Bricmont, D. Dürr, M. Galavotti, F. Petruccione, and N. Zanghi (eds.), *Chance in Physics: Foundations and Perspectives*, Berlin: Springer, p. 39. Available online at http://arxiv.org/abs/cond-mat/0105242.

Gour, G., M. P. Müller, V. Narasimhachar, R. W. Spekkens, and N. Yunger Halpern (2015). "The resource theory of informational nonequilibrium in thermodynamics." *Physics Reports 583*: 1–58.

Hahn, E. (1953). "Free nuclear induction." *Physics Today 6*: 4–9.

Hemmo, M., and O. Shenker (2012). *The Road to Maxwell's Demon: Conceptual Foundations of Statistical Mechanics*. Cambridge: Cambridge University Press.

Ismael, J. (2009). "Probability in deterministic physics." *Journal of Philosophy 106*: 89–108.

Jarzynski, C. (1997). "Nonequilibrium equality for free energy differences." *Physical Review Letters 78*: 2690–2693.

Kittel, C., and H. Kroemer (1980). *Thermal Physics* (2nd ed.). New York: W. H. Freeman.

Ladyman, J., S. Presnell, A. Short, and B. Groisman (2007). "The connection between logical and thermodynamic irreversibility." *Studies in History and Philosophy of Modern Physics 38* (1): 58–79.

Ladyman, J., and K. Robertson (2013). "Landauer defended: reply to Norton." *Studies in History and Philosophy of Modern Physics 44*: 263–271.

Ladyman, J., and K. Robertson (2014). "Going round in circles: Landauer vs. Norton on the thermodynamics of computation." *Entropy 16*: 2278–2290.

Landau, L., and E. Lifshitz (1980). *Statistical Physics* (3rd ed.). Amsterdam: Elsevier. English translation by J. B. Sykes and M. J. Kearsley.

Leff, H., and A. F. Rex (2002). *Maxwell's Demon: Entropy, Information, Computing* (2nd ed.). Bristol: Institute of Physics Publishing.

Lewis, D. (1980). "A subjectivist's guide to objective chance." In R. C. Jeffrey (ed.), *Studies in Inductive Logic and Probability, volume II*. Berkeley: University of California Press. Reprinted, with postscripts, in David Lewis, *Philosophical Papers*, Volume II (Oxford: Oxford University Press, 1986); page numbers refer to this version.

Loewer, B. (2002). "Determinism and chance." *Studies in the History and Philosophy of Modern Physics 32*: 609–620.

Maroney, O. (2007). "The physical basis of the Gibbs–von Neumann entropy." https://arxiv.org/abs/quant-ph/0701127.

Maxwell, J. C. (1867). Letter to P. G. Tait, 11 December 1867. Reprinted in C. G. Knott, *Life and Scientific Work of Peter Guthrie Tait* (London: Cambridge University Press, 1911), 213–215.

Maxwell, J. C. (1871). *Theory of Heat*. London: Longmans, Green and Co.

Myrvold, W. C. (2011). "Statistical mechanics and thermodynamics: A Maxwellian view." *Studies in History and Philosophy of Modern Physics 42*: 237–243.

Myrvold, W. C. (2020). "The science of $\theta\Delta^{cs}$." *Foundations of Physics 50*: 1219–1251.

Myrvold, W. C. (2024). "Shakin' all over: Proving Landauer's principle without neglect of fluctuations." *The British Journal for the Philosophy of Science* 75 (3): 587–616.

Norton, J. D. (2005). "Eaters of the lotus: Landauer's principle and the return of Maxwell's demon." *Studies in the History and Philosophy of Modern Physics 36*: 375–411.

Norton, J. D. (2011). "Waiting for Landauer." *Studies in History and Philosophy of Modern Physics 42*: 184–198.

Norton, J. D. (2013a). "Author's reply to Landauer defended." *Studies in History and Philosophy of Modern Physics 44*: 272.

Norton, J. D. (2013b). "The end of the thermodynamics of computation: A no-go result." *Philosophy of Science 80*: 1182–1192.

Pusz, W., and S. Woronowicz (1978). "Passive states and KMS states for general quantum systems." *Communications in Mathematical Physics* 58: 273–290.

Rhim, W.-K., A. Pines, and J. Waugh (1971). "Time-reversal experiments in dipolar-coupled spin systems." *Physical Review B 3*: 684–696.

Saunders, S. (2018). "The Gibbs paradox." *Entropy 20*: 552.

Sklar, L. (1993). *Physics and Chance: Philosophical Issues in the Foundations of Statistical Mechanics*. Cambridge: Cambridge University Press.

Throne, K. S., and R. D. Blandford (2017). *Modern Classical Physics: Optics, Fluids, Plasmas, Electricity, Relativity, and Statistical Physics*. Princeton: Princeton University Press.

Tolman, R. C. (1938). *The Principles of Statistical Mechanics*. Oxford: Oxford University Press.

Wallace, D. (2012). *The Emergent Multiverse: Quantum Theory According to the Everett Interpretation*. Oxford: Oxford University Press.

Wallace, D. (2014). "Thermodynamics as control theory." *Entropy 16*: 699–725.

Wallace, D. (2016). "Probability and irreversibility in modern statistical mechanics: Classical and quantum." To appear in D. Bedingham, O. Maroney, and C. Timpson (eds.), *Quantum Foundations of Statistical Mechanics* (Oxford University Press, forthcoming). Preprint at https://arxiv.org/abs/2104.11223.

Wallace, D. (2020). "The necessity of Gibbsian statistical mechanics." In V. Allori (ed.), *Statistical Mechanics and Scientific Explanation: Determinism, Indeterminism and Laws of Nature*. Singapore: World Scientific.

Yunger Halpern, N., and J. M. Renes (2016). "Beyond heat baths: Generalized resource theories for small-scale thermodynamics." *Physical Review E 93*: 022126.

3

Thermodynamics without Time

FRANCESCA VIDOTTO

Thermodynamics developed on the basis of the classical prerelativistic notions of absolute space and absolute time. With the revolution brought about by general relativity, these absolute notions have been understood not to be general enough to describe the entire world accessible to our experience, and a novel way conceptualizing space and time has developed. This development, and especially the revision of the notion of time, has an impact on the foundation of the thermodynamics. It raises the question: to what extent can thermodynamics be extended or modified to remain applicable in regimes where the classical, prerelativistic conception of absolute time loses its coherence?

This question is only partially addressed in the theoretical physics literature. The extension of thermodynamics to special relativity has been developed by numerous authors, and is not particularly troublesome, besides perhaps on terminological disputes regarding how temperature transforms under Lorentz transformation. The extension of thermodynamics to the case of the thermal behavior of material systems on a fixed gravitational field (the gravitational analog of the thermodynamics of charged matter in a given electromagnetic field) also is quite well understood (Tolman 1934).

What instead remains largely open is the thermodynamics of the gravitational field itself (the gravitational analog of the thermodynamics of the electromagnetic field, namely the black-body radiation problem). In nature, indeed, there is no reason for thermal energy not to dissipate also into gravitational degrees of freedom. The very spacetime geometry in which we live is thus a dynamical field with thermal properties. For these, we do not have yet a theoretical understanding. Part of the difficulty lies in the fact that gravity is attractive and long-range. Objects that interact gravitationally, even in the nonrelativistic approximation, are challenging to describe with standard techniques. But the root of the difficulty is much deeper than this, as discussed in this chapter. In the physics literature, the thermodynamics of the gravitational field are largely treated haphazardly in special cases only, such as for black holes, in the complete lack of a foundation or a general theory.

Here I illustrate the main reasons underpinning the current confused state in the foundation of relativistic thermodynamics, and some ideas that have been put forward to resolve them. The core of the problem is that the notion of absolute time used in the foundation of classical thermodynamics is not anymore available in general relativity. Therefore, I first briefly recall the reasons why this notion is not available in the context of relativistic gravitation (Section 3.1). I then illustrate a specific physical phenomenon, the Tolman effect, that clearly shows that conventional thermodynamics need a foundational revision and provide several useful hints for possible alternative foundational ideas (Section 3.2). I then discuss the notion of thermal time, which could be a key idea pointing to a possible solution of the problem (Section 3.3) and the peculiar way in which the notion of equilibrium could be recovered in a relativistic context, not as a property of two systems, but rather a property of three systems (Section 3.4). I close in Section 3.5 with some general remarks on the conceptual novelties needed for gravitational thermodynamics.

3.1 Notions of Time

Time enters in both the first and the second principles of thermodynamics: energy is conserved in time and entropy does not decrease in time. When thermodynamics were developed, time was understood as Newton's absolute time. This conception is characterized by its uniformity in space – meaning it is the same at all locations – and in time. This is a key aspect that distinguishes it from the time of general relativity, which is dynamical, affected by matter, and which flows at different rates in different locations, as witnessed by the fact that in a gravitational field in general two clocks that meet twice measure two different amounts of time lapsed between the two encounters.

It is perhaps interesting to observe, here, that the Newtonian notion of time, so ingrained in our modern way of thinking, is not necessarily the most primitively intuitive, nor was it common before Newton. Aristotle defined time as a measure of change, and such a way of understanding time remained dominant until Newton. If time is defined as a measure of change, it does not (unlike Newtonian time) lapse when nothing changes. This definition of time, as we shall see, remains consistent within relativistic physics; it is compatible with a relational notion of change through the use of clocks, and even gets a very concrete physical interpretation in the properties of thermal time, illustrated in this chapter.

In general relativity, the notion of time is radically altered in a number of ways. First, time cannot be separated from space; it is conceived as an aspect of the unified entity called *spacetime*. The radical conceptual novelty introduced by Einstein with general relativity is the identification of this spacetime with the gravitational field. Since all fields in physics have a dynamic nature, this time is also in a sense dynamical. The rate at which it lapses (as defined by a clock) is affected by these dynamics. For instance, it is modified by the vicinity of masses.

In this rich conceptual scheme, the Newtonian monolithic notion of time breaks up into a variety of notions, only loosely connected to one another: coordinate time, proper time along world line, clock time, asymptotic time . . . In the nonrelativistic flat-space limit, all these notions agree, but in general they are quite distinct notions, none of which retains the full properties of Newtonian time. Newtonian time is today understood as an approximation: in reality, there are no preferential variables that constitute an absolute time; at most there are particular physical situations where one variable plays the role of Newtonian time, but with a limited scope.

The Hamiltonian formulation of the theory makes this situation particularly vivid. A nonrelativistic system admits a natural Hamiltonian formulation in which, in phase space, the evolution of the system in time is governed by (is the flux of) a Hamiltonian. Not so for general relativistic systems. These can only be naturally formulated in a generalized Hamiltonian formalism, where the dynamics are expressed by a constraint on the phase space: on the subspace where it vanishes, the constraint generates (its flux defines) a relation between variables.

If dynamics are understood as a relation between variables, then all variables are a priori on the same footing and many different clocks can play the role of independent variable that Newtonian time used to play. This relational understanding of temporality is perfectly fine for describing the world, and in fact general relativity does describe the world very well. But how do we do thermodynamics in a theory that refuses to provide us a preferred time variable?

Consider in particular the statistical underpinning of thermodynamics. We understand thermodynamic states as statistical distributions of microstates. In a gravitational context, this appears to imply that a thermodynamic state of the gravitational field (say the gravitational analog of the electromagnetic black-body distribution) is actually a statistical mixture of spacetimes. In each of these clocks, it moves at a different rate. So, how to identify the global time variable needed for classical thermodynamics? Thermal fluctuations in the gravitational field affect the very rate of the clocks. Extending statistical mechanics to effectively describe the thermal fluctuations of spacetime is an open problem in physics (Rovelli 2013, Chirco and Josset 2017). Even neglecting quantum phenomena.

Quantum physics introduces further complications. In the quantum theory, clocks become quantum systems themselves (Rovelli 2015). They can be in superposition. Superposition of clocks implies superposition of the causal structures they define. In what follows I shall ignore any quantum phenomenon and just focus on the foundation of thermodynamics and statistical mechanics in the classical (nonquantum) relativistic gravitational context.

Perhaps surprisingly, this is a problem whose solution is even further off than quantizing gravity! We have consistent and reasonable tentative theories of quantum gravity (Rovelli and Vidotto 2014), that have not been confirmed empirically, but have a clear conceptual foundation (Rovelli and Vidotto 2022) and might be right. We do not have a consistent and reasonable tentative theory of statistical gravity with a clear conceptual foundation.

3.2 Tolman–Ehrenfest effect

At the very core of thermodynamics, there is the notion of equilibrium. This notion is based on the observation that if two bodies, A and B, at different temperature can exchange energy, they do so in such a way as to converge to a common temperature. They equilibrate. At equilibrium, the temperature of the two bodies is the same: $T_A = T_B$. Some authors have emphasized this basic fact by giving it the name of "Zeroth Principle of Thermodynamics." It is a basic fact stated at the beginning of any thermodynamics textbook, and it is very clear.

Except that in a gravitational field this is not true. Consider a concrete situation: take a vertical cylinder filled with gas or a fluid. Call the lower part of the cylinder A and the upper part B, then move a thermometer at the bottom and at the top, and check whether it registers the same temperature. It does not. Rather, it measures a tiny temperature difference $\Delta T/T = gh/c^2$, where g is the Galileo's gravitational acceleration ($\sim 9.8\ \mathrm{m/s^2}$), h is the height difference, and c is the speed of light.

This phenomenon is called the Tolman effect. It was first explained by Richard Tolman's 1930 paper "On the Weight of Heat and Thermal Equilibrium in General Relativity" (Tolman 1930) and then generalized to wider contexts in collaboration with Ehrenfest (Tolman and Ehrenfest 1930). The phenomenon has never been directly measured because on Earth it is very small, but it has been derived in a variety of ways and is universally considered credible.

It is a genuinely relativistic phenomenon (it disappears in the c to infinity limit) and it is a genuine gravitational phenomenon: it disappears if the limit to which Newton's constant (hence g) goes is taken to zero.

The derivations of the phenomenon are numerous, but none relies on first principles. Each of them is, so to say, a patchwork of different chunks of physics. Here is one: Temperature is inversely proportional to the derivative to the entropy with respect to the energy: $\frac{dS(E)}{dE} = \frac{1}{kT}$, where κ is the Boltzmann constant. This allows us to understand temperature in terms of the microstates of a given system, since the entropy $S(E)$ is the logarithm of the number of microstates $N(E)$ that have energy E. Say two bodies can exchange energy so that a small amount of energy dE can exit one, and the same amount of energy enter the other. When this happens, the entropy of each changes. Say equilibrium is when the total entropy, which is the sum of the two, is at a maximum; this immediately implies that the two temperatures must be the same, because the two derivatives with respect to dE (see Figure 3.1) must be the same.

However, in a gravitational field the energy that exits one body is not the same as the energy that enters the other! The reason is that energy is like mass: it is attracted by gravity, hence it falls. In a relativistic context, that is, energy has weight. Hence it acquires energy by falling, like any mass does. Hence if some energy dE exits the upper body, the energy that enters the lower body would be larger. Hence the maximal entropy is not attained when the two derivatives are equal. The energy increase in falling by a bit of energy dE can be easily estimated; for instance, by thinking that it is carried by a photon. A photon is blue-shifted in falling (this is a general relativistic effect) with a change in frequency $\Delta\nu/\nu = gh/c^2$. Since the energy of a photon is proportional to its frequency, it changes its energy in falling (see Figure 3.2), and therefore the two temperatures have to be off by this same factor in order for entropy to be maximal and entropy to be reached.

All this is clear and fully convincing. Yet, the conclusion is radical: In a gravitational field, equilibrium does not mean that temperature (as measured by the same thermometer) is uniform. The zeroth principle of thermodynamics collapses.

A more precise statement of Tolman law is that in a static gravitational field described by a Newton potential $\Phi(x)$, equilibrium is given by

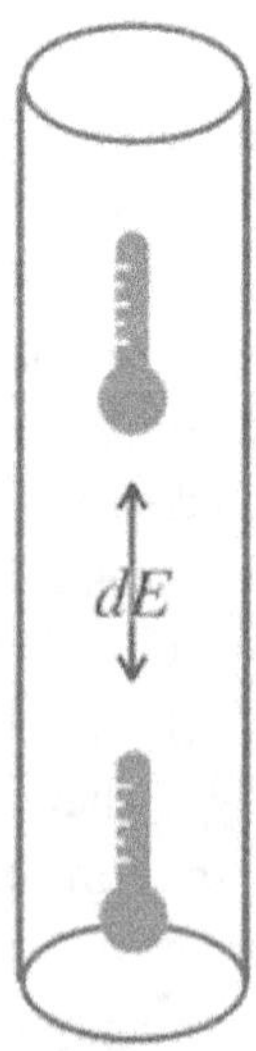

Figure 3.1

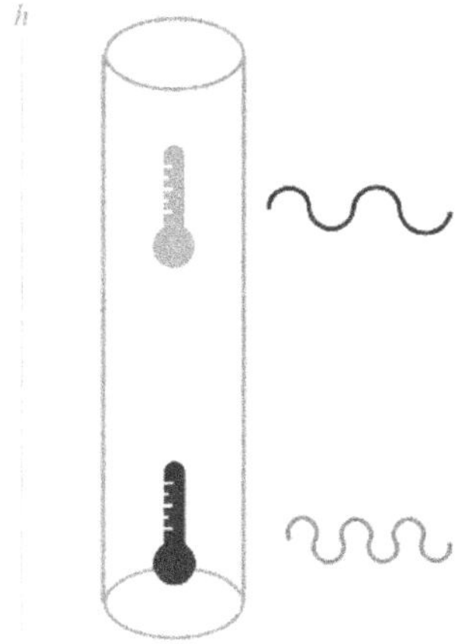

Figure 3.2

$$T_A\left(1+\frac{\Phi(x_A)}{c^2}\right)=T_B\left(1+\frac{\Phi(x_B)}{c^2}\right), \tag{3.1}$$

an equilibrium that is given by equality of the temperature at those positions multiplied by the red-shift factor determined by the gravitational potential.

This observation can be extended (Chirco et al. 2013). In a generic gravitational field, we can talk about equilibrium only if there is a sense in which spacetime is stationary. This is when there is a time-like Killing field ξ. (A Killing field is a vector field along whose integral lines the metric remains the same.) In this case, the Tolman equilibrium equation generalizes to $T|\xi| = const$, namely: what is constant at equilibrium is the product of the temperature times the norm of the Killing field.

In the light of this, one could of course define a notion of *proper temperature*, which is the temperature measured by a thermometer scaled by the red-shift factor. This is the actual *temperature* whose uniformity signals equilibrium. But it is not the quantity measured by the devices we call thermometers.

The Tolman relation, thus, shows an important property of the temperature (measured by thermometers): temperature is defined in a way that depends on the local time flow. Since in relativity Newtonian time is replaced by multifinger time, that is, by clocks, temperature becomes dependent on these, hence depends on space. The exact way in which this works is illustrated in the next section.

In any case, the Tolman effect makes clear that the standard nonrelativistic way of thinking about thermodynamics does not generalize to relativistic gravity. However, it also points us toward an intriguing direction for a different foundation of thermodynamics. This is illustrated in the following section.

3.3 Thermal Time

The concept of thermal time was introduced by Carlo Rovelli in 1993 (Rovelli 1993) and extended to the full context of quantum field by him in collaboration with Alain Connes (Connes and Rovelli 1994) using the Tomita–Takesaky modular theory. In this discussion, we are not interested in the extension to the quantum domain, hence we only briefly mention the definition, in order then to focus on the foundations of classical thermodynamics. Let's illustrate the notion of thermal time, first in the classical case.

Given any statistical state ρ, represented as a distribution over phase space, we can define its *thermal Hamiltonian* to be the $H_\rho = -ln\rho$ and thermal time τ as the parameter of the flow that the thermal Hamiltonian generates on phase space. For later convenience, let us rather define it as a dimensionless parameter by dividing it by $\hbar$. Here $\hbar$ is only used as a unit of action, so far with no physical reason. That is, given any observable function A, we consider the flow

$$\frac{dA(\tau)}{d\tau} = \hbar\{A, -ln\rho\}. \tag{3.2}$$

(In the general quantum case, the thermal time is the parameter of the Tomita flow of the state, seen as a function in the algebra.) In the special case in which the state is the Boltzmann–Gibbs state $\rho \sim e^{-\beta H}$, where $\beta = 1/kT$, we have clearly that

$$\tau = \frac{k}{\hbar} Tt. \tag{3.3}$$

That is, in this case the thermal time is the usual time scaled by the temperature.

Now, let's see what the thermal time in the case of the Tolman effect is. Locally, it is proportional to clock time, but the proportionality constant changes with the temperature, namely, it changes from point to point. But it does so, and precisely in the same manner, as shown by the Tolman low, the parameter that labels the flow generated by the Killing field. In other words, the thermal time is precisely the time flow along which the system is in equilibrium.

This is already a nice result, but there is much more: temperature is the ratio between the thermal time (namely the time along which there is equilibrium) and the clock time! This surprising result suggests a very compelling interpretation of temperature: it is the ratio between thermal time and proper time. Equivalently, it is the speed at which clocks move, in the time along which there is equilibrium. Temperature is, we can say, the speed of time (Rovelli and Smerlak 2011). This will become clearer in a moment.

In fact, there is even more, because the thermal time can be given a very concrete physical interpretation. For this, fix a unit of action. For instance, in view of quantum theory and the Heisenberg principle, consider units of action of size $\hbar$. Accordingly, consider cells in phase-space of size $\hbar$ per degree of freedom. Now consider a statistical state concentrated in one such phase-space cell. Call t_o the average time it takes this state, under the dynamical flow, to move from one such cell to the next one. The same quantity can be defined in quantum theory by asking what is the time t_o that it takes a quantum state $\psi(0)$ to evolve to a state $\psi(t_o)$ such that $|\langle\psi(0)|\psi(t_o)\rangle| \ll 1$. In Haggard and Rovelli (2013), this quantity is computed in the case of a thermal context and shown to be universal and only depending on the temperature:

$$t_o = \frac{\hbar}{kT}. \tag{3.4}$$

This in turn implies that the thermal time lapsed along a motion is nothing else than the number of such elementary steps t_o along such motion:

$$\tau = \frac{t}{t_o} = \frac{k}{\hbar}Tt. \tag{3.5}$$

Let's summarize. The system jumps from cell to cell in phase-space. If we take the cells as being of size $\hbar$, the system on average jumps τ cells during the clock time t. Thermal time counts the jumps. Temperature is the number (in natural units) of cells the system jumps per unit of clock time.

Notice how all of this fits with the Tolman phenomenology: thermal time is defined globally. Not so clock time, because its rate is affected locally by gravity. Hence the ratio between the two, namely temperature, changes locally.

And finally, most spectacularly: two systems are in equilibrium along a motion, in general, not if the temperature is the same, but if the number of elementary cells they have jumped along the motion is the same!

Expressing this in terms of quantum information, we discover that thermal time counts a number of states. Here, the fundamental quantum discreteness plays a central role: thermal time counts how many cells of size $\hbar$ are present in phase space.

This way of viewing equilibrium shifts the language from that of energy to that of information, in the sense of number of states. We can say that equilibrium is when the information flow between the two systems vanished, where the information flow is the difference between the number of states of one system "seen" by the other during the common motion. We can then switch to a fully informational language. The general covariant condition for equilibrium is given by a zero-information flux. The two subsystems are in interaction with one another: each system has access to the amount of information, in the sense of Shannon given by the number of states, of the other system it is exposed to during a process. Equilibrium is achieved when each of the two systems "sees" the same number of states in the other system, that is, when the value of the thermal time coincides for the two systems. Hence, equilibration is a process that selects a preferred common time. Notice that, conceived in this manner, equilibrium is not defined among states: it is defined among *processes*.

This is a hint for generalizing thermodynamics. While energy is a concept that in general relativity becomes ambiguous, information, counting of jumps, might provide a viable way to express thermodynamical quantities in a general relativist context with no preferred time variable.

3.4 Equilibrium in (Dynamical) General Covariant Systems

The previous two sections dealt with equilibrium of matter in a given gravitational field. They illustrated some concepts and ideas that play a role in relativistic thermodynamics in the presence of gravity, but they did not address the problem of the thermodynamics of the gravitational field.

Let's get back to the question of why this is difficult. For this, let's backtrack again to the statistical underpinning of thermodynamics. In standard thermodynamics, dynamics is given by a Hamiltonian on a phase-space generating time evolution. Statistical mechanics can be derived from an assumption about a probability distribution on phase-space invariant under these dynamics.

As mentioned, general relativity does not fit into this Hamiltonian formalism. Rather it fits into a *covariant* extension of this formalism, which I sketch in what follows.

The (so-called *extended*) phase-space is again a symplectic space, but the dynamics are not given by a Hamiltonian. Rather, they are given by a function C on phase-space called constraint. On the surface $C = 0$, the motions generated by C itself are one-dimensional lines. Each of these lines is a motion. A motion can be arbitrarily parameterized with a parameter t, but this parameter is (indeed) arbitrary and in general does not correspond to anything physical or observable. Along the motion, functions $(A, B, C, \ldots)$ on phase space take values $\big(A(t), B(t), C(t), \ldots\big)$. Each such function $(A, B, C, \ldots)$ is associated with an operational procedure and can be observed. They are denoted "partial observables" (Rovelli 2002). What the theory predicts, and can be compared with experiments, is the relative dependence of these partial observables from one another, for instance $A(B)$, obtained by solving t away from $\big(A(t), B(t)\big)$. The space of the motions is sometimes called the *physical phase-space*.

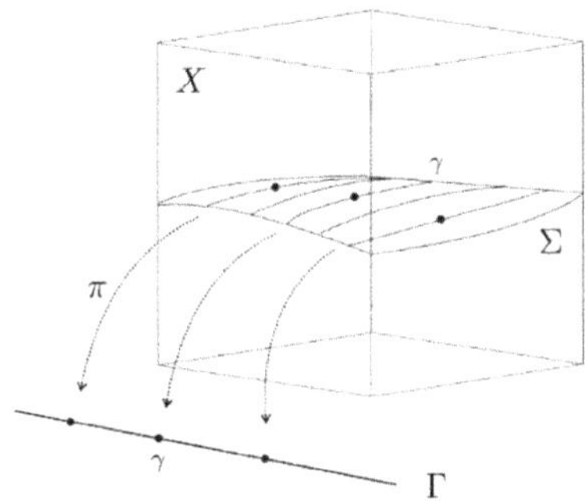

Figure 3.3 From Chirco et al. (2016) with permission from IOP Publishing Ltd

The connection between this general formalism and the special case of nonrelativistic Hamiltonian mechanics is simple. The special case is obtained when one of the partial observables is identified with the Newtonian time T and C has the special form $C = p_T + H$, where p_T is the variable canonical conjugate to T, and H is the usual Hamiltonian. In the general case, different partial variables can be identified with different kinds of clocks, as is the case in general relativity, where different real clocks measure different time lapses between the same couple of events.

The open question of general relativistic thermodynamics is how to define a useful statistical mechanics and thermodynamics, given only the generalized "covariant" Hamiltonian formalism sketched earlier. Some conceptual steps in this direction have been taken in Chirco et al. (2016); see Figure 3.3. I illustrate here the central idea. Consider a given partition of the system into two subsystems, such that the extended phase-space can be seen as the Cartesian product of two phase-spaces. One of the two is interpreted as a clock, the other is a physical system evolving in terms of the clock time. The resulting construction, illustrated in the following, depends on the choice of this partition.

In nonrelativistic statistical mechanics, we can consider ergodic systems. For these, we can write an equilibrium measure v on phase-space by requiring that the time average of an observable f along any motion $s(t)$ is equal to the phase-space average of the observable computed with this measure:

$$\lim_{T\to+\infty}\frac{1}{2T}\int_{-T}^{+T} f\Big(s(t)\Big)\,dt = \int_V f dv \tag{3.6}$$

This notion of ergodicity can be generalized to the relativistic case as follows. First, choose a one-form $d\theta$ on the phase-space of the clock system. Then, given a finite motion γ, we can define the time average of an observable f along this motion by

$$\bar{f}(\theta,\gamma) \underset{def}{=} \frac{\int_\gamma f\theta}{\int_\gamma \theta} \tag{3.7}$$

and say that a measure on the phase-space is an equilibrium measure if this approaches the phase-space of the observable with such measure, for every sufficiently long γ.

Now consider the case in which the system can be split into two subsystems, S_a and S_b, such that the overall phase-space is the Cartesian product of the two respective phase-spaces and the constraint has the form $C = C_a + C_b$, where the first (rest second) term depends only on variables of the first (response second) system. Then, it is easy to see that this split determines a foliation of the constraint surface where the leaves are labeled by the value I of $C_a = -C_b$, and each leaf, in turn, is the Cartesian product of one surface per system.

Interestingly, the total number of degrees of freedom of the system is the sum of the numbers of degrees of freedom of each subsystem – seen as an isolated constrained system defined by C_a (rest C_b) – plus one. The single additional degree of freedom is a relative variable connecting the two systems, and we can view I as the value of the generator of its evolution. Now, say that we are interested in measuring partial observables of the subsystem S_a only, as it evolves with respect to a clock defined by a variable in the subsystem, S_b that is, by a one-form $d\theta$ depending only on the variables of the second (clock) system. Then assuming ergodicity, it is proven in Chirco et al. (2016) that the equilibrium states on the phase-space of the first system are independent from the choice of $d\theta$ and can be labeled by I and written explicitly as

$$d\mu_{\Sigma^a_{I^a}} = \frac{\delta(C^a - I^a)d\mu_a}{\int \delta(C^a - I^a)d\mu_a}, \tag{3.8}$$

where μ_a is the Liouville measure on the phase-space of the first system. This result permits us to define the full structure of statistical mechanics in a generally covariant system, on the sole basis of the choice of a (weakly interacting or noninteracting) clock subsystem.

Given now a split of the total system into *three* components, the preceding construction allows us to say when two of these are in equilibrium *with respect to the third*, namely, with respect to the clock system. This shows that in a generally covariant system, the notion of equilibrium is defined when three systems are given, not just two. Hence the subtitle of the paper Chirco et al. (2016): *It Takes Three to Tango.* I refer the reader to that paper for rigorous derivations and extended details on the resulting form of thermodynamics.

3.5 Conclusions

A fundamental lesson of general relativity is the dissolution of time as a unique and absolute entity. It is nonetheless possible to generalize thermodynamics to the wider covariant context of relativistic gravity, where there is no preferred time. Earlier I illustrated some ideas on the direction to go in order to do so. Here are some of the resulting possible lessons:

- In a covariant context, equilibrium between two systems needs to be defined *with respect to a third system*. ("It takes three to tango.")
- Generically, every state defines its own thermal time. The thermal time of equilibrium states is a global notion. On a fixed background spacetime, it is globally proportional to the flow of the (timelike) Killing field along which equilibrium is established.
- Temperature is then not uniform, in general. Rather, it is defined locally, and it is the ratio between proper time and clock time. It is the *speed of time,* namely, the speed at which systems change, measured in proper time.
- This rate of change can be defined as the rate at which the microphysics moves the system from a cell in space to the next one. Or, in quantum theory, a quantum state moves to an orthogonal state.
- At equilibrium, two systems may not have the same temperature. Rather, their rate of change is the same. This gets an informational interpretation by noticing that the number of states of one system to which the other system is exposed can be called quantity of information, or flux that can be acquired, or information flux. At equilibrium, the net information flux (the difference between the fluxes in the two directions) vanishes.
- Equilibrium is then defined between processes, not states.

I close with some general observations. The use of the language of information theory should not be taken here as a reduction of everything to information. That would be the same mistake as going from the observation that energy is a very useful concept in numerous situations to the misleading and empty rhetoric of "everything is energy."

Still, this use of information language reinforces the idea that different conceptual domains often converge in forming a coherent vision of reality. When comparing theories with different metaphysical frameworks and the same empirical content, it is useful to consider the explanatory power they hold in light of their ability to offer a conceptual means for transitioning toward more fundamental and more explanatory theories. A second major observation is that *relational* thinking plays a major role in this context. Rovelli's work on the foundations of covariant thermodynamics is independent from his work on the interpretation of quantum mechanics and his introduction of the relational interpretation of quantum theory in the 1990s (Rovelli 1996). Yet, in both cases, the interpretation of quantum mechanics and the formulation of relativistic gravitational statistical mechanics, problems are solved by bringing into view the underlying relational character of entities previously mistakenly taken as absolute (Vidotto 2022). Information plays a major role in both, as well.

The results presented here for covariant thermodynamics, much of which stems from Rovelli himself, show the influence of relational thinking that emerges from general relativity and (relational) quantum mechanics.

In the definition of equilibrium discussed earlier, a fundamental role is played by the partition of the system and the resulting relational degree of freedom that this partition singles out. It is this variable that ends up playing the role most similar to the Newtonian global time. Here, a partition defines a relation (Rovelli 2014). Once again, the useful focus is on the interactions between systems, and on processes, rather than on the systems and their states.

Finally, let me emphasize again that these results can be seen at best as the opening of a very intriguing research direction, certainly not yet as a complete theory. Also, the extension of these ideas to quantum theory is still entirely lacking.

Yet, it seems to me that this is where some of the more philosophically and conceptually interesting aspects of modern physics are nested: we have discovered that time works radically differently than in nonrelativistic physics, and it is still taking a long time to integrate this profound discovery about the structure of reality into our worldview. Rethinking physics without time in a fully relational way, prioritizing processes, may even require a deep conceptual discussion of, and perhaps even openness to models of thought different from those traditionally employed in the development of Western science (Kohl 2012, Latten 2022).

References

Chirco, G., Haggard, H. M., and Rovelli, C. (2013). "Coupling and thermal equilibrium in general-covariant systems." *Physical Review D*, 88(8): 084027.

Chirco, G., Josset, T., and Rovelli, C. (2016). "Statistical mechanics of reparametrization-invariant systems: Takes Three to Tango." *Classical and Quantum Gravity*, 33(4): 045005.

Chirco, G., and Josset, T. (2017). "Statistical mechanics of covariant systems with multi-fingered time." *Classical and Quantum Gravity*, 34(3): 035011.

Connes, A., and Rovelli, C. (1994). "Von Neumann algebra automorphisms and time thermodynamics relation in general covariant quantum theories." *Classical and Quantum Gravity*, 11(12): 2899–2918.

Haggard, H. M., and Rovelli, C. (2013). "Death and resurrection of the zeroth principle of thermodynamics." *Physical Review D*, 87(8): 084001.

Kohl, Christian Thomas (ed.). (2012). *Nagarjuna and Quantum Physics*. Saarbrücken: AV Akademikerverlag.
Latten, T. (2022). "Levels of Relationality: Understanding relational quantum mechanics through Nāgārjuna." Masters Thesis.
Rovelli, C. (1993). "Statistical mechanics of gravity and the thermodynamical origin of time." *Classical and Quantum Gravity*, 10(8): 1549–1566.
Rovelli, C. (1996). "Relational quantum mechanics." *International Journal of Theoretical Physics*, vol. 35: 1637–1678.
Rovelli, C. (2002). "Partial observables." *Physical Review D*, 65: 124013.
Rovelli, C. (2013). "General relativistic statistical mechanics." *Entropy*, 15(1): 158–171.
Rovelli, C. (2014). "Why gauge?" *Foundations of Physics*, 44(1): 91–104.
Rovelli, C. (2015). "Is time's arrow perspectival?" *Studies in History and Philosophy of Modern Physics*, 52(B): 5–9.
Rovelli, C., and Smerlak, M. (2011). "Thermal time and the Tolman–Ehrenfest effect: temperature as the 'speed of time'." *Classical and Quantum Gravity*, 28(1): 015007.
Rovelli, C., and Vidotto, F. (2014). "Covariant Loop Quantum Gravity." Cambridge: Cambridge University Press.
Rovelli, C., and Vidotto, F. (2022). "Philosophical foundations of loop quantum gravity." In C. Bambi, L. Modesto, and I. Shapiro (eds.), *Handbook of Quantum Gravity*, Springer. *arXiv preprint* arXiv:2211.06718.
Tolman, R. C. (1934). *Relativity, Thermodynamics and Cosmology*. Oxford: Clarendon Press.
Tolman, R. C. (1930). "On the weight of heat and thermal equilibrium in general relativity." *Physical Review*, 35(8), 904–924.
Tolman, R. C., and Ehrenfest, P. (1930). "Temperature equilibrium in a static gravitational field." *Physical Review*, 36(12): 1791–1798.
Vidotto, F. (2022). "The relational ontology of contemporary physics." In V. Allori (ed.), *Quantum Mechanics and Fundamentality: Naturalizing Quantum Theory between Scientific Realism and Ontological Indeterminacy*, Synthese (SYLI, volume 460), Cham: Springer International Publishing.

Part II

The Arrow of Time and Philosophical Issues

Part II

4

Explaining the Psychological Arrow of Time in Time-Symmetric Physics

MEIR HEMMO AND ORLY SHENKER

4.1 Introduction

> The problem of time has always baffled the human mind. Not only the events of the external world but even all our subjective experiences occur in time. It appears as though the flow of time, which orders the events of the physical world, passes through human consciousness and compels it to adjust itself to the same order. Our observations of physical things, our feelings and emotions, and our thinking processes extend through time and cannot escape the steady current that flows unhaltingly from the past by way of the present to the future. What we experience in one moment, glides, in the next moment, into the past. There it remains forever, irretrievable, exempt from further change, inaccessible to further control by anything that the future will bring us – and yet enshrined in our memory as something that once filled our experience as an immediate present. Will it never come back? Why can it not be with us a second time?
>
> *(Reichenbach 1956: 1)*

In these famous words Reichenbach addresses several different problems. In this chapter, we focus on only some of them. It is useful to start by distinguishing between them.

First of all, Reichenbach mentions the *direction* of time (which is the title of his book) and the *flow* or *passage* of time. The directionality means that there is a difference between past events and future events, in that past events are (according to the prevalent intuitions) remembered and unchangeable, while the future ones are unknown and open to influence. The idea of passage is, roughly, that time somehow flows such that events that are in the future gradually become present and then past. The ideas of flow or passage and the present are famously opaque (especially since the conceptual difficulties pointed out by McTaggart 1908). Also as pointed out by Rietdijk (1966, 1976) and Putnam (1967a) they seem to be in tension with both determinism and the relativity of simultaneity in the special theory of relativity. In this paper we don't address the flow or passage of time; we discuss *only* the *direction* of time. Of course, the two issues are closely related and may implicate each other in various ways, but nevertheless it seems to us that by distinguishing between them analytically we may be able to make some progress in understanding the question of the direction of time and also in connecting it with other issues in philosophy of physics and mind.[1]

Secondly, Reichenbach says that directionality and passage are facts in the world that are, somehow, *epistemically accessible* to us. Our consciousness is compelled, he says, to adjust itself to the temporal nature of the physical world. In this chapter we shall be interested in both parts of this claim: (i) whether and in what sense it is a fact, and (ii) whether and in what sense it is epistemically accessible to us.

We shall investigate what we shall call "*the psychological arrow of time,*" which is the very fact that our *experience* has temporal features, in particular, as we said, that we *experience* temporal directionality.[2] Our starting point is Flat Physicalism, which is an identity theory implying *reductive physicalism about the mind*. That is, we *assume* that our mind, including our experience and, in particular, the psychological arrow of time is (i.e., is *identical* with) a physical feature of the world, possibly of our brains.[3] This is a minority view: in contemporary philosophy of mind the prevalent view is so-called *nonreductive physicalism*, according to which the mind *supervenes* on the physical, presumably on the brain, but is not identical with it. We have shown elsewhere (Hemmo and Shenker 2023) that nonreductive physicalism, including computational functionalism, is a form of mind–body dualism in the sense that it entails that nonphysical facts play an indispensable causal or nomological or other effective role in the evolution of the universe, and moreover that in some cases these nonphysical facts are epistemically and psychologically accessible to us. Therefore, when we say that the psychological arrow of time is physical, we mean it in a reductive physicalist sense of the term. And then the main question we address is this: how is the psychological fact of a temporal direction to be accounted for by contemporary physics?

Let us state right away that we are not about to propose any specific answer to this question, neither in terms of physics nor in terms of brain processes: we will only discuss where one might look for such an answer. The main reason for this is

[1] Buonomano and Rovelli (2021) address the flow of time and *assume* (as is standard in similar discussion in physics and in neurosciences) that there is a temporal arrow as a matter of fact. In their view, "the arrow of time is not part of the basic grammar of nature, rather it is a complex probabilistic macroscopic phenomenon." We reject the latter part of this statement and argue that the arrow of time *cannot* be accounted for in this way, for reasons that we describe in Section 4.3. For this reason, we do not engage with other elements of their argument in this chapter. For similar reasons, we do not engage here with proposals such as that of Gruber, Block and Montemayor (2022), since their very notion of information gathering presupposes a direction of time, which is the very notion we seek to explain by physics.

[2] For *other* aspects of temporal experience (e.g., passage, present, order, duration), see Ismael (2011, 2017); Maudlin (2007: ch. 4); Paul (2015); Dieks (2016). For overviews of the various issues in philosophy of time, see, e.g., Le Poidevin (2019).

[3] For an overview of the state of art concerning physicalism (and references), see Stoljar (2024).

that in contemporary science, including all the achievements of brain science, there is no explanation (in our view, not even a beginning of an explanation) for how the mental comes about from the physical or the physiological. There is, to be sure, some knowledge concerning how brain processes bring about other bodily processes (including behavior, perhaps even some linguistic behaviour); but this is a far cry from explaining the mind. Here science is still in the dark. Given this general state of affairs, the reader might wonder, what is the point of asking about the physical basis of the psychological arrow of time. Our answer is that, interestingly, we may be able to say some things about this particular kind of experience despite this situation. We explain why as we proceed.

In this chapter we examine three kinds of approaches to the psychological arrow of time.

Reflected arrow: The first approach says that an arrow of time is a feature of the physical world "out there" as it were, and is *reflected* in our experience *of* the world "out there." Here an important idea is that this temporal feature of the world "out there" can be accounted for by contemporary physics, in particular according to the mainstream view, by the temporally asymmetric Second Law of thermodynamics, which in turn can be grounded in time-symmetric fundamental physics together with some assumptions about a measure (of typicality or of probability) over the macro-state in the remote (low-entropy) past of the universe.[4]

As we show (in Section 4.3), *all* the arguments along these lines are *circular* (or a priori) since they depend on our experience already being temporally directed. Additionally, if an "external" (as it were) arrow is to give rise to the psychological arrow of time, then it has to be "sensed" in order to be reflected in the degrees of freedom of our brains which form the experience. As we show (in Section 4.3.1), this cannot be done, and this impossibility is a matter of *principle*. Finally, we show (in Section 4.3), that even if one could (counterfactually) find a way to account for the Second Law of thermodynamics in a noncircular way *and* to offer a "sensor" of this arrow, this line of thinking would lead to a notion of an arrow of time that is radically different from the one people (presumably) have in mind when starting this explanatory project; and moreover, endorsing it ("biting the bullet") leads to inconsistencies.

We conclude that contemporary physics does not support this mainstream view. If one insists on there being an arrow of time in the world "out there" (as it were)

[4] For various approaches to the foundations of classical statistical mechanics and overviews of attempted proofs, see Sklar (1993), Uffink (2007), Frigg (2008), Werndl and Frigg (2015), Shenker (2017a), Shenker (2017b), Goldstein, Lebowitz, Tumulka and Zanghi (2020), Myrvold (2021). For quantum statistical mechanics, see Linden, Popescu, Short and Winter (2009), Goldstein, Huse, Lebowitz and Tumulka (2016). A universal proof of the second law cannot be attained, since a Maxwellian Demon is provably compatible with the principles of classical statistical mechanics; see Hemmo and Shenker (2010, 2011, 2012a, 2013, 2016a); for the quantum mechanical case, see Hemmo and Shenker (2020). There are, however, important special cases, for example, Lanford's theorem, but these are subject to the minimum entropy problem; see Uffink and Valente (2015).

that is reflected in our experience, then one needs to change the fundamental principles of physics in a way that will render them temporally *asymmetric*, and such that the relevant brain processes will give rise to the psychological arrow. This requires a radical change of our contemporary physical theories and we do not explore this route here.

Additionally, to keep this discussion focused, we do not discuss here either the possibility of grounding the psychological arrow in the cosmological arrow, nor in the arrow stemming from the violation of parity symmetry. It is noncontroversial that despite the temporal symmetry of *fundamental* physics, there are several kinds of physical facts that are time *asymmetric*. One is the various facts described by the Second Law. Those are usually taken to be observable facts, that is, to be a generalization from *experience* (we don't challenge this here). The second is the cosmological arrow, which is part of a *theory* that explains, for example, the observed red shift, in terms of general relativity, where the latter is itself a time-symmetric theory.[5] The third is a result in the electro-weak *theory* of the CPT (Charge-Parity-Time) symmetry together with the *observed* violation of the parity and time symmetries in experiments.[6] We conjecture that the conclusions of this chapter, in which we focus on the first arrow, would be hold (mutatis mutandi) for the other two, but we don't argue for this here.

Created arrow: The second approach (discussed in Section 4.4) says that some feature of our brain, which is *asymmetric* in some *non*temporal sense, and which brings about mentality in general, *creates* in us the feeling that we *call* "time directionality." We describe (in this section) some attempts to account for an in-brain asymmetry that may account for the psychological arrow. Recently, Smith (2014) proposed that this in-brain asymmetry is an in-brain entropy gradient. We show (in Section 4.4) that an in-brain entropy gradient cannot play this role. In a reductive Flat Physicalist approach, the psychological arrow must be identified with an instantaneous state of the brain (even taking into account the "extended present" idea, known in contemporary psychology as "integration window"[7]), and therefore entropy gradient is not even a possible candidate for accounting for the psychological arrow. A possible account may be some nontemporal (i.e., spatial or other) effective asymmetry in the brain. Such an account has, however, the radical implication that the psychological arrow of time, which is physically nontemporal (!), is the fundamental arrow,

5 See a summary of the state of art of the standard model of cosmology and its experimental testing in Kunze (2015); see also Carroll and Chen (2004) for a different connection between spontaneous inflation and the second law.

6 Atkinson (2006) proposed to explain the CPT symmetry (and the observed violations of parity and time symmetry) by the physical features of the low-entropy state immediately after the Big Bang. But the hypothesis that as a matter of fact the universe was in such a state is required to account for the temporal direction of entropy increase according to the Second Law of thermodynamics!

7 Following Dennett and Kinsbourne (1992).

and it is this arrow that underpins what *appear* to be other "arrows of time" "out there," rather than the other way around (which is the intuition and mainstream view; see Hemmo and Shenker 2019b).

No arrow: The third approach, which we shall discuss only very briefly (in Section 4.5), denies that we have a psychological arrow of time to begin with, and turns to explain why we think that we do. This option is a version of "denialism" in the philosophy of mind, which is the view that first-person reports concerning one's own mental states are not always reliable (see Shenker 2021). Since this last approach is only weakly connected to what is usually taken to be part of the foundations of physics, we shall not expand on it; we mention it only to complete the picture and because, given the state of art so far, this option should not be ruled out.

The case of the psychological arrow of time turns out, then, to be extremely important in contemporary science in two respects. If the first option of the preceding three is the case, then this exploration brings to the surface a huge lacuna in contemporary physics and the need for a radical change of it. If the second or third option is the case, then the arrow of time offers a unique case for studying the physics of the mind.

Since our analysis builds on the assumption that the mental is physical, and since our view on this matter deviates somewhat from some mainstream ideas, we begin with a brief explanation of what physicalism means.

4.2 What Is a Physicalist Theory of the Mind?

Our starting point here is (as we said) that the mental and, in particular, the psychological arrow of time, the feeling of directionality, *is physical*. It may be, according to contemporary science, something in the brain. This special case of mind-brain identity is part of our approach called Flat Physicalism, according to which everything (!) is physical. The idea that the mental is physical is a minority view in contemporary philosophy; the majority endorses so-called nonreductive physicalism. These thinkers believe that their view is physicalist, as its name suggests. However, *this is a mistake*. So-called nonreductive physicalism is a form of mind–body dualism. We will only hint at why this is so in order to motivate our endorsement of a full-blown reductive physical approach to the mind (for more details and arguments, see Hemmo and Shenker 2023 and references therein).

According to *all* forms of physicalism (reductive as well as nonreductive) the individual facts in the world (objects or states of affairs or properties) are (fundamentally) physical. Denote the individuals by the numerals 1, 2, 3, and so on.[8]

[8] In physicalism, both reductive and nonreductive, the individuation conditions are physical. The difference between reductive and nonreductive physicalism allegedly comes about with respect to the high-level facts only. In what follows we challenge this view and show that in fact the view concerning the status of high-level facts has repercussions with respect to individuals, and also with respect to the low level (or fundamental level).

Additionally, according to *all* forms of physicalism (reductive as well as nonreductive), (all) the individuals belong to *physical kinds*: individuals belong to the same physical kind if they share a physical fact or fall under the same physical description. Examples are all the objects made of pure carbon; all the cases of fission of two hydrogen nuclei into one helium nucleus; all the electric circuits in which the current is of 2A; etc. Suppose that individuals 1 and 2 fall under the physical kind P1, 3 and 4 fall under P2, and 5 and 6 fall under P3. And assume – and this is important for our present discussion – that P1 and P2 and P3 are all equally mutually *physically heterogeneous*, that is, individuals that fall under them share *no relevant* physical property. The idea that two physical individuals may share no physical property whatsoever is, of course, dubious, since for example all the objects are made of the elementary particles described by the standard model of quantum field theory. This is why we stress the notion of *relevant* difference or relevant similarity, where by "relevant" facts here we mean those physical facts that can account for the so-called high-level facts.

By "high-level" facts one means the facts described by the so-called *special sciences*, such as biology or psychology (Fodor 1974). Individuals can belong not only to physical kinds but also to (so-called) *high-level kinds*, that is, the kinds described by the special sciences. Denote the high-level kinds by M1, M2, and so on. According to the thesis of *reductive physicalism* (RP), high-level kinds are physical kinds. The reason why the individuals 1 and 2 fall under M1 is that they belong to P1; and the reason why 3 and 4 don't fall under M1 but under a *different* kind M2 is that they belong to a different physical kind, P2, which is heterogeneous relative to P1. According to RP, physics fully explains how individuals are partitioned into kinds. Even if we are not aware of all the details of the different kinds, this is the physical ontology of the kinds in nature.

By contrast, according to the thesis of so-called *nonreductive physicalism* (NRP)[9], individuals that fall under the same physical kind *always* (or even *must*) fall under the same high-level kind (this is called *supervenience* of a special science kinds on physical kinds). This means that if 1 and 2 both fall under the same physical kind (say, P1), they (must) both fall under the same high-level kind (say, M1). But *not vice versa*: individuals that fall under the same high-level kind can belong to *different heterogeneous* physical kinds (this is often called *multiple realization* or *realizability* of the special sciences kinds by physical kinds). In our example, on this view, it may be that 1, 2, 3, and 4 fall under M1, although 1 and 2

[9] Examples of nonreductive views are Putnam (1967), Fodor (1974), Davidson (1970), Lewis (1994). All contemporary versions of functionalism (causal, computational, and others) in the foundations of the special sciences (including psychology) are nonreductive; they standardly assume supervenience of high-level kinds on physical kinds and imply multiple realization. They are all subject to our argument in this section.

fall under P1 and 3 and 4 fall under P2. (Individuals 5 and 6, which fall under P3, may fall under M2, but this is of less interest to us at the moment.)

There is a debate in the literature on whether our world is such that RP obtains or such that NRP obtains, that is, whether in our world there are cases of multiple realization, and both sides bring empirical support for their claims (see, e.g., Polger and Shapiro 2015; Maimon and Hemmo 2022).

Consider the individuals and physical kinds mentioned earlier. In our example, the three physical kinds P1, P2, and P3 are heterogeneous, so that 1 and 2 do not share any (relevant) physical property with 3 and 4 or with 5 and 6. Nevertheless, according to NRP, 1 and 2 and 3 and 4 are of the same high-level kind, but 5 and 6 aren't. *What explains this fact? What fixes the partition of the individuals to these sets?* Here are four options, which we think cover everything that is being said about this in the literature.

Option 1: Individuals 1, 2, 3, and 4 but not 5 and 6 share a physical fact. This is not the case in our example, in which P1 and P2 are physically heterogeneous. Were that the case, this would have meant that P1 and P2 are subkinds of the same physical kind (call it P12), and we would have here an example of RP, not of NRP. If we want an example of NRP, we need to try something else.

Option 2: There is an external system, say a device D, such that in the presence of 1 or 2 or 3 or 4, D's pointer reads "M1," while in the presence of 5 or 6, D's pointer reads "M2."[10] This case teaches us that we didn't partition the universe correctly into individuals: those should be 1+D, 2+D, and so on. And then, the individuals 1+D, 2+D, 3+D, and 4+D fall under the same kind M1, just because they share a physical property, namely, the position of D's pointer. And once again, this is an example of RP, not of NRP. If we want an example of NRP, we need to try something else.

By the way, option 2 is open to us if our aim is to explain why some fact in our environment *seems* to us to be multiply realized: we are D. It is a good (albeit reductive!) explanation of this case. But it is not open to us if our aim is to explain our mind, that is, if 1, 2, and so on are individual states of (say) our brain, and M1 and M2 are our mental states, since here there is no device D that can do the trick of uniting physically heterogeneous facts into one mental kind.

Option 3: Individuals1 and 2 are physically heterogeneous relative to 3 and 4, but all four of them are *functionally* identical. They occupy the same position in, for example, a structure of a causal network or of an implementation of a computation. And this accounts for their all being cases of M1 despite their physical heterogeneity. Of course, a "function" is not a physical notion, since it involves a target as

[10] Interestingly, D is *not* a measuring device that measures physical properties of the individuals, since P1 and P2 are (assumed to be) physically heterogeneous.

well as a preference for a certain perspective of looking at a system; so, we are already in a nonphysical framework; but forget this for a minute. An individual case of a "function," from the point of view of physics, is a sequence of physical states. So, we have here, in this view, four sequences of physical states. The same question immediately arises: In virtue of which fact do *sequences* 1, 2, 3 and 4, but not sequences 5 and 6, fall under the same (functional) kind? If it is in virtue of a physical fact that they all share, then, again, this is a case of RP and not NRP. If we want an example of NRP, we need to try some other explanation for the partition of the individuals to the high-level kinds. Notice that here functionalism of all sorts (causal, computational, etc.), as a special case of NRP (applied to sequences of physical states) *repeats the problem rather than solves it.*

Option 4: Brute fact. It is a brute fact that 1 and 2 and 3 and 4 but not 5 and 6 fall under M1. What kind of fact is a brute fact? We needn't answer this question; all we need to notice is that a brute fact is, by its very definition in this context, *not a physical fact*. In other words, *it is a nonphysical fact.* In this case, then, individuals 1, 2, 3, and 4 share a nonphysical fact (call it B1) which is different from the brute fact shared by 5 and 6 (call it B2). So, there is a nonphysical element *within each and every individual fact. This is token dualism.* We stress: it is sometimes said that according to NRP every individual is "fundamentally" physical, so that each one of them, taken by itself, is fully physical, and the high-level facts are only a matter of their partitioning into sets. But this claim ignores the question of how individuals are partitioned into sets, namely, what kind of fact they share such that they come to be of the same set or to fall under the same kind. Once this question is addressed, the nonphysical element of each individual is exposed.

Notice that if we are *given* (by some supernatural oracle) the partition to high-level facts that satisfies both supervenience and multiple realization, then, given a physical kind of an individual we can know its high-level kind. Often people take this to mean that the physical kind fixes or determines the high-level (e.g., the mental) kind. But to see that NRP is a version of dualism we need only ask the deeper question: in virtue of which facts do individuals fall under a kind in the first place? It is here that the brute facts appear as clearly nonphysical elements within each and every individual.

It is noncontroversial that supervenience is the hallmark of physicalism, since its negation (a case where the brain state doesn't change when the mental state changes in kind) straightforwardly entails dualism. But our argument shows that supervenience is not sufficient for physicalism because it is compatible with multiple realization, which implies mind-body (token-)dualism. Therefore, the only option for physicalism is a type–type identity of physical facts with mental facts. QED.

Interestingly, the nonphysical facts in the world (that exist according to NRP) are epistemically *accessible* to us, since we are able to distinguish, say, between two individuals that belong to heterogeneous physical kinds, but nevertheless belong to the same psychological kind. And if this account is to explain not only the appearance of multiple realization in our environment but also multiple realization of our mental kinds by physical kinds, then this dualism is good old mind–body dualism.

Soon after it was proposed (by, e.g., Putnam 1967b and Fodor 1974), NRP became extremely popular. This is understandable, since it promised to save us from having to decide between materialism (or physicalism) and dualism, that is, between the naturalist intuition, and the intuition according to which there is something special about the mind. It appeared to these thinkers as if there is a magical third option. But it turns out that NRP cannot stand up to these expectations, and we have no choice but to make up our minds between the old two options of physicalism (or materialism) and dualism.

In this chapter our starting point is physicalism, and we shall not argue for it any further. We now turn to ask how the psychological arrow can be explained in this framework, going through the three kinds of approaches sketched earlier: reflected arrow, created arrow, and no arrow.

4.3 Reflected Arrow

Let us start by supposing that the psychological arrow of time reflects or represents some arrow in the environment outside the brain. (Of course, the brain is physical and, as such, it is subject to the same temporal laws of nature. This point will be discussed in the next section.) Here the intuition is twofold: (i) There is a physical arrow of time which is "out there" (independently of us) in our environment; and (ii) our feeling of temporal directionality is *of* that arrow, that is, this external arrow somehow *brings about* in us, given the appropriate interactions, the experience of temporal directionality. Notice that Reichenbach (1956) seems to be in this direction, when he says (in the preceding quote): "[T]he flow of time, which orders the events of the physical world, passes through human consciousness and compels it to adjust itself to the same order."

We begin by analyzing the preceding twofold claim along the lines standardly used to account for (e.g.) the mundane situation in which there is a hot glass of water in our environment which, by way of some appropriate interactions (say, if we put our finger in the water), brings about in us the sensation of heat. In the first part of our discussion (in this section), we make the analogy between heat out there (in the water or the air around us) and the external thermodynamic arrow of time (out there around us). In the second part (Section 4.3.1), we make the analogy between our *sensation* of heat (by measurement) and our *sensation* of the entropic arrow out there in our environment.[11] We begin with the first part, that is, with the very claim that there *is* an external arrow of time. Which facts in our environment constitute an arrow of time?

When we say "constitute," we mean it in the sense of *identity*: there is only one thing, which may have two descriptions given in two kinds of circumstances. Although this idea may seem trivial, let us explain in some detail what it means in our context. According to the most prevalent view, the facts reflected in the psychological arrow are those described by the Second Law of thermodynamics. This claim is an *identity* statement: the direction of the future is *identified* with the direction in which entropy increases. Just like the natural kind term "heat" refers to the same thing as the scientific kind term "motion of molecules"; or the natural kind term "water" refers to the same thing as the scientific kind term "H_2O"; or the natural kind term "light" refers to the scientific kind "electromagnetic radiation"; the term "arrow of time" and the term "arrow of entropy increase" refer to the *same* arrow. What we call "temporal arrow" *just is* an entropic arrow, so that there is only one (kind of) thing, which is described using two terms: "the direction of the future" and "the direction in which entropy increases." These two terms describe (or refer to) the same thing, namely, the thing that triggers in us (somehow; see the next section) the psychological arrow. On this view, *all there is*, is an entropic arrow; and sometimes, for whatever reason, we call the entropic arrow by the name "temporal arrow." (It is not the case that the direction of entropy increase *brings about* or *creates* or *causes* the temporal direction, nor that the direction of entropy increase is correlated with the direction of the future or is indicative of it, because all these cases would imply that there are two things, the direction of time and the direction of entropy increase, and this would leave unanswered the question of what is the temporal direction that is brought about in this way.)

As is well known, the *laws* of contemporary fundamental physics are *time symmetric* (by this we mean invariant under the reversal of the direction of time; we use these two terms interchangeably).[12] Let us clarify this feature a bit. Consider the simplest kinematic equation of classical mechanics, something like $x = vt$. The parameter t appears here twice: it appears in both terms on the right-hand side, since v describes the change of position with time: dx/dt. There are three ways to insert a minus sign to this equation that will yield the same result for x, each having a different meaning: replacing dx with $-dx$ means *velocity reversal* (relative to the *same* direction of time); replacing dt with $-dt$ (not a well-formed formula, admittedly) means *retrodiction* (relative to the same direction of time; i.e., inferences about the past as opposed to prediction); and replacing t with $-t$ means the *reversal*

[11] See Kripke (1980) for the distinction between heat and the sensation of heat.

[12] For a discussion of the notion of time-reversal invariance, and more references, see Allori (2015).

of the direction of time itself (whether or not this is metaphysically meaningful is a question we leave open). (For more details on distinguishing between the three cases, see Hemmo and Shenker 2012a: ch. 4, 2012b, and 2016b.) The psychological arrow of time is about the direction of time, that is, it is about the *third* sense, and in particular, it is about experiencing the direction of *time itself*, relative to which we can reverse velocity or carry out retrodictions. By contrast, the temporal asymmetry of the Second Law of thermodynamics concerns the *first* sense, namely, it concerns *velocity* reversal *given* a direction of time: the second law says that some velocity reversals (given a direction of time itself) that are compatible with fundamental physics are not found in nature or are not likely to be found. And the question is whether some connection between the first and third kinds of reversals might explain the psychological arrow of time: does the *velocity* irreversibility described by the Second Law explain the *time* asymmetry of our experience? Many think that it does;[13] we think that it doesn't, and here we explain why and offer an alternative.

With respect to the thermodynamic arrow, our starting point is the (by and large) noncontroversial idea that it is a result of the underlying physics, and (moreover) that it is the task of statistical (classical or quantum) mechanics to settle between the temporal symmetry of fundamental mechanics and the thermodynamic temporal asymmetry.[14]

Notice, first, that if the direction of the future is accounted for by the *reduction* of the arrow of time to the direction of entropy increase, then the Second Law of thermodynamics (if true) and its statistical mechanical counterparts are all *tautologies*. Entropy cannot decrease because the direction of decrease would then be identified as past. We stress: If the arrow of time and the arrow of entropy increase are two different facts, then the Second Law describes a nontrivial and perhaps explanatory fact about the universe. Not so in the reductive view, in which the temporal direction is identified with the entropic arrow. In this latter case, the Second Law has no empirical content. Let us show this point in a different way.

Consider **Figure 4.1**, which is a variation of the famous illustration presented by Boltzmann and Schuetz. Boltzmann and Schuetz used a similar figure to illustrate the statistical mechanical account of the probabilistic counterpart of the Second Law of thermodynamics in a world without a beginning or end. We use it for a different purpose, and therefore include in it an adaptation to a universe that conforms to the Big Bang model and assume that the universe was in a low-entropy state just after the Big Bang (t_0 in Figure 4.1). The graph in this figure describes

13 Approaches in this direction are, e.g., Albert (2000); Loewer (2012); Mlodinov and Brun (2014). Hawking's (1988) approach to the psychological arrow on the basis of thermodynamics of memory systems relies on mistaken ideas about the relation between entropy and information; see Hemmo and Shenker (2012, 2013, 2019c).

14 Strictly speaking, classical mechanics is not true according to contemporary physics; we take it that it retains the relevant features of the true fundamental theory in this context; see on this point, e.g., Wallace (2001).

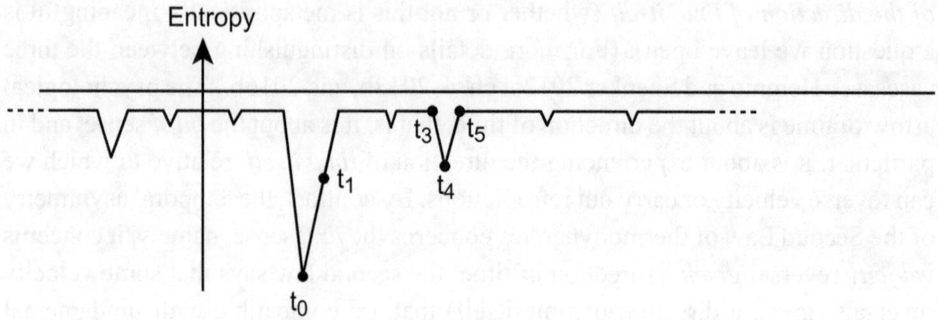

Figure 4.1 Change of entropy along a nondirected time. It is standardly assumed that deviations from equilibrium (i.e., maximum entropy) are rarer, as they are bigger. To the extent that this assumption is based on typicality arguments (and not on mere induction from local experience), we reject it and therefore do not assume here that this is the case.

entropy changes along some given mechanical trajectory of the universe,[15] in which "most" of the "time," that is, along most of the length of the trajectory according to an appropriate *measure*, the trajectory is in the state space region in which it has properties corresponding to thermodynamic equilibrium, and as the deviation from equilibrium is larger, the rarer it is. Boltzmann conjectured that the trajectory of our universe behaves in this way. (This idea involves conjectures about the dynamics and a choice of measure, which we do not discuss here.) The Boltzmannian conjecture is that, given this picture, if we find the universe in some nonequilibrium state (e.g., our own state, which is perhaps t_1 in the figure), it is likelier to climb back to equilibrium rather than to evolve farther away from it.

As we said, if the arrow of time is taken to be the direction in which entropy increases, that is, if the higher entropy state is *defined* as "future," then the Second Law (if true) is either tautological, or simply presupposes an arrow of time which is *not* entropic, a primitive arrow of time perhaps, or an arrow that comes from a different theory, which is *not* time symmetric.[16]

Consider again Figure 4.1. According to statistical mechanics, entropy fluctuations are a feature of the universe, even if they are rarer as they get bigger (and cosmological "small" may be very big and long in human scales). Suppose that the trajectory of the universe passes through the states $t_3 - t_4 - t_5$. At t_3 the "past" is in

[15] According to statistical mechanics, there are trajectories of the universe along which entropy changes in many other different ways. We abstain from describing this trajectory as "typical," since the notion of "typicality" involves circularity; see what follows and Hemmo and Shenker (2012): ch. 8 and Hemmo and Shenker (2015).

[16] If there is a collapse of the quantum state (as in the theory of spontaneous collapse by Ghirardi, Rimini and Weber (GRW, 1986)), then physics is fundamentally not time-symmetric; in which case one should examine whether and how this quantum arrow may account for the psychological arrow of time. See some related points in the last section.

the direction of t_4, and then t_5 is farther into the past. But at t_5 the "past" is, again, in the direction of t_4, and then t_3 is farther into the past. This means that if the "arrow of time" and "arrow of entropy increase" refer to the same thing, then there are fluctuations also of the direction of time, such that the open future and the fixed past are flipped. This case is neither imaginary nor rare: small fluctuations are frequent, and on the view we examine here, they entail small but frequent fluctuations in the direction of time.[17]

Suppose that one were to take the "past" to be the direction along the time axis in which there are (statistically or probabilistically) more lower entropy states (within a Big-Bang-style universe, not an eternal one). Would this solve the problem? The answer is in the negative because this characterization of the "past" would provide a direction only to an observer that looks at the world's history from the outside; it would not explain the *experience* of observers like us, who are immersed in fluctuating environments. Similarly, taking the "past" to be the direction in which the low-entropy state of the *entire* universe is placed, and arguing that we (somehow) experience the increase of the total entropy of the universe from this state will not work either, because the Boltzmann–Schuetz fluctuations (see **Figure 4.1**) apply to the behavior of the total entropy of the universe. Finally, one may argue that fluctuations of the total entropy of the universe are unlikely to occur along typical (or probable) trajectories of the universe and, therefore (contrary to our argument earlier), our psychological arrow is unlikely to flip. But this argument is wanting because typicality (and probability measure) are circular (insofar as they are meant to account for our experience, since they assume, by construction, that our experience is typical or probable).[18]

4.3.1 Measuring the Direction of Time

Assume that the arrow of time is indeed a feature of the physical world in our environment, "out there" as it were, and the psychological arrow is its reflection in our experience. This means that this arrow of time "out there" has to be somehow "sensed" by us and eventually come to bring about something in the degrees of freedom of our brains, which form the psychological arrow of time (in accordance with our physicalist framework). Given our discussion so far, the time arrow in our

[17] For the past hypothesis of low entropy, see Feynman (1965). For various applications of Feynman's hypothesis, see Sklar (1973); Albert (2000); Callender (2004); Callender (2010); Wallace (2023). For a different hypothesis about the reliability of our memories, see Hemmo and Shenker (2012). Price (1996) argues that the past hypothesis implicitly presupposes time-asymmetry. Earman (2006) criticizes Feynman's dismissal of Boltzmann's remarks on this problem and also argues that the past hypothesis is not well supported by contemporary cosmology; see also Goldstein, Tumulka and Zanghi (2016).

[18] The choice of measure of probability (or typicality) presupposes that the actual trajectory of the universe in which our experience arises is typical (or probable) amongst all possible (counterfactual) trajectories. For the typicality approach, see Goldstein (2012) and references therein. For the circularity argument, see Hemmo and Shenker (2012a: ch. 8; 2012c, 2015, 2019b).

environment, which is to be sensed, is *not* identical with the temporally asymmetric facts described by the Second Law of thermodynamics. Therefore, assuming, as we said, that there is a temporal arrow "out there," it must be something which is not the entropic arrow described by the Second Law. It can be either identical with some other fact described by our (present or future) theories of physics, for example, the cosmological arrow, or perhaps the arrow associated with the violation of parity symmetry, or a primitive arrow. And here we explore two options: either we sense that arrow of time *directly*, or it may be the case that the entropic arrow is somehow correlated with that temporal arrow, and we sense the entropic arrow and thus sense the temporal arrow *indirectly*. In all these cases, we need to have suitable sense organs that would be sensitive to the respective arrow and would then map it to the suitable brain state in which we have the experience of a psychological arrow. Boltzmann (1964) proposes to illustrate this idea using the case of the experience of distinguishing between the spatial direction called "up" and the direction called "down" near the surface of Earth (see Sklar 1981). The distinction between up and down is not fundamental in physics, and in outer space there is no such distinction; it comes about only near the surface of Earth, where it refers to the local gradient of the gravitational field. This local direction is measured, for example, by a sense organ of human beings that consists of the vestibular system, which is sensitive to this local direction and sends electrochemical signals to the brain, leading to the "up-down" feeling. So, the idea is this: if one argues that the psychological arrow reflects some external arrow, one is obliged not only to say what that external arrow is, but also to make the case that this external arrow can be sensed. Of course, there is no requirement for showing how precisely the sense organ in question works; the question is not technical or one of precise details; but the requirement is that this sensing needs to be possible as a matter of principle, according to the theories of physics.

One challenge here is that the external temporal arrow needs to be mapped, by way of physical interactions to other degrees of freedom in the brain, which are (according to the identity–physicalist theory we assume here) the experience of the psychological arrow. To avoid confusion, notice that we assume here that the feature of the brain which is the psychological arrow is not *itself* the physical origin of the temporal arrow. We explore this latter option, namely that a temporal arrow in the brain itself is responsible for the psychological arrow, in the next section. Here we shall assume that this is *not* the case and that the trigger for the psychological arrow is a temporal arrow in our external environment. How, then, can the temporally directed facts "out there" be mapped to nontemporal features of the brain in which we form the mental state (or the experience) of the psychological arrow?

Boltzmann (1964) thought that the temporal arrow "out there" is the entropic one (notwithstanding our argument earlier):

> [I]n the universe, which is in thermal equilibrium throughout and therefore dead, there will occur here and there relatively small regions of the same size as our galaxy (we call them single worlds) which, during the relatively short time of eons, fluctuate noticeably from thermal equilibrium, and indeed the state probability in such cases will be equally likely to increase or decrease. For the universe, the two directions of time are indistinguishable, just as in space there is no up and down. However, just as at a particular place on the earth's surface we call "down" the direction toward the center of the earth, so will a living being in a particular time interval of such a single world distinguish the direction of time toward the less probable state from the opposite direction (the former toward the past, the latter toward the future). *(Boltzmann 1964: 446–447)*

Our question now is this: *pace* our argument above, let us assume (for the sake of the argument) that there is an entropic arrow of time "out there." How can it be sensed? We shall argue that it cannot be sensed, and so this argument is an additional one against the idea that the Second Law of thermodynamics plays a role in explaining the psychological arrow of time.

For Boltzmann, the "past–future" distinction refers to (what we might call) the local gradient of an "entropy field" described by the Second Law of thermodynamics.[19] Boltzmann *doesn't* worry about the way the local "entropy field" can be measured or sensed by us, in a way analogous to the vestibular system, which is the sensor for the external gravitational field, in order to bring about the psychological arrow. How can this be done?

In Boltzmann's theory, entropy is a function of some *measure* over the state space (standardly, the Lebesgue measure[20]): the entropy of a microstate at a time is a function of the measure of a continuous set of microstates (of which the present microstate of the universe is actual and the others are counterfactual at that time) that share with it some macro-variable (state function).[21] That is, entropy is *not* a macro-variable (this is expressed by the fact that it is a number without units), and therefore it is not a (directly) measurable property. There isn't, and there *cannot* be, an "entropy meter" nor an "entropy change meter." Consequently, despite the

[19] The idea of a temporally orientable spacetime in which the direction of time is defined at every spacetime point is consistent with general relativity; see Earman (1974) and Weingard (1977).

[20] A choice of measure is required to determine the size of sets of microstates (or regions) in the phase space; which facts determine this choice is an open question; see Goldstein (2012) for the so-called typicality approach and references; see Hemmo and Shenker (2012, 2015) for a different view. Regarding the connection between entropy and probability: it is not analytic that the function for calculating entropy should be the Lebesgue measure, even if the Lebesgue measure turns out to fit the generalization of the relative frequencies of the thermodynamic macro-variables in our experience, i.e., the probability of statistical mechanics; see Hemmo and Shenker (2012): ch. 6 and 7 on the questions of the measure of probability and entropy, respectively.

[21] See Hemmo and Shenker (2016a, 2019) for the role of *macro-variables* in statistical mechanics.

importance of this quantity, our physiology doesn't contain and cannot contain a sensor of entropy nor of entropy change.[22]

Can entropy or entropy change be measured *indirectly*, to provide a temporal arrow? The second law says that quantities which are macro-variables change in a specified way that is generally expressed in terms of entropy. For example, temperature distribution or pressure distribution tend to become uniform with time. Could we perhaps measure those thermodynamic properties and their change over time? Obviously, strictly speaking, one cannot measure the instantaneous temporal gradient of such properties, since the time derivative is a limit concept which raises the Zeno Paradox (see Arntzenius 2000). But this by itself is not a problem since it is good enough to measure these thermodynamic properties at two (or more) times and then see at which of them the temperature distribution (say) is more uniform. Can we measure the temporal arrow "out there" in this way? The answer is no, for several reasons.

First, following our earlier discussion, since we do not identify the temporal direction with entropy increase, and we take the Second Law of thermodynamics to express a contingent fact and not an a priori truth, the question of whether entropy increases or decreases is contingent, to be decided empirically. And the problem here is that in order to decide on this empirical question we need to know, beforehand or as a precondition, which of the two states, the more uniform or less uniform distribution, is *later*. Therefore, the very difference between these quantities, which is an indirect indication of the entropy change, cannot be used as an indirect measurement of the temporal arrow "out there" in a noncircular way.

Another problem that arises here is this. All the macro-variables described by statistical mechanics, associated with the thermodynamic magnitudes, are invariant (or symmetric) under time reversal or under velocity reversal (since they are functions of either position or velocity-squared). This implies that within each set of microstates that share the same thermodynamic macro-variables, some microstates sit on trajectories that are entropy-increasing (at the time of measurement or sensation) and others sit on trajectories that are entropy-decreasing (at the same time). So, a sensor intended to distinguish between entropy-increasing and entropy-decreasing microstates would have to be sensitive to the difference between subsets of the thermodynamic sets, that is, the sensor would have to be sensitive to macro-variables that are *not* those appearing in statistical mechanics and thermodynamics, but rather to some different, perhaps "finer grained," macro-variables. How exactly this could be done is, by the very nature of these properties, beyond the scope of standard thermodynamics and statistical mechanics.

[22] We thank Yemima Ben-Menahem for discussions on this point.

And so, the idea that the psychological arrow of time is a reflection of an entropic arrow in our environment described by the second law of thermodynamics is to be rejected.

As we remarked, we do not discuss other physical arrows like the cosmological arrow or the arrow stemming in high-energy scales from the parity symmetry-violation (or the CPT symmetry). As long as physics remains time symmetric, we believe that the above arguments will apply to them as well, mutatis mutandis.

4.4 Created Arrow: An In-Brain Arrow of Time

Up to now, we explored whether in time-symmetric physics there can be an external arrow of time that is sensed and thus reflected in the mind, bringing about the psychological arrow; and our conclusion was that there can't be such an arrow. How, then, does the psychological arrow come about, in a physicalist approach to the mind? The next option to explore is this: since (1, lemma from the preceding) there is *no external* sensed arrow, and since (2, datum from experience) there is a *psychological* temporal arrow, and since (3, according to the physicalist framework in which we work here) the psychological arrow is a physical fact about the brain, it must be the case that (4, conclusion) the psychological arrow must somehow be a physical state of the brain. On this conclusion, the psychological arrow is *created* within the brain, *not* reflecting anything about the (otherwise) objective matter of facts in the world. This is the option we explore in the present section.

According to the physicalist view, mental facts are physical facts. *Which* facts about our brains could be identical with the psychological arrow? As we remarked in the introduction, we do not purport to make any specific proposals concerning these facts; we shall only point out some conditions that should guide the search in brain science or in physics. We shall look for characteristics of the physical state of the brain that are *necessary* for being the psychological arrow of time. We hope that these features would guide the research concerning the details of the brain state which is the psychological arrow of time.

When looking for the facts in the brain which *are* the psychological arrow, it is of utmost importance to bear in mind that we are after discovering *identity* statements: that is, we are looking for physical facts which *are* (*identical* with) the mental fact of the psychological arrow. We are not looking for facts that bring about (or cause) the psychological arrow, since this would mean that there are two kinds of facts: the sought physical facts in the brain, and the facts that they bring about, the latter (but not the former) being the psychological arrow. This case would leave open the question of the nature of the psychological arrow itself, and it is this latter fact that we are after here. Additionally, we are not after facts that merely *realize* (or implement or ground etc., depending on one's metaphysical theory) the psychological arrow, but after facts that are *identical* with it. In our view, notions of realization (etc.) are dualistic (see Section 4.2) and (we reiterate) leave open the question in which we are interested, namely, what is the physical nature of the psychological arrow of time itself.

Let us digress for a moment on the notion of *identity* statements. We employ an understanding of identity in the spirit of the influential analysis of Kripke (1980) (but without arguing for it nor going into the details of its many implications).[23] Kripke distinguished between what he called "heat" (molecular motion) and "sensation of heat" (a mental state), the former being the state of affairs in the world which (causally or otherwise) brings about the latter. In this analysis, in the present section we are interested in the physical nature of the "sensation of heat," not of "heat." Kripke (1980), famously, used his analysis to argue for mind–body dualism; for example, for him "sensation of heat," and (we assume, similarly, the psychological arrow of time) is not a physical fact. We don't discuss Kripke's dualistic argument, and we mention his analysis only to clarify that we are after a kind of fact that Kripke thought doesn't exist. We shall not argue here for our reductive physicalist view; it is our starting point, and we seek an account of the psychological arrow of time in this framework.

Given all of this, what we seek is something about the physics of the brain that will have two features: one is that it will bring about a brain state which is a *mental* state; and the other is that this physical-mental state will involve some *symmetry-breaking* in the brain.[24] The "mental" feature is the very fact that it is of the kind of physical facts that are mental facts. We have no idea what this feature might be, and (as we have already remarked) this is still an open question in contemporary brain science. The "asymmetry" feature is that, since this mental fact involves some asymmetry, we conjecture that in a reductive physicalist approach it must be associated with a *physical asymmetry* in the brain.[25] Thus, our guiding hypothesis is that there must be some physical symmetry-breaking in the degrees of freedom that instantiate this particular mental condition. It is this condition on which we focus in this section, from now on.

We start off with the assumption that *all* the fundamental theories of physics are time symmetric.[26] And therefore, the explanation of the psychological arrow must involve some other suitable symmetry-breaking. One line of thinking may be Boltzmann's idea of a *local* symmetry-breaking (see the earlier quotation). In our

[23] For a recent Kripke-style analysis of identity, see Hoefer and Marti (2019).

[24] These two conditions are called "Mental" and "Asymmetry" in Hemmo and Shenker (2019b).

[25] Obviously, the mental asymmetry expressed by the psychological arrow of time cannot be derived from a symmetric physical theory without some symmetry-breaking amendments; see van Fraassen (1989).

[26] As we said, the cosmological arrow and the parity symmetry breaking are, in this view, empirical facts that are not derived from the fundamental laws alone but involve some additional assumptions. We don't discuss this point here.

approach, we are interested in a local symmetry-breaking which takes place in the brain. Here, by "brain" we mean the system which is, according to science, associated with the mental, and of which some degrees of freedom *are* (on the physicalist view) mental. Let's explore some proposals along this line.

A recent proposal (by Smith 2014) appeals to processes of diffusion in memory formation and gene expression in the brain, which are (arguably) locally entropy-increasing. Here the conjecture is that such processes in the brain produce in us the psychological arrow of time. The brain, however, is an open system, and its total entropy decreases as memories are formed, so in order to support Smith's approach, brain science must identify the relevant features of the brain for accounting for the psychological arrow and show that these involve entropy increase.[27] This proposal brings out a number of important issues.

Smith relies on generalizations of empirical findings concerning temporally asymmetric phenomena of dissipation (local entropy-increasing diffusion processes or gene expression) in the brain, rather than on general statistical-mechanical (alleged or attempted) proofs of the Second Law of thermodynamics. While this approach has the advantage that it is not subject to the well-known difficulties in the attempts to prove the Second Law in statistical mechanics, it does not escape from the *circularity* charge described in Section 4.3. The reason for this is that the very claim that entropy "increases" (rather than "decreases") already assumes a direction of time, and therefore the direction of time in this proposal is either identified with the direction of entropy increase or is sensed via the direction of entropy increase. Our circularity charge (from the previous section) applies in both cases. Therefore, this line of thinking cannot explain the psychological arrow of time.

To avoid the circularity, one must appeal to some local asymmetry in the brain which is *not temporal*, that is, an asymmetry that does not involve velocity (or other such first derivatives) or notions like "increase." Given contemporary physics, this asymmetry may be in spatial or spin degrees of freedom. For example, Hemmo and Shenker (2019b) put forward a caricature proposal, just to make the point, according to which whether an event is felt as past or future is determined by whether a certain molecule in the brain is closer to our left or right ear, a property that is well defined on the face of Earth (much like the notion of up and down mentioned by Boltzmann). Needless to say, we do not mean to suggest that the brain features associated with the temporal arrow are anything like this; but what we do want is to illustrate what it would mean to have a brain feature that is not temporally oriented and nevertheless accounts for the psychological arrow of time. In this case, the mental experience that we *call* "the psychological arrow

[27] We thank Ronen Golan for discussions on this issue.

of time" has nothing to do with anything associated with "time" as we intuitively think of it: it is the name of a spatial (nontemporal) feature of the brain, which is identified with a certain mental state. The discovery of this local nontemporal asymmetry in the brain, and the fact that it gives rise to the psychological arrow, are to be parts of the study of the brain.

Is this conceptually possible? It is known in contemporary brain research that some spatial experiences are accounted for in terms of closely related spatial features of the brain itself. This is a surprising discovery, which is by no means a priori trivial: it might have turned out, from all we know about the brain, that the neural structure responsible for spatial experiences would not be organized in a spatially similar way. After all, the experience of the color red is not accounted for by anything red in the brain, and the experience of heat is not accounted for by anything hot in the brain. Still, it is a prevalent intuition that has its supporters among philosophers, that temporal experiences need to be accounted for by temporal features of the brain (e.g., Mellor 1998; Phillips 2017). Our preceding *proposal* goes against this intuition.

However, we must add a caveat here. Our discussion addresses the arrow of time only, and (as we said) we try to avoid discussing the *passage* of time. The intuition that time needs to be accounted for by time is of special strength in the context of the passage of time, mainly because contemporary science and philosophy are unable to account for the passage of time in a reductive way. It seems to us that our proposal of the "distance from the left ear" can apply to the arrow of time only, not to passage. But we will not go further into the question of passage.

The idea that the psychological arrow is created by a physical nontemporal asymmetry in the brain has the following implication: standard descriptions of the world in physics include sequences of states, parameterized by something called "time" and are somehow directed. In the present proposal, the directionality of the processes described by the laws of physics is a reflection of the psychological arrow (created by a nontemporal symmetry-breaking in the brain), rather than the other way around (which is the standard way of thinking about the arrow, as in Section 4.3). Fundamental physics describes the sequence of instantaneous states, and the appearance of a temporal direction in this sequence reflects the *nontemporal asymmetry* in the brain, which we call the "psychological arrow of time."

In this proposal, the Second Law of thermodynamics is not tautological, since it makes the nonanalytic and noncircular statement that the sequence of states is such that the "side" of higher entropy states in the sequence coincides with a certain "side" of the states of the asymmetric (nontemporal) degrees of freedom (for example, according to our preceding caricature proposal, states in which the indicative molecule is closer to the left ear). In particular, if we consider the direction of entropy increase in the Second Law of thermodynamics, the "side" on which we place the past hypothesis of low entropy is derivative and determined by the degrees of freedom in the brain that satisfy our mental and asymmetry conditions and in which the psychological arrow of time comes about. If it is in fact the case that some local asymmetry gives rise to the psychological arrow, this means that we have a direction of time *before* postulating the past hypothesis of law entropy in statistical mechanics and regardless of its truth. The past hypothesis is placed on the "side" of the sequence of states which we already experience as "past." The psychological arrow of time fixes the temporal direction of the past hypothesis, rather than the other way around. Therefore, if there were no creatures like us with this local in-brain asymmetry, the sequence of states of the universe would simply be temporally directionless, as described by our time-symmetric physics.

And so, it turns out that (on our proposal) the *past* hypothesis – the fact that the low entropy is "in the past" – is derivative, and not primitive; it is explained by the psychological arrow and doesn't explain it. Thus, our way of solving the tension between the temporal symmetry of fundamental physics and the temporal asymmetry of our experience is by reversing the explanatory direction: a local nontemporal asymmetry in the brain is the origin of our experience, as if there is a temporal direction out there; not the other way around. The psychological arrow is the basis for all other so-called "temporal" arrows (inside and outside of physics).

4.5 No Arrow: The Unreliability of Introspection

This section is somewhat different from the previous ones, as it describes an argument that does not take us to physics in any direct way. We bring it in order to complete the picture, and because it sheds light on some assumptions that underlie the cases explored in the previous sections. In both Sections 4.3 and 4.4, when discussing the reflected arrow and the created arrow, we assumed that it is a fact about us, that we *have* a psychological arrow of time (see, e.g., the argument given in the first paragraph of Section 4.4). The assumption that this is a fact is usually taken to be supported empirically, where this support is taken to be so trivial, so noncontroversial, that we need not actually carry out the experiments that test it. The experiments in question would, presumably, be in the form of asking people about their experiences, that is: they would be based on first-person *reports*, which are grounded in what is often called "introspection." The assumption that, when asked about their introspection, people (under normal circumstances) report experiencing a psychological arrow, and the assumption that these reports are reliable, are both usually taken to be noncontroversial, even truisms.

The reliability of first-person reports based on alleged introspections has, however, been challenged in the literature in cognitive studies. Those who challenge these assumptions are called "Denialists" (see Shenker 2021).[28] Denialists emphasize that there is a *gap* between an experience (of any kind) and its report (by any means), so that these are two distinct events that take place at two distinct moments, and this gap makes it possible that the two are not as strongly correlated as we usually take them to be. The intuition is that an experience causally (or otherwise) brings about its report, so that from the report one can infallibly infer the experience. But denialists describe examples in which this is not the case. This is taken to suggest more generally that introspection is not always veridical and also that the reports of experiences are not veridical (see, e.g., Jacobson-Horowitz 2008 and references therein).

Applying denialism to the case at hand, it may be that also the content of our experience of the psychological arrow of time is not veridical. Both approaches of reflected arrow (discussed in Section 4.3) and created arrow (Section 4.4) attempt to provide a physicalist explanation of the psychological arrow, and thus start off by assuming that the psychological arrow is a fact. The no-arrow approach denies this assumption or at least raises doubts concerning it. On the denialist view, "the psychological arrow" is the name of a mental state here and now, that need not reflect or represent or even be correlated with anything "out there" nor "inside" (introspectively). Perhaps it is a kind of a mental state that is *not "about"* anything; it is a purely phenomenal noncognitive state. If so, this feeling does not entail that the description of our experience in introspective reports is veridical. Be that as it may, it is still the job of physics and, in particular, of brain science which builds on physics, to explain what in the physics of the brain makes it the case that we believe (mistakenly, perhaps) that we have a psychological arrow of time, even though we don't in fact have this experience.[29]

4.6 Conclusion

The psychological arrow of time is a unique case study. Our attempt to understand it in a physicalist framework, that is, to account for the arrow by finding the physical features of the brain that are the psychological arrow of time brings to the surface ideas in physics and in brain science, and no less in the philosophy of mind, and offers an intriguing angle from which to explore them.

If, however, the psychological arrow is a *reflection* of a fact about the world "out there" as it were, then it turns out that there is a huge lacuna in contemporary physics, which calls for a radical change in it. The *new* physics will have to be time asymmetric at the fundamental level.

[28] On the invasion of introspection, see Chirimuuta (2014). For various versions of denialist views, see, e.g., Dennett (2015, 2016), Frankish (2016), Graziano (2019).

[29] See in this context Chalmers' (2018) meta-problem of consciousness.

The asymmetries known in *contemporary* physics are not fundamental and involve theories that *postulate* asymmetries that seem to be contingent, and it is hard to see what (in the relevant theories) brings them about. For example, the *postulation* of the time-asymmetric Big-Bang theory (including the idea of baby universes) explains the observed red-shift and is compatible with many other observations (though perhaps not with all observations, as was recently suggested). But, first, the interpretation of observations like the red-shift already presupposes a background direction of time. Second, it is not clear whether the cosmological arrow is to be *identified* with the temporal arrow, which is then somehow "sensed" and reflected in our experience, or whether it is contingently *correlated* in that particular way with a *primitive* arrow of time and serves as an indirect way of sensing it. In both cases, one would have to explain how we sense the cosmological arrow. Similar considerations (mutatis mutandis) arise with respect to the violation of parity- and time-symmetry in high-energy physics, which leads to the temporal arrow associated with the CPT theorem. As we saw, since the entropic arrow described by the (classical or quantum statistical mechanical counterpart of the) Second Law of thermodynamics cannot be identified with the temporal arrow, nor even give a contingent indication of it (on pain of circularity or of being a tautology), associating the cosmological arrow and the CPT arrow with the entropic arrow (as suggested by, e.g., Atkinson 2006) is of no help.

We didn't discuss in this chapter theories which are *not* time-symmetric like some quantum mechanical theories that involve a collapse of the quantum state. Von Neumann (1932) famously tried to explain the temporal direction of his "Process 1," in which a superposition is to be formally replaced by one of its components (in some basis), in circumstances sometimes called "collapse" or "measurement" (or something similar), by placing it against the background of the thermodynamic entropic arrow. This attempt is problematic since it is based on the entropic arrow, which (as we argued) cannot explain the psychological arrow of time. And so, the attempt to explain the *psychological* arrow of time as a mental state that reflects a fact "out there" (as it were), suggests that we need some new physics, presumably time-asymmetric physics or at least a substantial addition to contemporary physics, that will introduce nontautologous temporal asymmetry.

However, there is a way to connect von Neumann's Process 1 approach and our created arrow of time approach by understanding von Neumann's outlook about the relationship between the entropic arrow and the quantum arrow in the "reversed" reductionist direction: that is, taking the thermodynamic arrow to be accounted for by the quantum arrow which is assumed by his Process 1. Since for von Neumann it seems that Process 1 is ultimately grounded in the *experience* of an observer, such a "reversed" approach may coincide with our approach (in Section 4.4) of an in-brain created arrow. Here is what von Neumann (1932) says in his description of Process 1:

"[I]t is a fundamental requirement of the scientific viewpoint – the so-called principle of the psycho-physical parallelism – that it must be possible so to describe the extra physical process of the subjective perception as if it were in reality in the physical world – i.e., to assign to its parts equivalent physical processes in the objective environment, in ordinary space . . . In a simple example . . . : We wish to measure a temperature. If we want, we can pursue this process numerically until we have the temperature of the environment of the mercury container of the thermometer and then say: this temperature is measured by the thermometer. But we can carry the calculation further, and from the properties of the mercury, which can be explained in kinetic and molecular terms, we can calculate its heating, expansion, and the resultant length of the mercury column, and then say: this length is seen by the observer . . . [W]e could go still further, tracing the chemical reactions which produce the impression of this image on the retina, in the optic nerve tract and in the brain, and then in the end say: these chemical changes of his brain cells are perceived by the observer. But in any case, no matter how far we calculate – to the mercury vessel, to the scale of the thermometer, to the retina, or into the brain, at some time we must say: *and this is perceived by the observer*. That is, we must always divide the world into two parts, the one being the observed system, the other the observer. In the former, we follow up all physical processes (in principle at least) arbitrarily precisely. In the latter, this is meaningless. The boundary between the two is arbitrary to a very large extent . . . [B]ut this does not change the fact that in each method of description the boundary must be put somewhere, if the method is not to proceed vacuously, i.e., if a comparison with experiment is to be possible. Indeed, experience only makes statements of this type: an observer has made a certain (subjective) observation; and never any like this: a physical quantity has a certain value." (1932: 418–420)[30]

In his so-called psycho-physical parallelism, at least at first sight, von Neumann does not seem to have embraced a reductive physicalist approach. For him, it seems that Process 1 ultimately occurs in the observer's experience, which lies outside physics (where the physics of the world "out there" (as it were) is described by "Process 2" only, namely in nonrelativistic quantum mechanics the unitary and deterministic Schrödinger equation). However, the coincidence of von Neumann's reversed approach with our created arrow approach may make sense if one makes another "small" jump in von Neumann's outlook and adopts a physicalist theory of mind. In this case, von Neumann's attempted account of the thermodynamic arrow by the quantum arrow would have to pass via a *time-asymmetric* collapse theory in which collapses occur inside the brain precisely and only in the nontemporal degrees of freedom in which our psychological arrow of time comes about.

[30] In the preceding quote von Neumann claims quite mysteriously that unlike Process 2, it is meaningless to follow up Process 1 (in principle at least) arbitrarily precisely. Let us just say that the reason for this (we think) is that in von Neumann's approach Process 1 is genuinely (quantum mechanically) stochastic! But this issue goes beyond our discussion of the arrow of time, so we set it aside.

To sum up. We showed that in contemporary physics, which is time-symmetric, the probabilistic Second Law of thermodynamics in statistical mechanics cannot account for the psychological arrow of time. We further saw that an approach in which the psychological arrow of time (or the illusion that we have a psychological arrow) is created by some physical degrees of freedom in the brain offers a unique case for studying the physics of the mind, by considering two conditions that are necessary (and we think also sufficient) for such an account: the condition that these degrees of freedom are *mental* and the condition that they involve some *nontemporal asymmetry.*

Of course, these lessons about physics are relevant only if one starts off straightforwardly with a full-blown reductive physicalist theory of mind–brain identity. This is another case in which the philosophy of mind turns out to play a major role in the foundations of physics.

Acknowledgments and Funding

This research was supported by the Israel Science Foundation, grant numbers 1148/2018 and 690/2021.

References

Albert, D. (2000). *Time and Chance*. Boston, MA: Harvard University Press.

Allori, V. (2015). "Maxwell's Paradox: The metaphysics of classical electrodynamics and its time-reversal invariance." *Analytica*, 1:1–19.

Arntzenius, F. (2000). "Are there really instantaneous velocities?" *The Monist*, 83: 187–208.

Atkinson, D. (2006). "Does quantum electrodynamics have an arrow of time?" *Studies in History and Philosophy of Modern Physics*, 37: 528–554.

Boltzmann, L. (1964). *Lectures on Gas Theory*. Berkeley, CA: University of California Press.

Buonomano, D. V., and Rovelli, C. (2021). "Bridging the neuroscience and physics of time." DOI: https://doi.org/10.48550/arXiv.2110.01976

Callender, C. (2004). "There Is No Puzzle about the Low Entropy Past." In C. Hitchcock (ed.), *Contemporary Debates in Philosophy of Science*. Malden, MA: Blackwell, 240–255.

Callender, C. (2010). "The Past Hypothesis meets gravity." In G. Ernst and A. Hüttemann (eds.), *Time, Chance and Reduction: Philosophical Aspects of Statistical Mechanics*, Cambridge: Cambridge University Press, 34–58.

Carroll, S., and Chen, J. (2004). "Spontaneous inflation and the origin of the Arrow of Time." unpublished, http://arxiv.org/abs/hep-th/0410270.

Chalmers, D. (2018). "The meta-problem of consciousness." *Journal of Consciousness Studies* 25: 6–61.

Chirimuuta, M. (2014). "Psychophysical methods and the evasion of introspection." *Philosophy of Science* 81 (5): 914–926.

Dennett, D. (2015). "Why and how does consciousness seem the way it seems?" In T. Metzinger and J. Windt (eds.), *Open MIND*. Frankfurt am Main: MIND Group. DOI: https://doi.org/10.15502/9783958570245.

Dennett, D. (2016). "Illusionism as the obvious default theory of consciousness." *Journal of Consciousness Studies*, 23, 11–12: 65–72.

Dennett, C. D., and Kinsbourne, M. (1992). "Time and the observer: The where and when of consciousness in the brain." *Behavioral and Brain Sciences*, 15: 183–201.

Dieks, D. (2016) "Physical time and experienced time." In Yuval Dolev and Michael Roubach (eds.), *Cosmological and Psychological Time, Boston Studies in Philosophy of Science*, vol. 285, 3–20.

Earman, J. (1974) "An attempt to add a little direction to 'the problem of the direction of time'." *Philosophy of Science*, 41: 15–47.

Earman, J. (2006). "The Past Hypothesis: Not even false." *Studies in History and Philosophy of Modern Physics*, 37: 399–430.

Feynman, R. (1965), "The distinction of past and future," in *The Character of Physical Law*, Cambridge, MA: MIT Press.

Fodor, J. (1974) "Special sciences: Or the disunity of science as a working hypothesis," *Synthese* 28: 97–115.

Frankish, K. (ed.) (2016). *Illusionism. Special issue of Journal of Consciousness Studies* 23 (11–12). Reprinted as a volume by Imprint Academic in 2017.

Frigg, R. (2008). "A field guide to recent work on the foundations of statistical mechanics," in D. Rickles (ed.), *The Ashgate Companion to Contemporary Philosophy of Physics*. London: Ashgate, 99–196.

Ghirardi, G., Rimini, A., and Weber, T. (1986) "Unified dynamics for microscopic and macroscopic systems." *Physical Review D*, 34: 470–479.

Goldstein, S. (2012). "Typicality and notions of probability in physics." In Y. Ben-Menahem and M. Hemmo (eds.), *Probability in Physics*, Berlin: Springer, 59–72.

Goldstein, S., Lebowitz, L. J., Tumulka, R., and Zanghi, N. (2020). "Gibbs and Boltzmann entropy in classical and quantum mechanics," in V. Allori (ed.), *Statistical Mechanics and Scientific Explanation: Determinism, Indeterminism and Laws of Nature*, Singapore: World Scientific, 519–582.

Graziano, S. A. M. (2019). *Rethinking Consciousness: A Scientific Theory of Subjective Experience*. W. W. Norton & Company.

Gruber, R. P., Block, R. A., and Montemayor, C. (2022). "Physical time within human time." *Frontiers in Psychology*, 13, 718505. DOI: https://doi.org/10.3389/fpsyg.2022.718505.

Hemmo, M., and Shenker, O. (2010). "Maxwell's Demon." *The Journal of Philosophy* 107(8): 389–411.

Hemmo, M., and Shenker, O. (2011). "Szilard's Perpetuum Mobile." *Philosophy of Science* 78(2): 264–283.

Hemmo, M., and Shenker, O. (2012a) The road to Maxwell's Demon, Cambridge: Cambridge University Press.

Hemmo, M., and Shenker, O. (2012b). "The mathematical representation of the arrow of time." *Iyyun. The Jerusalem Philosophical Quarterly*, 61: 167–192.

Hemmo, M., and Shenker, O. (2012c). "Measures over initial conditions." In Y. Ben-Menahem and M. Hemmo (eds.) *Probability in Physics*, The Frontiers Collection, Berlin: Springer, 87–98.

Hemmo, M., and Shenker, O. (2013) "Entropy and computation: The Landauer–Bennett thesis reexamined." *Entropy* 2013, 15: 3297–3311.

Hemmo, M., and Shenker, O. (2015). "Probability and typicality in deterministic physics." *Erkenntnis*, 80: 575–586.

Hemmo, M., and Shenker, O. (2016a) "Maxwell's Demon." Oxford Handbooks Online. www.oxfordhandbooks.com/view/10.1093/oxfordhb/9780199935314.001.0001/oxfordhb-9780199935314-e-63.

Hemmo, M., and Shenker, O. (2016b). "The arrow of time." In Y. Dolev and M. Roubach (eds.), *Cosmological and Psychological Time, Boston Studies in Philosophy of Science*, vol. 285, Springer, 155–164.

Hemmo, M., and Shenker, O. (2019a) "Two kinds of high-level probabilities." *The Monist* 102(4): 458–477.

Hemmo, M., and Shenker, O. (2019b) "The Second Law of thermodynamics and the psychological arrow of time." *The British Journal for the Philosophy of Science*, axz038, DOI: https://doi.org/10.1093/bjps/axz038.

Hemmo, M., and Shenker, O. (2019c) "The physics of implementing logic: Landauer's principle and the multiple-computations theorem." *Studies in History and Philosophy of Modern Physics*, 68: 90–105.

Hemmo, M., and Shenker, O. (2020). "Maxwell's Demon in quantum mechanics." *Entropy*, 22: 269.

Hemmo, M., and Shenker, O. (2023). "Is the mind in the brain in contemporary computational neuroscience?" *Studies in History and Philosophy of Science*, 100: 64–80.

Hoefer, C., and Marti, G. (2019). "Water has a microstructural essence after all." *European Journal for Philosophy of Science* 9 (12), DOI: https://doi.org/10.1007/s13194-018-0236-2

Ismael, J. (2011). "Temporal experience." In C. Callender (ed.), *Oxford Handbook on the Philosophy of Time*, Oxford: Oxford University Press, 460–482.

Ismael, J. (2017). "Passage, flow, and the logic of temporal perspectives." In C. Bouton and P. Huneman (eds.), *Time of Nature and the Nature of Time*, Cham: Springer, 23–38.

Jacobson-Horowitz, H. (2008). "The scientific untraceability of phenomenal consciousness." *Philosophia*, 36: 509–529.

Kripke, A. S. (1980). *Naming and Necessity*, Boston, MA: Harvard University Press.

Kunze, K. E. (2015) "An introduction to cosmology." In M. Mulders and G. Zanderighi (eds.) *Proceedings of the 2015 CERN*–Latin-American School of High-Energy Physics, Ibarra, Ecuador, 4–17 March 2015, CERN-2016-005 (Geneva: CERN, 2016). Available at arXiv:1604.07817.pdf (2016); see also summary of experimental results in https://cds.cern.ch/record/2148792/plots?ln=en.

Le Poidevin, R. (2019). "The experience and perception of time." *The Stanford Encyclopedia of Philosophy* (Summer 2019 Edition), Edward N. Zalta (ed.), https://plato.stanford.edu/archives/sum2019/entries/time-experience/.

Lewis, D. (1994) "Reduction of mind." In S. Guttenplan (ed.), *A Companion to the Philosophy of Mind*, Oxford: Blackwell, 412–431.

Linden, N., Popescu, S., Short, A. J., and Winter, A. (2009). "Quantum mechanical evolution towards thermal equilibrium." *Physical Review E*, 79: 061103.

Loewer, B. (2012). "Two accounts of laws and time." *Philosophical Studies*, 160: 115–137.

Maimon, A., and Hemmo, M. (2022). "Does neuroplasticity support the hypothesis of multiple realizability?" *Philosophy of Science*, 89 (1): 107–127.

Maudlin, T. (2007). *The Metaphysics within Physics*. Oxford: Oxford University Press.

Mctaggart, J. E. (1908). "The unreality of time." *Mind*, New Series, Vol. 17, No. 68: 457–474. http://www.jstor.org/stable/2248314.

Mellor, D. H. (1998). *Real Time II*. New York: Routledge.

Mlodinow, L., and Brun, T. (2014). "On the relation between the psychological and the thermodynamic arrows of time." *Physical Review E*, 89: 052102.

Myrvold, C. W. (2021) *Beyond Chance and Credence: A Theory of Hybrid Probabilities*. Oxford: Oxford University Press.

Paul, L. A. (2015). "Experience and the arrow," in A. Wilson (ed.), *Chance and Temporal Asymmetry*, Oxford: Oxford University Press.

Phillips, I. (2017). *The Routledge Handbook of Philosophy of Temporal Experience*. London: Routledge.

Polger, T., and Shapiro, L. (2016). *The Multiple-Realization Book*. Oxford University Press.

Price, H. (1996). *Time's Arrow and Archimedes' Point*. Oxford: Oxford University Press.

Putnam, H. (1967a). "Time and physical geometry." *Journal of Philosophy*, 64: 240–247. Reprinted in Putnam's Collected Papers (Volume I), Cambridge: Cambridge University Press, 1975.

Putnam, H. (1967b). "Psychological predicates." In W. H. Capitan and D. D. Merrill (eds.), *Art, Mind and Religion*, 37–48, Pittsburgh, PA: University of Pittsburgh Press. Reprinted under the title: "The nature of mental states," in Putnam, H., *Mind, Language and Reality*, 429–440, Cambridge: Cambridge University Press (1975).

Reichenbach, H. (1956). *The Direction of Time*. Berkeley, CA: University of California Press.

Rietdijk, C. W. (1966). "A rigorous proof of determinism derived from the special theory of relativity." *Philosophy of Science*, 33: 341–344.

Rietdijk, C. W. (1976). "Special relativity and determinism." *Philosophy of Science*, 43: 598–609.

Shenker, O. (2017a). "Foundations of statistical mechanics: Mechanics by itself," *Philosophy Compass*. DOI: http://doi.org/10.1111/phc3.12465.

Shenker, O. (2017b). "Foundations of statistical mechanics: The auxiliary hypotheses," *Philosophy Compass*. DOI: http://doi.org/10.1111/phc3.12464.

Shenker, O. (2021). "Denialism: What do the so-called consciousness deniers deny?" *Iyyun: The Jerusalem Philosophical Quarterly*, 68: 307–337.

Sklar, L. (1973). "Statistical explanation and ergodic theory." *Philosophy of Science*, 40: 194–212.

Sklar, L. (1981). "Up and down, right and left, past and future." *Noûs*, 15: 111–129.

Sklar, L. (1993). *Physics and Chance*. Cambridge: Cambridge University Press.

Smith, R. (2014). "Do brains have an arrow of time?" *Philosophy of Science*, 81: 265–275.

Stoljar, D. (2024). "Physicalism," *The Stanford Encyclopedia of Philosophy* (Spring 2024 Edition), Edward N. Zalta & Uri Nodelman (eds.), https://plato.stanford.edu/archives/spr2024/entries/physicalism/.

Uffink, J. (2007). "Compendium to the foundations of classical statistical physics." In J. Butterfield and John Earman (eds.), *Handbook for the Philosophy of Physics*, Part B, Amsterdam: Elsevier, 923–1074.

Uffink, J., and Valente, G. (2015). "Lanford's theorem and the emergence of irreversibility," *Foundations of Physics*, 45: 404–438.

Van Fraassen, B. (1989). *Laws and Symmetry*. Oxford: Clarendon.

Von Neumann, J. (1955). *Mathematical Foundations of Quantum Mechanics*, Princeton, NJ: Princeton University Press.

Wallace, D. (2001). "Implications of quantum theory in the foundations of statistical mechanics." Unpublished manuscript. Preprint available at http://philsci-archive.pitt.edu/410/.

Wallace, D. (2023). "The logic of the Past Hypothesis." In B. Lowever, E. Winsbert, and B. Weslake (eds.), *The Probability Map of the Universe: Essays on David Albert's Time and Chance*, Cambridge, MA: Harvard University Press, 76–109.

Weingard, R. (1977). "Space-time and the direction of time." *Nous*, 11(2): 119–131.

Werndl, C., and Frigg, R. (2015). "Rethinking Boltzmannian equilibrium." *Philosophy of Science*, 82:1224–1235.

5

The Consequence Argument Meets the Mentaculus

BARRY LOEWER

> It is as impossible for the subtlest philosophy as for the commonest reasoning to argue freedom away. Philosophy must therefore assume that no true contradiction will be found between freedom and natural necessity in the same human actions, for it cannot give up the idea of nature any more than that of freedom.
>
> *(Kant,* Fundamental Principles of the Metaphysics of Morals, *75–76).*

Kant feared that freedom may be in conflict with natural necessity. This fear has been supported by well-known philosophical arguments that aim to show that if all events, including decisions and actions, conform to laws of nature, then none of our decisions or actions are free. Chief among these is the "Consequence Argument." Its more recent formulation and defense by Peter van Inwagen has spawned an enormous literature in response.[1] The most notable of these is David Lewis' response based on his account of counterfactuals. My reason for adding to this literature is that while Lewis' diagnosis of the argument is on the right track, his account of counterfactuals that he relies on to rebut the argument is defective and so consequently is his reply to the argument. I will develop a response that is in some ways similar to Lewis' but relies on a different and better account of counterfactuals based on an approach to statistical mechanics that goes back to Boltzmann and has recently been developed by David Albert in his book, *Time and Chance*. This account, which Albert and Loewer refer to as "the Mentaculus," provides a framework for explaining and connecting the various so-called arrows of time, including those of thermodynamics, causation, knowledge, and influence. It is the last of these arrows that is key to my response to the Consequence Argument.[2] If my response is effective, then it will turn out that physics (together

[1] Van Inwagen (1983) is the locus classicus for the argument. Van Inwagen subsequently reformulated the Consequence Argument in various ways in response to criticisms.

[2] The response to the Consequence Argument that I will defend is novel but has similarities to discussions by Hoefer (2002), Loewer (2007), Dorr (2016), and Ismael (2016).

with some philosophy), rather than conflicting with freedom, is able to rescue it, at least, from the Consequence Argument.

Digging more deeply, I will argue that metaphysical views about the nature of time and laws underlie the arguments for the incompatibility of free will and determinism and more generally for the difficulty in seeing how there can be free will in a world in which the motions of material bodies conform to fundamental laws of physics. I will conclude by showing why this is so and how the Mentaculus response to the Consequence Argument involves relacing these metaphysical views with an alternative account of laws and time more in tune with Hume-friendly metaphysics. If this is correct, then a proper metaphysical account of laws and time removes further obstacles to the compatibility of free will with laws of nature. Kant's assumption is vindicated.

A preliminary version of the Consequence Argument goes like this:

(1) Determinism: The fundamental laws of nature and the past entail the future.
(2) The fundamental laws are not up to me.
(3) The past is not up to me.

Therefore,

(4) The future is not up to me.

The upshot of this argument is that if determinism is true, then the future is not up to me; since my exercise of free will requires that the future, at least to some extent, be up to me, it follows that I lack free will. Compatibilists about free will reject this conclusion. But to be convincing, they need to say what is wrong with the argument and show why it is wrong. This is the primary aim of this chapter. A second and related aim is to explain the Mentaculus account of the foundations of statistical mechanics and show how it explains the second law of thermodynamics and also accounts for why we cannot have control over past events but can, to a limited extent, control future events. It is this account that is used to disable van Inwagen's argument. The third and also related aim is to show how certain views about laws and time make the Consequence Argument persuasive and how the Mentaculus account dispenses with these metaphysical views. The result is a view about laws that is compatible with free will.

Determinism means that a proposition specifying the entire fundamental physical state of the universe at any time *t* and the fundamental dynamical laws implies a proposition describing the entire fundamental physical state at every other time. It is usually thought that classical mechanics is deterministic.[3] In classical mechanics,

[3] There are complications that have been pointed out by John Earman and John Norton who describe worlds compatible with classical mechanical laws where determinism fails. (Earman 1986) But these don't affect the Consequence Argument.

the state of the world at t consists of the positions (or relative positions) and momenta of all particles, and the laws are the dynamical laws of classical mechanics together with the usual force laws. Since the positions of my hands at t_1 supervene on the state of the universe at t_1, it follows that the state of the universe at times before my birth entails the positions of my hands at t_1. This seems to mean that the positions of my hands at t_1 are not up to me now considering it had already been "decided" by the world's laws and its physical state at a time before I was born. The Consequence Argument is supposed to make this reasoning persuasive.

In discussions of free will, it is often pointed out that classical mechanics has been superseded by quantum mechanics as the fundamental theory of the world. Because quantum mechanical predictions are probabilistic, the theory is said to not be deterministic, and so the conflict between physics and free will is supposedly moot. But this is not so. While quantum theory as presented in physics textbooks is a fantastically accurate instrument for making probabilistic predictions, its ontology and the dynamics that the ontology obeys are notoriously obscure. The usual textbook account (the so-called Copenhagen interpretation) makes reference to "measurement collapsing the wave function" in formulating its laws without a clear account either of measurement or the ontological status of the wave function. There are both deterministic and indeterministic interpretations or reformulations of quantum mechanics that address these issues.[4] Since I don't want to get entangled in these issues, I will restrict attention to the argument that claims to show that free will is incompatible with determinism. In doing so, I will pretend that classical mechanics is true.[5]

A decision or action is up to me only if I am able to choose between two or more incompatible actions that are, in some sense, open or possible for me to choose. There are differing views about the pertinent sense of "possibility" at issue regarding the ability to choose. If one thinks that the relevant notion of possibility is physical possibility – namely compatibility with the laws and the microphysical state at a time t prior to my choosing – then the incompatibility between free will and determinism is immediate. No further argument is needed. But there are other notions of possibility that have been proposed to account for the ability involved in free choice. For example, Christian List suggests that the relevant notion of possibility involves compatibility not with the microphysical state prior to making the choice and fundamental laws but with what he calls "the agential state" and

[4] Bohmian mechanics and the "many worlds" understanding of quantum theory both have deterministic dynamics. There are so-called spontaneous collapse theories, like GRW, that modify the quantum formalism and introduce indeterministic dynamics. While these theories are not subject to the Consequence Argument, they pose other challenges to free will.

[5] There are arguments analogous to the Consequence Argument that claim to expose the incompatibility between free will and probabilistic theories, like GRW (see Loewer 1996). Van Inwagen's "rollback argument" is also designed to show that indeterminism is incompatible with free will (2000). The account in this chapter refuting the Consequence Argument also refutes the rollback argument.

whatever laws govern an agent's psychology (List 2019). Since it is plausible that an agent's psychological state together with the macroscopic physical state of her body and environment are not linked by deterministic law to her subsequent decisions and actions, there may be more than one action that is agent-possible at *t*.[6] Van Inwagen's argument aims at showing that even if one grants to the compatibilist that, in some sense, an agent is able to choose among alternative decisions, free will and determinism are incompatible.

My initial formulation of the Consequence Argument employed the expression "up to me." I understand this to mean at least that there is more than one alternative decision that I am able to make (i.e., that is possible for me) such that my decisions *influence* which alternative states of affairs actually occur.[7] For my decisions to influence what states of affairs occur is for there to be dependence of the states of affairs on these decisions.[8] This dependence is expressed by subjunctive conditionals or conditional probabilities. If the antecedent is false, then the subjunctive conditional is a counterfactual. For example, my decision at *t* to go to the university (or decision to stay home) influences whether or not I am later at the university; if I were to decide to go (stay home), I would later be at the university (not be at the university)."

A more precise definition of influence is:

A decision d influences a state of affairs o (or the objective probability of o) if and only if o counterfactually depends on d. That is, both $d > o$ and $\neg d > \neg o$ (i.e., $d > P(o) = x$, $\neg d > P(\neg o) = y$ and $x \neq y$) are true.

In other words, which state of affairs (or probability assignment to states of affairs)[9] obtains counterfactually depends on which decision I make. Note that this technical notion of influence is not exactly the same as causation.[10]

The Consequence Argument formulated in terms of influence proceeds as follows:

(1) Determinism: The total fundamental state of the world at time t and the fundamental laws imply the fundamental states of the world at all other times t'.

[6] The deterministic laws of physics (e.g., classical mechanics) connect the complete state of the universe at a given time with states at other times. I will return to this point later.

[7] The way we usually understand "*S* is up to me" requires not only that my decisions influence whether or not *S* obtains, but also that I know or have good reason to believe that my decisions influence whether ¬*S* obtains. Later, I will discuss a version of the argument that appeals to this stronger understanding of "up to me."

[8] Van Inwagen's initial formulations of the argument and ensuing discussion obscured the role of counterfactuals (1975). Lewis formulates a version of the argument in terms of counterfactuals, and Kadri Vihvelin (2013) convincingly argues a proper formulation of the argument must involve counterfactuals. Van Inwagen accepts the counterfactual formulation in response to Lewis.

[9] I will later discuss how probability should be understood in this context.

[10] David Lewis' account of causation in terms of counterfactuals (Lewis 1986) and probabilistic accounts (Williamson 2009) are both accounts of event causation. There may be counterfactual or probabilistic influence between states of affairs that are not events and so do not count as causation.

(2) LAWS: I have no influence over the laws at time t. There are no alternative decisions d_1 and d_2 that are open to me at t such that $d_1 > L_1$ and $d_2 > -L_1$, where L_1 are the laws.
(3) PAST: I have no influence over past states at time t. There are no alternative decisions d_1 and d_2 that are possible for me at t such that $d_1 > s_1$ and $d_2 > s_2$, where s_1 and s_2 are incompatible states of affairs that pertain to times prior to t.

Therefore,

(4) I have no influence over the future at t.

Since my having free will requires that my decisions influence the future, therefore:

(5) I lack free will.

Some philosophers reconcile themselves with the conclusion of this argument by arguing that a lack of free will doesn't interfere with our practice of holding people responsible for their actions.[11] Even if one finds this plausible, for those of us who think that we can influence and thereby, choose, at least to an extent, our futures, the Consequence Argument presents a challenge to discover where it goes wrong.[12] There are only four possibilities: (i) the argument is invalid; (ii) determinism is false; (iii) LAWS is false – one can influence the laws; or (iv) PAST is false – one can influence the past. I will discuss each in turn.

Rejecting the validity of this argument is not a plausible way out. This argument is not only intuitively valid, it is valid according to Lewis' or Stalnaker's framework for counterfactual logic and valid in probability theory.[13] Determinism may be false, but since we don't know that it is false and we are interested in whether it conflicts with free will, I will assume that it is true.[14] So that leaves LAWS and PAST (possibilities (iii) and (iv)) as potential targets. These premises strike many as obviously true. Laws are exceptionless and, on most views, possess some kind of necessity. How can we influence them? No one thinks that deciding to suspend the laws of gravitation while falling from a great height will be of any avail. The past,

[11] This is the route taken by John Fischer in *The Metaphysics of Free Will* (1995).

[12] Another response is that the argument presents a mystery. There is something wrong, but human beings are not capable of discovering it. This response is suggested by van Inwagen in a recent talk.

[13] Both Lewis' (1973) and Stalnaker's (1968) semantics are based on similarity orderings of possible worlds that have the following truth conditions: $A > B$ is true if and only if there is a no A and $\neg B$ world that is more similar to the actual world than an $A \wedge B$ world. For Stalnaker, the ordering is simple and discrete, while for Lewis, there may be ties and the ordering may be dense. I will assume that the ordering allows for ties but is discrete. These differences make no difference to my discussion of the Consequence Argument. The argument is valid when formulated in Lewis or Stalnaker logics for counterfactuals. Suppose I actually decide d_1 and o_1 obtains. I have influence over o_1 only if I am able to make a decision, d_2, such that $d_2 > \neg o_1$. Unless I have influence over the past or the laws, in the most similar worlds in which I decide d_2, the past and the laws are the same as they actually are. But since the actual past and the actual laws imply the actual outcome, if I have no influence over the past or the laws, I have no influence over whether o_1 occurs.

[14] In any case, as I mentioned earlier, there are other arguments lying in wait that claim to devour free will even if determinism is false (see Loewer 1996 and van Inwagen 2000).

moreover, is fixed, over and done with. How can we have any influence over it? No one thinks that they can now influence whether or not the milk spilled yesterday. Nevertheless, there have been attempts to show that either LAWS or PAST is false. I will discuss both.

Can we influence the laws? One thought which crops up in the literature on free will and laws is that on Humean accounts of laws, there is a sense in which laws metaphysically depend on events including decisions since which generalizations are laws supervene on the totality of events. So, if Humean accounts of laws can be made plausible, then this may seem to be a way to support rejecting LAWS. Helen Beebee and Al Mele (2002) call this response to the Consequence Argument, without endorsing it, "Humean Compatibilism." Here is how it is supposed to work. According to David Lewis' version of Humeanism, the Best Systems Account, laws are certain generalizations entailed by the axiom system that best systematizes the totality of fundamental events. Since my decisions and actions (and the more fundamental events that constitute them) are among the events that the best system systematizes, there is a sense in which the laws depend on my decisions and actions.

According to Beebee and Mele,

> [The] Humean perspective, then, imposes a modus tollens on van Inwagen's quick argument quoted above – and, a fortiori, on the modal version of the Consequence Argument. Since, in his deterministic world, it is up to Fred whether or not he skips breakfast, and it is manifestly not up to Fred what went on in the distant past, it is up to Fred what the laws of nature are. We admit that this may sound peculiar to some readers, but to Pat's Humean ears it is merely a way of saying that laws do not have the kind of ontological clout that would enable them genuinely to impede exercises of free will. According to the Humean view we have been assuming, there is no feature of the world that deprives Fred of the ability to skip breakfast tomorrow. *(Beebee and Mele 2002: 208–209)*

Non-Humean accounts of laws encourage the feeling that laws have ontological clout that constrain events, including our decisions and actions. Jenann Ismael suggests that that this is partially what underlies the worry that determinism conflicts with free will.

> When we adopt a globalist perspective, our activities become part of the pattern of events that make up history. Since our activities partly determine the pattern, and the pattern determines the laws, our activities partly determine the laws. But then something weird happens. We invert the order of determination and reify the laws, so that now it looks like the laws are not simply descriptions of patterns that is partly constituted by our actions but are instead iron rails built into the spatial and temporal landscape that won't let us act in any way not in accord with them. (Ismael 2016: 111)[15]

[15] Ismael's (2016) account of free will is subtle and full of insights about the nature of free will. I agree with her that reifying laws as certain non-Humean accounts do, making them "enforcers" of regularities, seems to threaten free will, but I don't think that rejecting a non-Humean account of laws is sufficient to disarm the

But whatever the merits or defects of Humean accounts of laws, Humean Compatibilism does not offer an escape from the Consequence Argument. The problem is that while on a Humean account, the laws do depend metaphysically on actual events, including my decisions and actions, this kind of dependence is not the kind that is relevant to the Consequence Argument. The kind of dependence relevant to the Consequence Argument is that my decisions influence the laws, namely that the laws are counterfactually dependent on my decisions. But the counterfactual dependence of laws on my decisions doesn't follow merely from a Humean account of laws. Humean accounts of laws are compatible with laws never counterfactually depending on my decisions. Whether or not decisions influence the laws depends on the truth conditions of counterfactual – not on the metaphysics of laws. Humeans can adopt the same recipe for evaluating counterfactuals as non-Humeans. There are accounts of counterfactuals on which if L is a law and A and $\neg A$ are compatible with the laws, then both $A > L$ and $\neg A > L$ are true whether the laws are Humean or non-Humean.[16] I will discuss such an account later.

But non-Humean accounts of laws and a companion account of time are I think behind why many find the Consequence Argument persuasive. The account I have in mind is one that understands fundamental dynamical laws as taking the state of the universe at a time and producing from it subsequent states. Some proponents of non-Humean accounts of laws, like Tim Maudlin, speak of time passing from past to future.[17] On his account, time, unlike space, possesses an intrinsic direction that laws follow in evolving states from past to future. If one has this view of law and time in mind it is not surprising that one finds that motions of my body are "produced" by the far distant past state of the universe and the laws.

I think that this view of laws is mistaken, but that is not the argument of this chapter. The response to the Consequence Argument that I will give does not depend on rejecting this non-Humean view but rather on a particular account of counterfactuals that I develop in what follows. However, that account of counterfactuals fits best with a Humean account of laws and chances, although it doesn't depend on it. If this is right, then Humean Compatibilism about free will is connected with Humeanism about laws and chances after all.

Although David Lewis is the most prominent proponent of a Humean account of laws, he does not rely on Humean Compatibilism to respond to the Consequence

Consequence Argument. Ismael argues against both LAWS and PAST. While I agree that it is PAST that needs to go, our reasons for rejecting it differ.

16 Humean accounts of laws entail that whether or not a proposition is a law supervenes on the totality of nonnomological facts, so they will endorse counterfactuals of the form, "If *A* were the case, then *L* would not be a law." But this is different from, "If *A* were the case, then *L* would not be true," and it is this latter form that is relevant to the Consequence Argument.

17 See Maudlin (2007), and for a discussion of how this view compares with Lewis' Humean account, see Loewer (2012).

Argument. He rejects LAWS not on the basis of his Humean views about laws but on the basis of his account of counterfactuals. Before discussing his response to the Consequence Argument, I will review his account of counterfactuals since it plays a central role in his response. Problems with his account of counterfactuals partly motivate my alternative account of counterfactuals and my response to the Consequence Argument.

According to Lewis, the counterfactual $A > B$ is true in the actual world, @, if and only if B is true if there is no possible world in which $A\& - B$ is true that is as similar to the actual world as a world in which $A\&B$ is true.[18] The truth conditions of counterfactuals depend on how "similarity" is understood. Kit Fine pointed out that if "similarity" is understood in an ordinary way, then the counterfactual, "If Nixon had pressed the button, there would have been a nuclear war," would turn out to be false since a world in which there is no nuclear war is surely more similar to the actual world than a world with a nuclear war. So, the relevant notion of "similarity" is not the ordinary one. What is it?

Lewis calls the type of counterfactuals related to influence and causation "non-backtrackers." Here is an example of a backtracking counterfactual: Suppose that I am on the sixth-floor ledge, and there is no safety net below me. Then the counterfactual, "If I had jumped, I would have landed safely since there would have been a safety net under me," understood as expressing a truth, is a backtracker. But "If I had jumped, I would have been badly hurt when I hit the ground," is a non-backtracker. For non-backtrackers, Lewis proposed an account of similarity that considers conformity to the actual laws as very important but allows for small, local violations of the actual laws in similar worlds. He calls these violations of @'s laws "small miracles." They are not literally miracles since they occur not in @ but in worlds similar to @ whose laws match @'s except for those of a small region. According to Lewis' well-known account, the considerations involved in determining world similarity are:

(1) Avoid big, widespread, diverse violations of @'s law. ("big miracles")
(2) Avoid even small, localized, simple violations of @'s laws. ("small miracles")
(3) Maximize the spatiotemporal region throughout which perfect match of particular fact with @ prevails.
(4) It is of little or no importance to secure approximate similarity of particular fact, even in matters that concern us greatly.

Lewis' proposal is that the truth conditions for non-backtracking counterfactuals are given by semantics in which the similarity relation satisfies 1–4. These

[18] More accurately, $A > B$ is true at @ if and only if there is a world W in which $A \wedge B$ is true and there is no world W^* that is just as or more similar than W to @ in which $A \wedge \neg B$ is true. This condition prohibits there being a most similar world to @.

conditions are intended to produce the result that the most similar worlds match the actual world up until a small miracle leads to a state in which the counterfactual antecedent, A, is true, which then evolves by the actual laws so that the counterfactual consequent, *B*, is true. For example, "If Nixon had decided to press the button at time *t*, there would have been a nuclear war" is true according to Lewis since the most similar world, *W*, in which Nixon decides to press the button at *t*, is a world that exactly matches @ up until a short time before *t* when there is a small, local violation that allows for Nixon's decision to occur and then proceeds in conformity with the actual world's laws, leading to a nuclear war. Therefore, even if the state of the universe long before Nixon was born and the actual deterministic laws entail that he doesn't decide to press the button, Nixon would have been able to press the button, and his doing so would have influenced whether there was a nuclear war.

Here are three quick observations about the truth conditions for counterfactuals based on Lewis' account of similarity that are relevant to my discussion of the Consequence Argument.

(1) If determinism is true in @, then it is clear that the most similar world to @, which is even slightly different from @ at some time, must be one in which either the history of the world is different or the laws are different or both.[19] Assuming that counterfactuals are not trivial, this is the case no matter how "similarity" is understood. On the account given by considerations 1–4, if *A* is false in @, the most similar world to @ differs from it by including a small violation of @'s laws prior to the time, *t*, at which *A occurs* and has a history that also differs from the time of the violation to *t*.

(2) Since the evaluation of typical counterfactuals with false antecedents, like the Nixon counterfactual, involves a world in which the laws of the actual world are violated in a small region, one might suppose that this account requires a Humean account of laws. But this would be a mistake. The "miracle" occurs not in the actual world but in counterfactual worlds where it doesn't violate those worlds' laws. Both the laws in the actual world and the counterfactual world can be either Humean or non-Humean laws. Nevertheless, a Humean account of laws does seem more amenable to describing a possible world in which the actual laws apply except for in a localized region for a brief period of time, from when the violation occurs to when the antecedent obtains.[20]

Lewis says that according to the notion of "similarity" characterized by conditions 1–4, a world in which Nixon presses the button, but there is no war, is less similar to

[19] This point is driven home by Vihvelin (2013).

[20] A Lewisian best system for the counterfactual world in which Nixon decides to press the button might be one that is just like the best system of the actual world with the exception that in a small region just prior to *t*, these laws don't hold, but an event leading to the decision occurs.

@ than one in which there is a war because a big miracle is required to eradicate the traces of the small miracle and subsequent button pressing. This conclusion is essential to Lewis' claim that the truth conditions determined by 1–4 entail the temporal asymmetry of counterfactuals. What is meant by "temporal asymmetry" is that if E is a relatively local event that occurs at time t, and if E had not occurred, then the world's history prior to t would have been pretty much the same while the future after t would have probably been very different from the actual future. We will later see that Lewis is mistaken that his account does capture temporal asymmetry and will propose an account that does.

Lewis uses his account of counterfactuals to respond to the Consequence Argument as follows: Assume that at the time at issue, Nixon was able to decide either to press or not to press the button. We think that had Nixon pressed the button, there would have been a nuclear holocaust, and Lewis' account seems, at first, to deliver this verdict. It would take only a small violation of the actual laws in a counterfactual world prior to Nixon's pressing the button – say, the firing of a few neurons in his brain – to lead by the actual laws to his decision to press the button, and hence, to a nuclear holocaust. So, Nixon was able to influence whether or not there would have been a nuclear holocaust. But Lewis' account also requires that whether or not there is a violation of actual laws counterfactually depends on his decision. In other words, if Nixon was able to decide either to press or not to press, he was also able to influence whether or not there would be a small violation of law prior to his deciding. Lewis rejects both LAWS and PAST, although his rejections are qualified. LAWS holds except for a very small region where there is a violation of the actual laws, and PAST holds except for the typically short period of time between the violation and the decision.

Lewis' claim that we are able to influence the laws seems incredible, at first. But he argues that it is simply a consequence of applying his account of counterfactuals to decision-making. Furthermore, once one recognizes exactly what the ability of choice consists in, the idea of decisions influencing laws can be seen as acceptable. It is the ability to counterfactually violate the actual laws in a small region of space-time prior to the decision (though they would then not be laws). He points out that possessing this ability doesn't mean that we are able to do something that causes a law-breaking event since the violation of law occurs prior to the decision. Further, the law is not an event. It doesn't give us the ability to suspend the laws of gravity or to build a rocket that travels superluminally. Even though we have the ability to *decide* to make a rocket that travels superluminally, we don't have the ability to make such a rocket. The ability to make alternative decisions doesn't violate deterministic laws. Of course, we never do make the decision that requires an actual violation of law since that decision is made not in the actual world but in another, similar world.

Recapitulating, Lewis claims that his account of counterfactuals captures the truth conditions of counterfactuals relevant to causation and influence. His response to the Consequence Argument is that it follows from his account of counterfactuals that LAWS is strictly false, so there is no additional cost to rejecting LAWS to refute the Consequence Argument. Van Inwagen agrees that this response succeeds in showing that the Consequence Argument is not a "slam-dunk," but he goes on to claim that the cost of responding to it as Lewis does is high since it involves accepting that our ability to freely choose requires "miracles."[21] Commenting on van Inwagen, Kadri Vihvelin agrees that the cost of evaluating counterfactuals with false antecedents, if determinism is true, must be either a violation of LAWS or PAST, so there is no additional cost of applying an account of counterfactuals to decisions.[22] Unless van Inwagen has an argument to show that determinism is incompatible with Lewis' account of counterfactuals and the truth of some counterfactuals with false antecedents – which would be a truly remarkable result – Vihvelin's point is sufficient to show that the Consequence Argument is unsound. Lewis' response shows one way in which van Inwagen's formulation of the argument might be unsound.

What does the work of disarming the Consequence Argument is not Lewis' particular account of counterfactuals but simply the observation that if determinism is true, then possible world accounts of counterfactuals with false antecedents require either that laws be violated or that the past differ from the actual past. Lewis picks both. While this is sufficient to defuse the Consequence Argument, I want to delve deeper into Lewis' account of counterfactuals since it turns out to be defective and this undermines his response. I will then sketch a different, better account of counterfactuals on which PAST is false while LAWS is true. This will provide, as I will argue, a more effective response to the argument.

There are two problems with Lewis' account of counterfactuals: one relatively small and the other devastating. The relatively minor one is this: it strikes many that the principle "If L is a law and A is logically compatible with L, then $A > L$" seems obviously correct.[23] It is because Lewis' account of counterfactuals violates this principle, and thus allows for "miracles," that van Inwagen thinks that Lewis' response "raises the cost" of compatibilism. In fact, it follows from Lewis' account

[21] Van Inwagen says that Lewis's paper is "the finest essay that has ever been written in defense of compatibilism – possibly the finest essay that has ever been written about any aspect of the free will problem" (2008).

[22] Vihvelin (2013) argues that the Consequence Argument doesn't raise the price of compatibilism since, assuming determinism, any account of counterfactuals requires that the most similar world to the actual in which a false antecedent is true is one in which either the laws of the actual world are violated or the past of the antecedent differs from the actual past (or both). She claims that this observation is sufficient to disarm the Consequence Argument.

[23] Marc Lange (2009) characterizes laws as propositions that satisfy the even stronger principle that if A is compatible with L and L is a law, then $A > L$ is a law. Of course, this is incompatible with a Humean account of laws even though $A > L$ is not.

that if I actually make decision d_1, then had I made decision d_2, determinism would have been false.[24] This conclusion does have an odd ring to it; it seems to say that even if both d_1 and d_2 are agent possible, there is some sort of incompatibility between determinism and my ability to choose.

The devastating problem is that Lewis' account of counterfactuals is not successful as an account of ordinary non-backtracking counterfactuals. It endorses obviously false counterfactuals, and it fails to capture the temporal asymmetry of counterfactuals as well as the influence typical of non-backtrackers. This was shown by Adam Elga who pointed out that Lewis' account of similarity delivers the verdict that if Nixon had pressed the button, there would not have been a nuclear war (Elga 2000; see also Loewer 2007).

Counterfactuals are temporally asymmetric in that counterfactual antecedents departing slightly from actuality can and often do lead to consequents departing significantly from the actual future but not the actual past. For example, if Nixon had pressed the button at time *t*, the course of events after *t* would have been radically different, but the course of events prior to *t* would have been pretty much as it actually was. For example, there would still have been a cold war, Kennedy would still have been assassinated, and so on. Lewis believed that the truth conditions embodied in 1–4 captured this asymmetry. But Elga showed that due to the temporal symmetry of the fundamental dynamical laws, Lewis' conclusion was wrong (Elga 2000; Loewer 2007). Under conditions 1–4, a world, W*, that satisfies the actual dynamical laws except for a "small miracle" after Nixon presses the button is at least as similar to @ as a world, W, in which the miracle occurs prior to Nixon pressing the button. In W*, the past of the button pressing is vastly different from the actual past while W*'s future matches that of the actual world. In W, it is the past that matches that of the actual world while its future differs.[25] The consequence is that on Lewis' account, since W* is as similar to @ as W is, the proposition "If Nixon had pressed the button, the past would have been pretty much as it actually was while the future would have been vastly different" turns out to be false.

Elga's result is fatal to Lewis' account of counterfactuals and so fatal to Lewis' response to the Consequence Argument. One might try to repair Lewis' account. One way is to alter the account of similarity so that match with respect to the history

[24] In fact, if I had made a decision other than the one I actually made, the actual fundamental laws would have been violated, and if I had done something that results in the violation of the laws of quantum mechanics, I should win a Noble Prize, it seems to follow, on Lewis's account, that if I had decided otherwise, I should have won a Noble Prize. Odd indeed!

[25] W* is constructed by temporally reversing the actual world from its end until a time shortly before Nixon presses the button; a "miracle" leads to him press the button, after which W* continues in accordance with the actual laws. W* differs from W and the actual world in that its entropy increases in both temporal directions from the time of the button pressing.

prior to the time of the antecedent is more important than the match after *t*.[26] This complicates the account and would mean giving up his hope of explaining time's arrows in terms of the asymmetry of counterfactuals. And we would still be left with the consequence that if we have the ability to freely influence the future, we also have the ability to freely falsify determinism. For these reasons, I now want to look at accounts of counterfactuals that respond to the Consequence Argument by taking the second horn of the dilemma. That horn is to reject the premise PAST instead of LAWS. If determinism is true, then this means that the most similar world to the actual world will have a different past and future than the actual world, but the deterministic laws are never violated, that is, there are no "miracles." Accounts along these lines have mostly been neglected because it is thought that a small departure from actuality at *t* can lead deterministically to very big differences in the past just as in the future. Lewis says,

> [T]here is no guarantee whatever that a world where the actual laws are true and where Nixon presses the button can be chosen so that the differences diminish and eventually become negligible in the more and more remote past. Indeed, it is hard to imagine how two deterministic worlds anything like ours could possibly remain just a little bit different for very long. There are altogether too many opportunities for little differences to give rise to bigger differences. *(Lewis 1979: 45)*

Lewis' worry is that if determinism is true, worlds which are macroscopically identical to the actual world at the time of the decision might have histories that greatly diverge from the actual history. In that case, it could turn out, for example, that if Nixon had pressed the button, Caesar might not (or would not) have crossed the Rubicon.[27] To avoid this, he devised criteria which he thought resulted in the most similar worlds being those whose pasts are almost identical to the actual world. But he needn't have worried. In classical mechanics (and in quantum mechanics), there are worlds in which Nixon pressed the button containing pasts that are macroscopically identical or almost identical to the actual past, although they differ greatly in their futures. The account of counterfactuals that I will now describe takes advantage of this fact. Instead of rejecting LAWS, my account rejects PAST.[28] Setting up the account requires a detour into physics.

The part of physics that plays a central role in rescuing free will from the Consequence Argument is statistical mechanics. Boltzmannian statistical mechanics was developed to explain thermodynamic regularities, for example, the gas laws

[26] If one does this and also adopts NP, then the result is an account of counterfactuals very close to the account that Jonathan Bennett develops in *A Philosophical Guide to Conditionals* (2003). Vihvelin employs it in her defense of compatibilism (2013).

[27] Following Albert (2000) and Loewer (2007), Cian Dorr (2016) points out that in possible worlds with the kinds of laws we think hold in our world (classical or quantum mechanical), small differences from actuality don't necessarily lead to big differences in the past.

[28] A number of other responders to the Consequence Argument also suggest rejecting PAST. Among these are Hoefer (2002), Ismael (2016), Dorr (2016), and Esfeld (2021).

and, in particular, the second law of thermodynamics. The second law says that the entropy of the universe and its suitably isolated subsystems increases over time until equilibrium (i.e., maximum entropy) is attained. The entropy of a system's macro state is measured by the size or volume (relative to a standard measure) of the set of microstates that realize it. There are arguments (due to Boltzmann and others) that make it plausible that (on the Lebesgue measure) most of the microstates compatible with the nonequilibrium macro states of an isolated system evolve in accordance with the fundamental dynamical laws to macro states with greater entropy. Boltzmann understood "most" as specifying a probability, so he argued that the second law should be understood as saying that the entropy of a system not yet at equilibrium very likely increases.[29] An instance of this law is that a half-melted ice cube sitting in a tub of warm water evolves to a higher entropy state in which the ice cube is melted. Good! But Boltzmann soon realized that, because of the temporal symmetry of the fundamental dynamical laws, it also follows that it's very likely that the system evolved from a state of higher entropy, namely., that the ice cube was melted prior to *t*. This is bad! The bad result can be avoided by positing that the entropy of a system to which statistical mechanics is applied starts out low. Applied to the entire universe, the proposal is that the entropy at the earliest time was very small. David Albert (2000) calls this assumption "the past hypothesis" (PH). He and Loewer have proposed a fundamental theory that includes the PH and the appropriate probability distribution as laws in addition to the fundamental dynamical laws that they call "the Mentaculus" that accounts for the second law and other temporal asymmetries. The Mentaculus has three ingredients:

(1) The fundamental dynamical laws,
(2) The past hypothesis (PH),
(3) An objective uniform probability distribution over the micro-histories compatible with the PH.[30]

Albert argued that the Mentaculus entails not only that the entropies of the universe and its suitably isolated subsystems increase (or are very likely to increase) from earlier to later until equilibrium is attained, but also that it plays a central role in grounding two other important temporal asymmetries – the asymmetry of records and the asymmetry of influence. Subsequently, it was shown how the Mentaculus can be

[29] "The initial state in most cases is bound to be highly improbable and from it the system will always rapidly approach a more probable state until it finally reaches the most probable state, i.e., that of the heat equilibrium. If we apply this to the second basic theorem, we will be able to identify that quantity which is usually called entropy with the probability of the particular state" (Sharp and Matschinsky 2015: 1, translation of Boltzmann 1877).

[30] It is sometimes said that if the fundamental dynamical laws are deterministic, then objective probabilities of events are trivial, either 0 or 1. But this is not true. Statistical mechanics involves objective probabilities even though dynamical laws may be deterministic and there are accounts of the nature of probability that support this. (see Loewer 2001).

used to characterize counterfactuals involving decisions (Loewer 2007). It is this account of counterfactuals that I will employ to respond to the Consequence Argument.

I won't repeat the arguments for Albert's claims here (see Albert 2000, 2015; Loewer 2007, 2020). However, I do want to emphasize one point. Despite its name, the PH does not presuppose the past–future distinction. Rather, it claims that at one time (around the time of the Big Bang), the entropy of the universe was very small. It doesn't say that this time is in our past. If the Mentaculus succeeds in explaining the temporal asymmetries of influence, records, and counterfactuals, then it implies that this time is in our past and *explains* time's arrows. In this way, the Past Hypothesis earns its name.[31] For most of the remainder of this chapter, I will show how the Mentaculus can be used to provide truth conditions for counterfactuals and how counterfactuals so characterized provide an effective rebuttal to the Consequence Argument. As a coda I will then show how the account fits into a Humean metaphysics of laws and probabilities.

The distinction between macro and microstates is central to understanding how the Mentaculus explains thermodynamics and also to how it grounds temporal asymmetries. The microstate of the world at time t is the complete, fundamental physical state at t. In classical mechanics, this consists of the positions and momenta at t of all the fundamental material particles. The thermodynamic macro state of a system S consists of the values of thermodynamic quantities – temperature, energy, pressure, frequency of radiation, mass density, and so on – averaged over small volumes of S. More generally, macroscopic properties are properties that can be expressed in language and realized by infinitely many microstates or histories. The macroscopic state of the universe relative to a coarse-graining or collection $\sum$ of macroscopic properties is the complete description of the universe in terms of the properties in $\sum$. A macro state is realizable by a continuous infinity of microstates. The entropy associated with a macro state is the measure of the set of microstates that realize it (given a uniform probability distribution, as stated in the third ingredient of the Mentaculus). The fundamental claim of Boltzmann's statistical mechanics is that if S is a suitably isolated system, whose macro state is M at time t is not one whose entropy is maximum, then the vast majority (on the usual measure) of microhistories that realize M at t realize states with lower entropy at times earlier (between t and the time of the PH microstate) and realize macro states at times after t (time after t) whose entropies are greater.[32]

[31] For elaborations and defenses of these claims, see Albert (2000, 2015) and Loewer (2001, 2007, 2020).

[32] The reason is that in the neighborhood of every abnormal, entropy-deceasing microstate, almost all states (on the standard measure) are normal, entropy-increasing states. It is also the case that every convex set of microstates contains some abnormal, entropy-decreasing states.

Given the Mentaculus' probability posit and PH, it follows that it is very likely that *S*'s entropy was less prior to *t* and greater after *t*.

The Mentaculus implies that while the evolution of the microstate of an isolated system is deterministic, the evolution of its macro state is indeterministic. There are micro-trajectories that are macroscopically identical until a time *t* when they macroscopically diverge and, in some cases, diverge drastically. The result is a branching structure for the evolution of macro-histories that looks like the illustration in Figure 5.1.

There are infinitely many micro-histories emanating from the PH that are macroscopically indiscernible until time *t* when they macroscopically diverge. There are also microstates that are macroscopically distinct for a period and then

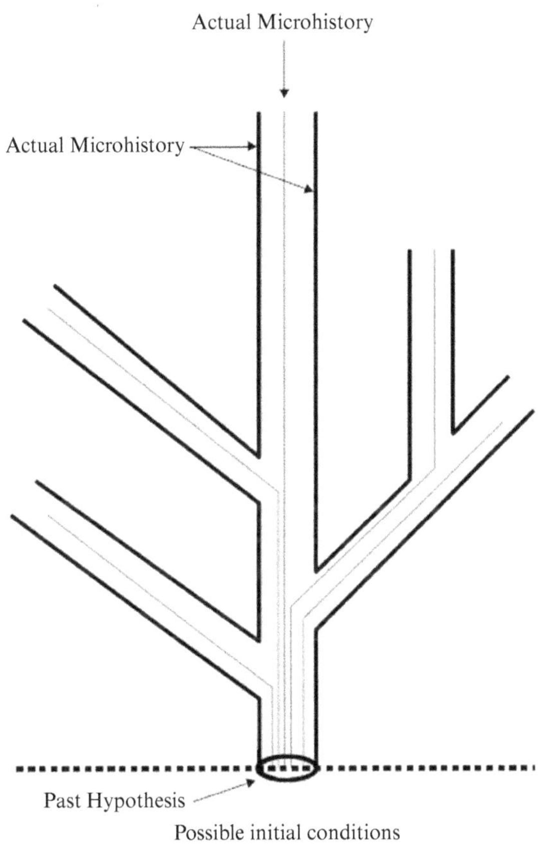

Figure 5.1 Illustration of deterministic micro histories (thin straight lines) and chancy macro histories (cylinders).

converge. But these are much fewer. As long as the time is not too long after the time of the PH, divergences greatly predominate over convergences. The overall picture is one in which, while determinism obtains on the microscopic level, indeterminism reigns at the macroscopic level.

What are the consequences for the influence of decisions? I will suppose, as seems plausible, that alternative decisions d_1, d_2 correspond to very small differences in brain states, say the firing of a few neurons. Decisions d_1 and d_2 each occur on mutually exclusive microstates that differ with respect to the firing of these neurons. I also assume that the macroscopic state at time t outside of the agent's skull is compatible with the agent choosing either d_1 or d_2. Two deterministic micro-histories h_1 and h_2 may be macroscopically indiscernible at a time t right before diverging; a moment after t, h_1 realizes decision d_1 and h_2 realizes decision d_1. Subsequent to t, the macroscopic histories H_1 and H_2, realized by h_1 and h_2 respectively, may be very different. For example, H_1 may include Nixon pressing the button and all that follows, whereas H_2 is the actual macro-history in which he didn't press the button. The macroscopic state M at t doesn't determine which decision an agent will make a moment after t, although the microscopic state that realizes M does determine the subsequent positions of Nixon's finger. This provides a sense of "possible" on which alternative decisions are available to Nixon at t. This is how the Mentaculus endorses Christian List's proposal that the ability to freely choose among distinct alternatives requires that alternative decisions and actions be possible. List thinks that his proposal is sufficient to defuse the Consequence Argument (List 2018).[33] Although freely choosing a decision is required to defuse the Consequence Argument, one must also show that either LAWS or PAST can be reasonably rejected. The Mentaculus shows why rejecting PAST is reasonable. To see how, we first need to see how decision counterfactuals can be characterized in terms of conditional probabilities grounded in the Mentaculus.

Decision counterfactuals express the probabilities of various outcomes when particular decisions are taken in current circumstances. For example, if I were now to decide to drive to the university, the probability that I would arrive at the university in 20 minutes is x. I propose that we evaluate decision counterfactuals as follows: $d_1 > P(A) = x$ is true if $P\left(A/\left(d_1 \wedge M(t)\right)\right) = x$, where $M(t)$ is the macroscopic state of the world a moment prior to the time when the decision is given by the Mentaculus.

For example, if d_1 is Nixon's decision of pressing the button, the probability of there being a nuclear war soon after d_1, given the macroscopic state

[33] List seems aware of this worry but tries to wriggle out of it by claiming that it is some kind of category mistake to combine agential and fundamental physical levels. But I don't see how this can show that the Consequence Argument is unsound.

a moment prior to d_1, will be very high. If d_2 is Nixon's decision of not pressing the button, then the probability of there being a nuclear war soon after d_2 is very low. These probabilities express the fact that Nixon wields influence over nuclear war. What about events prior to the decision? If E is a macroscopic event that occurred prior to the decision, then except under very unusual circumstances, $P\big(E/(d_1 \wedge M(t))\big) = P\big(A/(d_2 \wedge M(t))\big)$; E is probabilistically independent of d_1 and d_2 given $M(t)$. In other words, $M(t)$ screens off d_1 and d_2 from E. And if $M(t)$ contains records of E, then the probability of screening off is near 1.

Decision counterfactuals are temporally asymmetric. In this way, the Mentaculus explains the temporal asymmetry of influence – that is, why we can have some influence over the future but no (or almost no) influence over the past. "Almost no" because this account does entail that Nixon (and we) have influence over the world's past micro-history. Since determinism obtains, the alternative decisions, d_1 and d_2, occur on different micro-histories of the world, h_1 and h_2, so the counterfactuals $d_1 > H_1$ and $> H_2$ are true where H_1 and H_2 are disjunctions of the microhistories on which d_1 and d_2 occur that are macroscopically identical back to the Big Bang and differ macroscopically only subsequently after the decisions. If determinism and LAWS are true while Nixon has influence over the future, he has to have this much influence over the past. But Nixon (and our) influence over the past is useless to him. Neither he nor anyone else has preferences for which the past micro-histories realize the past macro-history. Nor does he know which microhistories realize his decisions.[34] Further, although the decision may influence H, it doesn't cause prior microstates since H is not an event but a disjunction of microhistories.

So far, my account applies only to decision counterfactuals. Showing how to extend it to counterfactuals in general is beyond the scope of this chapter.[35] But the basic idea is that in evaluating $A > B$, where A is not a decision, we look at states in which the antecedent is true that are as similar as possible to the actual macro state. The probability assigned by the Mentaculus to the macro-history leading to the state in which the antecedent is true should also be as high as possible. The result is an account that superficially gives results that are much like the results Lewis intended.[36] But it is

[34] If there is no record of an actual macroscopic event E, then the Mentaculus may assign a low probability to its occurring; therefore, the relevant counterfactual d_2 worlds are mostly ones in which E does not occur. But regardless of which decision is made, the probability of E is low, so the decision has little to no influence on whether *E obtains*. This observation dispenses with Elga's "Atlantis" objection to the Mentaculus account of counterfactuals (2000). Also see Dorr (2016).

[35] See Loewer (2007) and (2024) and for an account of counterfactuals based on statistical mechanics.

[36] This account closely connects counterfactuals and counterfactual accounts of causation with intervention along the lines of accounts like Woodward's (2003, 2007).

also significantly different from his account in a number of respects. The five most important differences are:

(1) Unlike Lewis' account, my account captures the temporal asymmetry of counterfactuals and is not subject to Elga's argument that devastates Lewis' account.
(2) Because on my account counterfactuals track conditional probabilities, all macroscopic consequents of true counterfactuals with contingent macroscopic antecedents are probabilistic. For example, the account endorses, "If Nixon had decided to press the button, there *very likely would* have been a nuclear war," but not, "If Nixon had decided to press the button, there *would* have been a nuclear war." This is because there will always be some "maverick" microstates that are macroscopically identical to the actual microstate which evolve aberrantly, for example, there is a macro state in which there is no subsequent war. But in the circumstances we believe to have obtained, "most" of the microstates in this macro state evolve so that there is a subsequent war. This is a virtue of, not a problem with the account. As Al Hajek has persuasively argued, most counterfactuals with nonprobabilistic consequents are literally false.[37]
(3) Instead of being a perfect match, most of the micro-histories in which the antecedent is true match only with respect to macroscopic events for which there are records. This has the consequence that the account violates PAST.
(4) Instead of relevant alternative histories violating the actual laws, they always conform to the fundamental dynamical laws of our world until the time of divergence. So unlike in Lewis' account, LAWS is preserved.[38] The account does not endorse "If I had decided to scratch my ear (when I didn't so decide), I should have won a Nobel Prize."
(5) The logic of counterfactuals on my account diverges from Lewis' logic in a number of ways since it is based on conditional probabilities. But developing that logic is beyond the scope of this chapter.[39]

To see exactly how my account of counterfactuals undermines the Consequence Argument, I will reformulate it in terms of conditional probabilities:[40]

(1) Determinism: if $S(t)$ and $S(t^*)$ are the complete states of the universe at any times t and t^*, then $S(t)$ and the complete set of dynamical laws entail $S(t^*)$.

[37] Hajek has several reasons for this claim, one of which is similar to mine (2014).
[38] In fact, not only is the Mentaculus upheld in these worlds, but it continues to be law even on Humean accounts of laws.
[39] See Leitgeb (2012) for an account of counterfactuals based on conditional probabilities in which a logic is developed.
[40] Although the logic of the conditional probability account of counterfactuals differs from the logic of Lewis' system, this argument is valid on the conditional probability account.

(2) LAWS: I cannot influence the laws. For all d_1, d_2 open to me at time t, if L is a law, then $P(E/(d_1 \wedge M(t))) = P(A/(d_2 \wedge M(t))) = 1$.
(3) PAST: I cannot influence the past. For all d_1, d_2 open to me at time t, if B is in the past of d_1, d_2, then $P(B/(d_1 \wedge M(t))) = P(B/(d_2 \wedge M(t)))$.

Therefore,

(4) I cannot influence the future at.

The Mentaculus response to the Consequence Argument can now be clearly stated. It is to reject the premise PAST by showing that given the physics of our world (i.e., the Mentaculus) and a proper understanding of influence, freely chosen decisions that influence the future also influence the past. But that influence over the past is negligible, involving only microscopic histories, and useless since it doesn't enable us to exert any control over past events that are of any interest to us. I am assuming that my decision can control a consequence only if I intend that my decision makes that consequence more likely, and I can tell whether I succeed. Clearly, I have no control over the microscopic past events I influence.

Like Lewis' account, my response has what may strike one as a counterintuitive consequence. Whereas Lewis' account implies that an agent has the ability to make a decision that violates laws prior to the decision, my account implies that she has the ability to make a decision that influences the past prior to the decision. Furthermore, whereas Lewis' account implies that if Newtonian dynamics correctly specifies the laws, I have the ability to falsify them and falsify determinism, my account doesn't have this consequence.

Both Lewis' account and my account have counterintuitive rings. Some philosophers respond to the suggestion that one has influence over the past, however meager, with incredulity.[41] But I think this reaction is mistaken given our views about influence and especially about time. It may be thought that one's having influence over a situation means that one can change it. But the counterfactual and probabilistic accounts of influence do not imply that one can change the past. I cannot change the future any more than I can change the past. They are what they were and will be. But I can have influence over both. The difference is that my influence over the future may involve macroscopic events that are significant and satisfy my desires while my influence over the past is negligible, noncausal, and uninteresting. If I spilled milk yesterday, there is no decision I can make now that will undo it, but I can avoid spilling milk later today.

[41] This was Peter van Inwagen's response to an earlier version of this chapter in an e-mail exchange.

Resistance to the claim that I have an ability to influence the past may also be due to adherence to an A-theory of time and its picture of time flowing. On this view of time as it flows, events are nailed down and cannot be influenced. This is encouraged by thinking about time in terms of the so-called "growing block" picture. But past events are no more "nailed down" than future events. Talk of "nailing down events," if it makes any sense at all, refers to the fact that we can influence future events in ways that we cannot influence past events.[42] All it means to say that "we have the ability to influence the past" is that the past would have been different had we made a different decision. Once that is realized, the worry that we can't influence the past evaporates and with it, the Consequence Argument.

Coda

Although the Mentaculus account of counterfactuals and my response to the Consequence Argument are compatible with non-Humean accounts of laws and probabilities, they fit better with a Humean metaphysics of both. One reason is that the Mentaculus involves a probability distribution over possible micro-histories of the universe even when the dynamical laws are deterministic. It is a widely held view that objective physical probabilities or chances require indeterministic dynamical laws and that, if dynamical laws are deterministic, then probabilities must be merely epistemic credences. Even David Lewis seems to have thought this (Lewis 2010). But this is not so. The probabilities that occur in deterministic statistical mechanics and in deterministic versions of quantum mechanics like Bohmian mechanics are objective features of the world. Like all objective physical probabilities, they guide epistemic credences, but they are distinct from them. The best account of laws and objective probabilities I know of is Lewis's "Best Systems Account" (BSA). It applies whether dynamical laws are deterministic or indeterministic.[43]

On the BSA, laws are certain generalizations entailed by the axiom system that optimally systematizes the full fundamental description of the world – what Lewis calls "the Humean mosaic."[44] Here is how Lewis adds probability to the BSA. A function term, "P(E,t)," denoting a function from propositions and times to real numbers between 0 and 1 is added to the language for candidates for best system to enhance informativeness while minimizing added complexity. Truth conditions for

[42] See Carl Hoefer's (2002) for an insightful discussion of how mistaken views about time underlie the alleged conflict between free will and determinism.

[43] This was first pointed out in Loewer (2001).

[44] The Humean mosaic is the distribution of fundamental perfectly natural properties throughout all of space-time. For a discussion, see Loewer (2024).

sentences containing the expression are then provided by specifying criteria for evaluating systems containing these sentences. The idea is that P(E,t) is true if and only if the best system of the world entails it. Here is an illustration of how this works. Consider a long sequence of the outcomes of coin tosses, for example, <HHTHTTTHHTTHTHHTTT . . .>. A very informative description of the sequence specifying each outcome would be very complicated. A simple description, for example, "approximately half Hs and half Ts" would be much less informative. But the description "outcomes of independent trials with probabilities P(H) = .5, P(T) = .5" is fairly simple and much more informative. If it is the best system for the sequence, then in the sequence P(H/T) = .5.

Understood this way, chances are not fundamental features of events. Rather, a function expression specifying chances is introduced into theories that are candidates for optimal scientific systems to provide information about the mosaic. The proposition "the chance of E at t is x" says that the actual mosaic is such that its scientifically optimal systematization(s) entail that the chance of E at t is x. In this way, chances are given by BSA laws that specify chances. Lewis's criteria for evaluating scientific optimality of a system are simplicity, informativeness, and fit. Fit is a measure of the likelihood of a history of the world or a certain select proposition on the system used to evaluate probabilistic laws.[45] The best system is the one that optimally balances satisfying these criteria.

Lewis applies this idea only to chances that are specified by a certain kind of indeterministic dynamical law. These are laws of the form "if *h* is the actual history of the universe or an isolated subsystem up until and including time *t*, the chance of *s* at *t* is *x*." But the account applies equally well when the probability distribution is over initial conditions as in the Mentaculus. In this case, fit is determined by the likelihood of macroscopic propositions on a system.

To summarize: The Mentaculus is a theory of the world that includes fundamental dynamical laws, a law specifying a low-entropy boundary condition, and a probability distribution over micro-histories compatible with it. This package accounts for the arrows of time – the second law of thermodynamics and the temporal asymmetries of records and influence. In this chapter, I applied the Mentaculus to show how free will can be reconciled with physics. First, the Mentaculus yields indeterminism at the macro-level despite being underwritten by determinism at the micro-level. This explains why our decisions appear to be undetermined and provides a sense in which we are able to choose among possible futures. Second, it shows how an agent can have significant influence over the future but not the past. It supports an account of decision counterfactuals that shows how an agent's decisions can have significant influence over the future

[45] There are issues concerning exactly how to characterize fit. For discussion, see Elga (2010) and Loewer (2012).

while having negligible influence over the past. This removes the threat that determinism and the Consequence Argument seemed to pose to free will. If my account is correct, then physics, rather than undermining freedom, is part of its explanation.[46]

References

Albert, D. (2000). *Time and Chance*. Cambridge: Harvard University Press.
Albert, D. (2015). *After Physics*. Cambridge: Harvard University Press.
Beebee, H., and Mele, A. (2002). "Human compatibilism." *Mind*, vol. 111 (442): 201–223.
Bennett, J. (2003). *A Philosophical Guide to Conditionals*. Oxford, UK: Clarendon Press.
Dorr, C. (2016). "Against counterfactual miracles." *Philosophical Review*, 125 (2): 241–286.
Earman, J. (2028) *A Primer on Determinism* Dordrecht: D. Reidel.
Elga, A. (2000). "Statistical mechanics and the asymmetry of counterfactual dependence." *Philosophy of Science*, 68 (3): 313–324.
Esfeld, M. (2021). "Super-Humeanism and free will." *Synthese*, 198: 6245–6258.
Fischer, J. M. (1995). *The Metaphysics of Free Will*. Malden: Blackwell Publishers Inc.
Hájek, A. (2014). "Probabilities of counterfactuals and counterfactual probabilities." *Journal of Applied Logic*, 12 (3): 235–251.
Hoefer, C. (2002). "Freedom from the inside out." In Craig Callender (ed.), *Time, Reality and Experience*. Cambridge, UK, Cambridge University Press, pp. 201–222.
Hoefer, C. (2019) *Chance in the World*. Oxford: Oxford University Press.
Ismael, J. (2016). *How Physics Makes Us Free*. Oxford: Oxford University Press.
Lange, M. (2009). *Laws and Lawmakers*. Oxford, UK: Oxford University Press.
Leitgeb, H., 2012, "A probabilistic semantics for counterfactuals: Part A," *The Review of Symbolic Logic*, 5(1): 26–84.
Lewis, D. (1973). *Counterfactuals*. Malden: Blackwell.
Lewis, D. (1979). "Counterfactual dependence and time's arrow." *Noûs*, 13 (4): 455–476.
Lewis, D. (1986). *Philosophical Papers (Volume II)*. Oxford: Oxford University Press.
Lewis, D. (2010). "A subjectivist's guide to objective chance." In A. Eagle (ed.), *Philosophy of Probability: Contemporary Readings*, New York: Routledge, pp. 263–293.
List, C. (2019). *Why Free Will Is Real*. Cambridge: Harvard University Press.
Loewer, B. (1996). "Freedom from physics." *Philosophical Topics*, Vol. 24 No. 2, Fall.
Loewer, B. (2001). "Determinism and chance." *Studies in History and Philosophy of Science Part B*, 32 (4): 609–620.
Loewer, B. (2007). "Counterfactuals and the second law." In Richard Corry and Huw Price (eds.), *Causation, Physics, and the Constitution of Reality (Russell's Republic Revisited)*. Oxford: Oxford University Press, pp. 293–326.
Loewer, B. (2012). "Two accounts of laws and time." *Philosophical Studies*, 160: 115–137.
Loewer, B. (2020). "The Mentaculus Vision." In Valia Allori (ed.), *Statistical Mechanics and Scientific Explanation: Determinism, Indeterminism and Laws of Nature*. Singapore: World Scientific Publishing, pp. 3–29.
Loewer, B. (2024) *The Mentaculus Vision*. New York: Oxford University Press.

[46] Of course, physics is only a part of the explanation of free will since at most, it makes free will physically possible. Free will involves making decisions on the basis of (rational) deliberation. Accounts of these aspects of decision-making are the business of psychology and rationality theory.

Maudlin, T. (2007). *The Metaphysics within Physics*. Oxford: Oxford University Press.

Sharp, K., and Matschinsky, F. (2015). "Translation of Ludwig Boltzmann's paper 'On the relationship between the second fundamental theorem of the mechanical theory of heat and probability calculations regarding the conditions for thermal equilibrium'." *Sitzungberichte Der Kaiserlichen Akademie Der Wissenschaften. Mathematisch-Naturwissen Classe*. Abt. II, LXXVI 1877, pp. 373–435 (Wien. Ber. 1877, 76:373–435). Reprinted in Wiss. Abhandlungen, Vol. II, Reprint 42, p. 164–223, Barth, Leipzig, 1909.

Stalnaker, R. (1968). "A theory of conditionals." In W. L. Harper, R. Stalnaker, and G. Pearce (eds.) *IFS. The University of Western Ontario Series in Philosophy of Science*, vol. 15. Dordrecht: Springer. DOI: https://doi.org/10.1007/978-94-009-9117-0_2.

Van Inwagen, P. (1975). "The incompatibility of free will and determinism." *Philosophical Studies*, 27 (3): 185–199.

Van Inwagen, P. (1983). *An Essay on Free Will*. Oxford: Oxford University.

Van Inwagen, P. (2000). "Free will remains a mystery." *Philosophical Perspectives*, 14: 1–20.

Van Inwagen, P. (2008). "How to think about the problem of free will." *Journal of Ethics*, 12: 337–341.

Vihvelin, K. (2013). *Causes, Laws, and Free Will: Why Determinism Doesn't Matter*. Oxford: Oxford University Press.

Williamson, J. (2009) "Probabilistic *theories* of *causality*" in *The Oxford Handbook of Causation*, pp. 185–212.

Woodward, J. (2003). *Making Things Happen: A Theory of Causal Explanation*. New York: Oxford University Press.

Woodward, J. (2007). "Causation with a human face." In H. Price and R. Corry (eds.), *Causation, Physics, and the Constitution of Reality: Russell's Republic Revisited*, Oxford: Oxford University Press, chapter 4, pp. 66–105.

6

"Freedom from the Inside Out": Revisited

CARL HOEFER*

6.1 Introduction

In a 2002 paper titled "Freedom from the Inside Out," I leveraged the so-called block universe view of time and ontology (Figure 6.1) to offer a new kind of compatibilist position in the free will *versus* determinism debate. The first key idea was that we should stop viewing the past → future direction of physical constraint and determination as somehow the "real" or natural one and stop thinking of the past as both "fixed" and entirely unaffected by our choices and actions now, or in the future. Instead, recognizing the time symmetry of physics and the ontological parity of future parts of the universe with "past" and "present" parts, we may be open to considering that future events put constraints on earlier facts as well as the more familiar *vice versa*. The second key idea was that we should embrace downward causation, namely, the idea that facts and events at higher ontological levels may be the causes and (at least partial) explainers of microphysical facts; as, for example, when my desire to finish typing this sentence in just the way you will soon see it end causes the atoms in my left middle finger to be pressing against the atoms of the 'd' key on my laptop keyboard just before typing the final period. As I type, the locations of my fingertips' atoms over time are *best* explained by my thoughts and intentions, not by anything expressible in purely physical language.

Putting the two ideas together, I argued that instead of feeling that we are held in unbreakable chains by the combination of deterministic laws plus some past full microstate of the world, we may instead regard the free choices of agents such as ourselves as the *partial* causes and explainers of the world's microstate – in both directions, toward the future *and* toward the past (see Figure 6.2). In this way we can turn the problem on its head. Our choices and actions are not what they are because of the past microstate, but rather the past microstate is what it is, in part, because of the free choices made by agents throughout space and time. This is

* 1. Department of Philosophy, University of Barcelona.
2. ICREA, Pg. Lluís Companys 23, 08010 Barcelona, Spain.

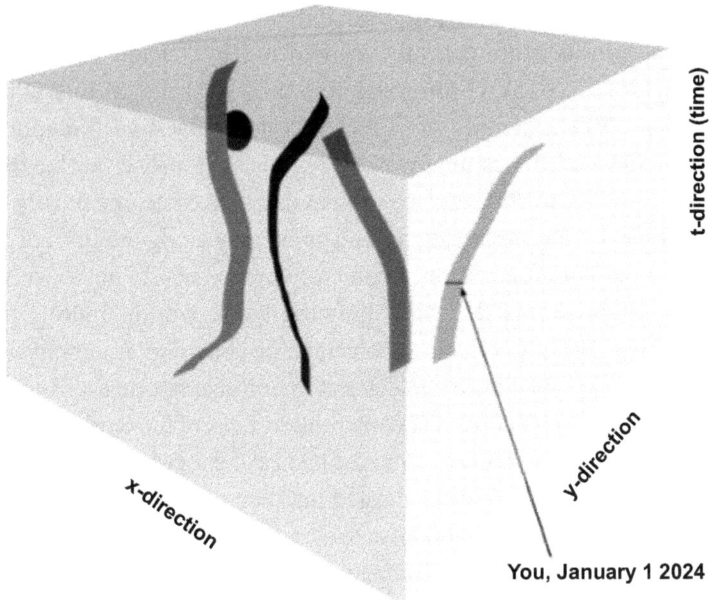

Figure 6.1 The block universe.

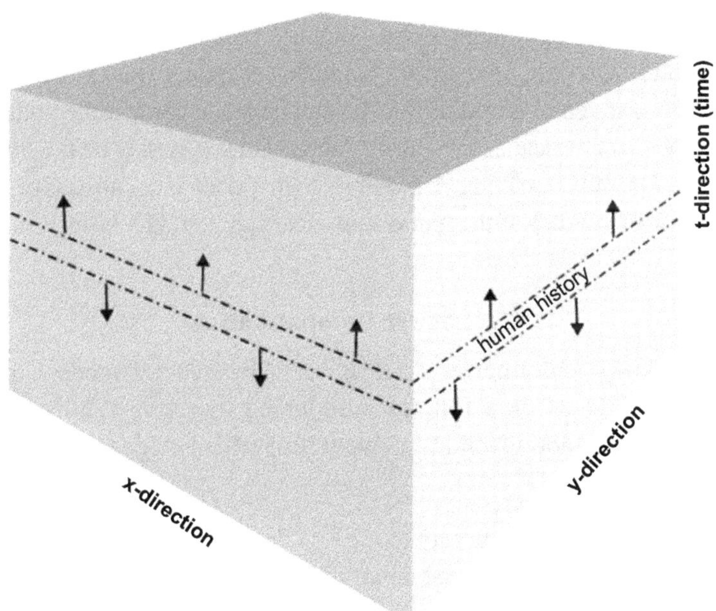

Figure 6.2 Constraint/determination relations flowing in both directions in the block.

a frankly rather idealist-sounding proposal, but one very close conceptually to other compatibilist proposals more recently worked out by Jenann Ismael (2016) and Barry Loewer (2022, 2025). (From here forward, I'll refer to this 2002 form of compatibilism as "FFIO," and refer to the 2002 paper itself as "Freedom.")

The first ingredient, the block universe ontology, is obviously crucial to the proposal, but unfortunately, between 2002 and the present day I have become mostly dissatisfied with that ontological view, preferring instead something along the lines of a "growing block" temporal ontology, according to which future events do not exist in the sense that past and present events do.[1] In a growing block world, future events, being nonexistent, cannot partially influence, constrain, or determine microphysical facts in the past or present. So, our future choices and actions cannot be ascribed freedom or explanatory priority *vis à vis* current or earlier microstates of the world in the manner of FFIO. Nor, obviously, can our current or past choices be ascribed freedom, since there were earlier times when the universe existed but they did not yet exist, and clearly – given determinism – the universal microstate at those earlier times was already real and complete, and was the unique state compatible with the way things later went (in full micro-detail). Turning the point around, our choices and actions at *any* time have been the unique ones compatible with how the universe was (i.e., its full microstate) at times when no agents yet existed. So FFIO is ruled out for all agents' actions at all times.

Now, to most of my friends and colleagues, the solution to this self-created problem will seem obvious: I should just reembrace the block universe and stop flirting with a growing block ontology, a temporal ontology widely considered to be deeply problematic in its own right. Perhaps so, but in this chapter, I want to explore whether something *close enough* to the idea of FFIO can be made compatible with a growing block ontology plus physical determinism. As we will see, it may be possible to effect this reconciliation, albeit paying a possibly high price: one must give up on the *determinacy* (or sharpness) of the world's physical states and its history.

6.2 Loewer Compatibilism

In both "Freedom" and Loewer (2025), van Inwagen's famous Consequence Argument (CA) is used as a foil for introducing the compatibilist solution on offer. In the quick version given in Loewer (2025), the CA goes:

1. The past is not up to me.
2. The fundamental laws are not up to me.
3. Determinism: The laws and past entail the future.

[1] I explain – without defending – the way I think of the growing block ontology in (2019), chapter 1. An illustration of a growing block world would be like Figure 6.2 but cutting off and discarding everything above the hypersurface corresponding to (say) the first dashed/dotted line in Figure 6.2. The half-block, representing everything that has happened so far, grows larger as time passes.

Therefore

4. The future is not up to me

Both Loewer and I reject premise 1. Supposing I actually choose A (yogurt for breakfast), despite determinism I could have chosen B (cereal for breakfast). And if I had chosen B, the past would have been different, assuming the laws stay true and as they are in actuality. Both Loewer and I think this is no problem. One can defend this way of defusing the CA without ascribing to ourselves either the ability to "change the past" or, indeed, any sort of *causal* control over anything in the past. Let's briefly see how this works.

Counterfactuals are key to our ordinary understanding of freely deciding and acting. One wants counterfactuals of this sort (which Loewer calls "decision counterfactuals") to be true: If I had decided to have cereal for breakfast, I would have eaten cereal.[2] This captures our *influence over* some aspects of the future: it's up to me what the future holds, in terms of facts about the late-morning contents of my stomach, what's available in my fridge, and so on. Some facts about the future counterfactually depend on our present (and past) decisions, fortunately. The truth of these decision counterfactuals emerges unproblematically from standard accounts of the semantics of counterfactuals.

But overcoming the CA by rejecting premise 1 also involves needing a certain sort of backtracking counterfactual to be true: for example, (DIFF): "If I had chosen to have cereal, the complete microphysical state of the world at earlier times would have been different." This in turn requires a semantics of counterfactuals able to deliver this, which Lewis' semantics at least is designed *not* to do. Loewer (2022) presents and defends just such a semantics, completing a task begun in his (2025).

Strictly speaking, Loewer's account gives a semantics just for counterfactuals with probabilistic consequents; instead of "If I had decided to have cereal for breakfast, I would have eaten cereal," what comes out as true in Loewer's account is (PROB): "If I had decided to have cereal for breakfast, the probability that I would in fact eat cereal for breakfast would have been x" (where x is in normal conditions something close to 1). In agreement with Hájek (2014), Loewer[3] thinks that most counterfactuals with contingent antecedents and nonprobabilistic consequents are false.[4] One then gets also nonspecific probabilistic consequents, as in (PROB*): "If I had decided to have cereal for breakfast, then I would most likely

[2] The truth of counterfactuals like this one is a necessary condition for freedom, although not sufficient: one wants to also have as true: "I could have chosen to eat cereal instead of yogurt, the world being as it was prior to my deliberating and deciding." We'll come back to this aspect shortly.

[3] But it seems clear that his account also entails the truth of counterfactuals like (DIFF), which has a non-probabilistic consequent.

[4] Roughly, Hájek argues that most ordinary counterfactuals with contingent antecedents and consequents are false because of indeterminism. Change the world's history so as to make it different in the way(s) described in the antecedent, and more than one possible consequence might subsequently happen. This applies to stochastic

have eaten cereal for breakfast." This is close enough to the ordinary, intuitive decision counterfactuals to be satisfying.

Telling a story about how suitable decision counterfactuals come out true, and explicitly rejecting premise 1 of CA, may formally be enough to answer the argument, but it is hardly a *fully* satisfying response! As noted in a footnote earlier, we also need for it to be the case that decision-power claims are true, such as (COUL): "I could have chosen to eat cereal instead of yogurt, the world being as it was prior to my deliberating and deciding." This has traditionally been where incompatibilists rest their case: they assume that if your story makes (DIFF) true, it cannot at the same time make (COUL) true. The past microphysical states of the world being different, all the way back to the beginning, seems hardly compatible with "the world being as it was prior to my deliberating and deciding"! Nevertheless, in both "Freedom" and Loewer (2025), it is argued that these two things are indeed compatible. Figure 6.3 (from Loewer [2025]) helps to see how this can be so.

Each thin line represents a dynamically possible, complete history of the world, with time increasing up the page.[5] Think of the circle labeled "Past Hypothesis" as containing all possible microscopic initial conditions of the universe (the square dots) that are compatible with the way we have observed the universe, *macroscopically*, to be in its early days. Obviously, our observations cannot reveal the detailed microstate of any object to us, whether it be the state of a small box of oxygen molecules, or the state of the world. So, at any time, what we can (maximally) observe about the world around us does nothing more than restrict the set of possibilities for what the complete, detailed microscopic state (now, and in the past) may be. It restricts the possibilities enormously compared to the full space of possible states for an object the size of our universe; but the possibilities remaining open are nonetheless still astronomical in number, perhaps infinite.[6]

Loewer's diagram is thus misleading in how few thin lines it contains, and one should instead imagine that every part of the tree is packed with an infinite number of thin lines. But all the time, as branching points are passed, bundles of lines are "lost" from the trunk representing actual macro-history. The evolution of macro-observable events as time unfolds looks indeterministic: there are always multiple possible future branches emerging from the trunk of what has happened up to now,

dynamics but also in case the indeterminism is only apparent and "at the macroscopic scale," as in Figure 6.3. To change the world's history so as to satisfy the antecedent involves picking one microstate from a large set of such microstates; but these microstates will not all evolve identically toward the future, in general, and so only *some* will make the consequent come to pass, not all of them. Therefore, there is no fact about what *would* (definitely) happen if the antecedent had been true.

5 Below I will sometimes refer to these lines as world-trajectories, because a more physical way to think about them is as trajectories in the configuration space of the physical theory being assumed.

6 Whether the possibilities are infinite or not depends upon the specific deterministic dynamics. From here onward I will assume that the possible states are infinite with at least the cardinality of the real numbers.

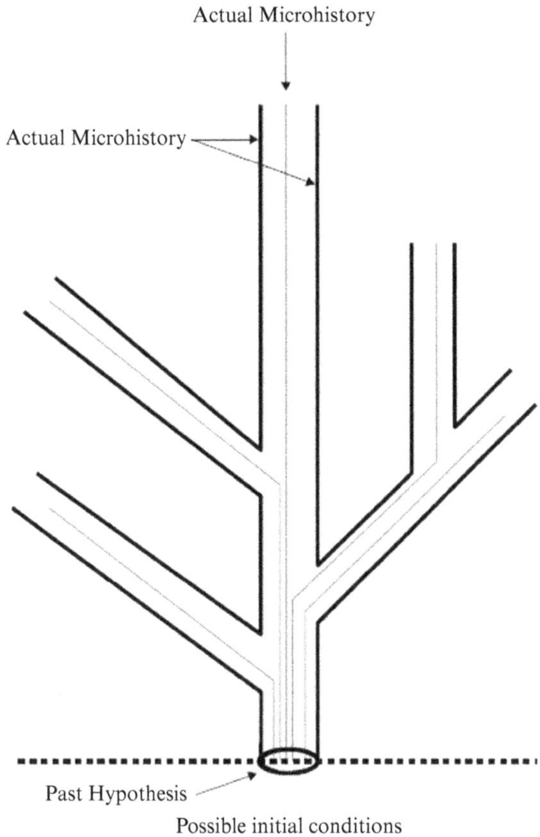

Figure 6.3 The branching-tree structure of possible macro-histories of the world.

and how things evolve is – at least sometimes – unpredictable, despite being in fact deterministic.

Some of the branching points will correspond to decisions for future action taken by rational agents such as ourselves: In the uppermost branching point, for example, I decide to have yogurt (central trunk) instead of cereal (leftward branch). Trace backward down the tree all the lines contained within these last two branches of the tree: they stay close, within a single backward trunk-path, all the way to the beginning. This is what gives us (COUL). I could have decided on cereal instead, with the world being *macroscopically, observably* exactly as it was in actuality; in this counterfactual scenario, the world's microstate history would have been one from the bundle that diverges leftward at the last branch point, instead of being one from the (disjoint) bundle that continues straight upward.

In both Loewer's version of compatibilism and FFIO, human agents thus have "influence" over the past microstate of the world: it's up to us whether it be one from *this* bundle or instead one from *that* bundle.[7] In "Freedom" I put it this way: the choices of free agents place *partial constraints* on what the universe's complete, detailed microhistory can be. These constraints go in both time directions, given determinism, but in a block universe with time-symmetric physics, so what? Importantly, we only have influence deserving the name "control" over how things evolve, at a macroscopic and observable level, in one direction: to the future. Loewer minimizes this influence over past facts by making it be *merely* the truth of certain backtracking counterfactuals (like (DIFF)). By contrast, in "Freedom" I more boldly suggested that we embrace our influence in both directions as being significant in an explanatorily important sense. Rather than treating microphysics as where the "real" action is going on, with the initial microstate "really" determining everything that happens later, I argued that rational agents impose important constraints on world history, from the inside "out" (both to the past and to the future). In a very real sense, that the universe's initial conditions are the precise way they are is properly explained, *in part*, by the fact that those conditions must be such as to accommodate the later free choices of rational agents.

Both Loewer's account and mine in "Freedom" agree that, in any event, this sort of compatibilism does not amount to saying that we can "change the past." In a block universe, the past is what it is, period; and so is the future. We cannot *change* either the past or the future, but that does not mean that our choices do not, in important ways, influence how both the past and the future are!

These sentences from Loewer nicely capture how resistance to allowing agents to have any influence at all over the past may be connected to temporal ontology:

> Resistance to the claim that I have an ability to influence the past may be due to the picture of time flowing and that as it flows events are nailed down and once nailed down cannot be influenced. This is encouraged by thinking about time in terms of the so called "growing block." (2025: 16)

This was one of the key insights of "Freedom," I thought, in 2002: the intuitive pull of incompatibilist arguments like CA rests largely on the commonsense view of time as passing and of the past as fixed.

But what if one becomes, perhaps perversely, attracted to that picture of time passing and starts seriously thinking of the world as a "growing block"? Must FFIO be then utterly abandoned?

[7] Up to us, in what sense? In the sense of counterfactual dependence, not in the sense of being something we can change, that is, make different than it has been so far.

6.3 Indeterminacy

While we tend indeed to think of past events as "nailed down" and hence not influenceable in any way, what exactly does it mean to say that they are nailed down? The most restrictive sense would be that, in terms of Figure 6.3, we should modify the figure by adding a "thin red line": choosing precisely *one* microstate trajectory from the bundle that makes it from the bottom to the very top, coloring it red, and saying: "This is the line representing our *actual* history in full detail (as opposed to the bundle containing it, which represents the possible world histories macroscopically/epistemically indistinguishable from actual history)." (The top of the diagram can then represent the present day, the edge of the growing block.) This is a natural and widely accepted assumption of analytic philosophers: that physical reality is some particular way that is sharp, precise, determinate, and complete. Vagueness, indeterminacy, and the like are aspects of our epistemic situation and our descriptions of the world using language, while reality in itself is perfectly determinate.

But Loewer's figure suggests a different way of understanding "nailed down": the past is nailed down in the sense that all the microstates in the backward-traced trunk *look the same*. Nothing we could possibly observe now (or have observed earlier) is left out or left indeterminate. But we do not have to pick a specific microstate from the bundle in that trunk and say, "*This* is how things are in full detail, though we can't know those details." Instead, we might explore the idea that physical reality itself – all that has yet happened in history – is *not* sharp, precise, determinate, or complete. The central trunk of Loewer's figure would then illustrate not just how world history looks to ideal observers with non-supernatural powers, but how the world *is*, in itself. As the block grows, reality (i.e., the past plus the present) gets in a sense *more* determinate. At every branching point, an enormous bundle of microstates ceases to be part of how the universe (indeterminately) *is*.[8]

In the rest of the chapter, I will explore this suggestion and argue that if we take it seriously, it allows a kind of restoration of FFIO. But first, some overdue remarks about physical determinism. What physics, exactly, have we been talking about?

Philosophical discussions of freedom and the threat determinism poses to it typically say nothing about what kind of physical theory is being assumed. If anything is said at all, it is usually a vague invocation of Newtonian point particle dynamics, or "something like it." More or less the same thing happens in Loewer's discussion, and in "Freedom." The Mentaculus account of statistical mechanical

[8] Loewer's figure is good for showing how things appear indeterministic at the observable level despite an underlying deterministic dynamics, and not so good at depicting just the growing block that we are after. The branches coming off the central trunk need to be "cut off" somehow, a point that we'll come back to below.

probabilities created by Albert and Loewer (Albert 2000) does likewise.[9] Often, in light of the evident fact that Newtonian physics is incapable of modeling anything like ordinary matter in all its glorious complexity (crucial for the organic chemistry that makes living agents possible), Bohmian Mechanics (BM) takes the place of Newtonian physics as the token representative of deterministic physics that involves the requisite ontological underpinnings: point particles with definite positions at all times, tracing out continuous paths in space over time. Bohmian Mechanics offers a well-defined configuration space in which a point represents the instantaneous positions of all material particles, and the history of a Bohmian universe is a trajectory (continuous path traced out in that space) suitable for being represented by the lines in Figure 6.3.

As a version of quantum mechanics (QM), BM is much closer than is Newtonian physics to being able to capture the complexity of real-world matter. But it is not able to do justice to *all* of current physics, unfortunately, even setting aside gravity: a full and satisfactory integration of quantum field theories (QFTs) into the Bohmian framework has not yet been achieved (see Wallace 2023).[10] But work is ongoing in the Bohmian camp, and it is far from clear that it is *impossible* that there be a future Bohmian QFT, still deterministic, able to capture the same empirical phenomena currently captured by standard QFT. For the purposes of the rest of this chapter, I'll pretend that this is indeed the case.

Besides the greater verisimilitude of gesturing toward BM rather than Newtonian particle physics as the assumed deterministic laws in the discussion, there is an additional virtue *vis à vis* the proposal being explored in this section. QM famously (or notoriously) is often presented as offering us a "blurry," indeterminate picture of the states of material particles. First introductions to QM frequently say things like: particles do not have determinate positions or velocities, instead only possessing "smeared out" versions of these properties. The famous Heisenberg uncertainty principle

$$\Delta x \Delta p \geq \frac{\hbar}{2} \tag{6.1}$$

also goes by the name "indeterminacy principle" at times, and standard quantum interpretations often make clear that it is not to be understood as merely epistemic, namely, merely expressing a limitation on how precisely we can know the values of

[9] In their writing related to the Mentaculus, Albert and Loewer do acknowledge that real physics is of course not Newtonian and give arguments for the claim that their approach is adaptable to quantum physics, whether under a deterministic interpretation such as BM, or a stochastic interpretation such as GRW. See Albert (2015).

[10] One sticking point is that QFTs involve particle creation and annihilation processes, whereas the basic framework of BM has a fixed number of particles that are never created nor destroyed. Bell (1984) proposed an extension of BM to include creation and annihilation processes, but at a cost: the resulting theory is stochastic rather than deterministic.

these (in themselves sharp, determinate) quantities. Instead, we are told that to the extent that particles even possess such things as positions and momenta, the possessed values are indeterminate ones, smeared out over some range of possible values in a way codified in the quantum wave function. Indeed, it is often pointed out that we need the positions of electrons and photons (for example) to be *genuinely* indeterminate, in order to have an intuitive understanding of how they can pass through *both* slits of a double-slit experiment and thus self-interfere in the way needed to explain the diffraction patterns observed on a screen.

Of course, BM – in its standard form – is designed precisely to overcome this ontic indeterminacy and restore the sharp determinacy of positions and velocities to quantum particles. The Heisenberg principle in BM is strictly an epistemic limitation that is explainable from the fundamental laws of BM.[11] So it might seem perverse to undo the core achievement of BM by suggesting that real particles' ontic states be represented not by a single sharp trajectory in the state space but instead by the bundle of such trajectories that are compatible with all current and past *observable* features of the world. But that is what I will do now – albeit only for the purpose of attempting a reconciliation of FFIO with a growing block world.

As mentioned at the beginning of this section, the idea here is to raise what is, in Figure 6.3, just an epistemic indeterminacy, to full ontic status: physical reality itself – all that has yet happened in history – is *not* sharp, precise, determinate, or complete. Rather it is objectively, ontically indeterminate in the precise sense that the bundle of possible world-trajectories remaining at the present day just is the full, complete representation of the way the world is *so far*. The dynamics of the world is just the presupposed deterministic, complete BM that is empirically indistinguishable from ordinary quantum theory.

Since the dynamical laws are deterministic, we are still *technically* engaged in the original task of responding to van Inwagen's consequence argument and trying to show how free action by agents is compatible with determinism. But notice that we have, in a different sense, turned the full dynamical situation into one of indeterminism. At any given time, multiple future branchings are included in *that which is physically possible*, namely, the panoply of distinct ways that the current complete ontic state leaves open as possible future evolutions. Given the way things are now, and have been in the past, multiple distinct future states are physically possible, which is one way of defining physical indeterminism. Our assumption here, consistent with the growing block picture, is that the world contains a single, unique history, unlike Everettian theory in which indefinitely many branches coexist in the past as well as the future (each one supposedly looking like

[11] The interference phenomena of the two-slit experiment are explained in terms of the quantum wave function, which *is* spread out in space and able to travel through both slits at once, although it is perfectly *determinate* qua mathematical object defined in configuration space.

a distinct world/history).[12] This raises a fundamental question: what is it that determines *which one* of the possible future branches becomes the actual one, at each branch point?

Notice that I don't need to worry about, or say anything about, what makes branching points occur when and where they do; that is determined by the underlying dynamics. When a subbundle of trajectories diverges from the rest sufficiently – that is, becomes such as to subvene an overall world-state that is observationally distinct from what the bundle "left behind" subvenes – we have branching. Like Everettian branching, this is ubiquitous and constantly happening. But what makes a certain branch be the one that becomes the continuation of world history, while all the others get cut off? Evidently, neither the dynamics nor the (indeterminate) physical state can supply an answer to this, so we have here (to repeat) a substantial form of indeterminism. But it is not the indeterminism of stochastic dynamics (although QM probabilities will *appear to* govern those branchings that correspond to a repeatable quantum measurement situation). Instead, we have the indeterminism of an indeterminate state evolving forward under deterministic dynamics. We have spontaneity, and a failure of physics to supply a *complete* story of why things happen as they do. This failure is part of what makes room for FFIO in this ontological picture.

Some branchings will correspond to choices freely made by rational agents, and as per the original FFIO, the proper (albeit partial) explanation for why the world's state is the way it is – why it evolved onto the branch that it did – is simply the choice freely made by the agent. Every choice freely made thus puts partial constraints on the state of the world, both toward the past and toward the future. Toward the future, we have the usual causal upshots of our choices, which amount to a limited but important amount of control. Toward the past, we can think of the world's history as getting progressively more *refined* as the flow of time advances. Every new branching point passed makes sharper the overall world-history that is developing. At each branching we have a lopping-off of trajectories that used to be in the bundle that constitutes the ontic state of the world: They are in effect *erased from history*, because from the branching moment onward one can say: the ontic state of the universe contains only those world-trajectories that evolve so as to lead to *this* observable state of affairs. The universe (and hence its history) becomes more and more definite, as time goes by. Agents' choices play a role in this process, making the path taken at certain branching points not a matter of mere spontaneity, but explicable as rational choice.

[12] Reminder: although history is singular, unique, it is not *sharp* – it is a bundle of Bohmian world trajectories, not a single trajectory. Without this uniqueness assumption – namely, taking the whole of Figure 6.3 as representing how the world is up to now (we should extend all branches until they have the same "height" on the diagram) – we would essentially have a combination of Everettian QM with BM.

In a sense, then, this very heterodox picture involves saying that we *do* change the past when we make a free choice: we refine it. Some physicists' interpretations of standard QM measurements have posited something similar, especially in discussions of delayed-choice experiments, where a position measurement of a particle taken after it has passed through a beam splitter is said to "make it the case that the particle went on the left branch of the interferometer," for example. When it comes to agential choices, in the picture I'm sketching, those choices "make it the case that the earlier physical state [of agent plus relevant environment] was such as to evolve into the later state in which that choice has been made."

6.4 Back to Loewer: FFIO-lite

The modified understanding of freedom from the inside out articulated just previously is, it seems, compatible with a growing block ontological picture of time, although it adds a twist: the increasing refinement or precision, as time marches forward, of the past state of the world. Both this twist, and the indeterminacy of the physical state, may seem like extravagant metaphysical commitments to take on, just in order to maintain something as close as possible (in a growing block world) to the original FFIO. As we saw in Section 6.1, Loewer's compatibilist response to the consequence argument is not too different from the FFIO of "Freedom," although it has a decidedly less idealistic flavor. So, it is worthwhile to take a moment to consider whether Loewer's story is compatible with a growing block ontology at a lower metaphysical cost.

As far as I can see, the answer is a clear "yes": it carries over directly and without need of modification, and it does so equally well whether one assumes a single determinate trajectory as the past history of the world, or an indeterminate bundle of such trajectories as in the approach I have explored. Because Loewer's compatibilism requires only his semantic account of counterfactuals (involving a limited form of backtracking), it is unaffected by temporal ontology as long as the ontology has room for the Mentaculus account of objective probabilities. A growing block ontology does accommodate the Mentaculus account, whether the past is determinate or indeterminate; what matters is the existence of a "Past Hypothesis" initial macrostate, and a suitable deterministic dynamics, which I have also assumed throughout this chapter.

This being the situation, a natural thought for the reader will be: why bother with FFIO, either the original or the new version, when Loewer's compatibilism overcomes the CA at the low, low price of simply adopting the Mentaculus account of probabilities plus the associated semantics of counterfactuals?

Discussing this question properly would involve getting into details about the Mentaculus story, including whether it is physically well grounded and whether it offers an acceptable account of objective probabilities; this is a huge topic, not something that can be done here, of course.[13] Moreover, while Loewer's (2022) account of counterfactuals fills in many details that are left open in his (2025), it is fair to say that it is not yet a complete semantic account of counterfactuals, nor (to my knowledge) has it been tested by other philosophers who work on counterfactuals and related topics. So, the approach of Loewer may have weaknesses that are not evident on the surface. But with that said, I have to say that it has great attractions for its elegance and simplicity, and I am happy to recommend it to those who have less idealistic leanings than me.

6.5 Conclusion

In this chapter I have tackled an inconsistency that arose in my views between 2002 and 2022, looking to see whether the ideas in my "Freedom from the Inside Out" could be modified in such a way as to be compatible with a growing block view of temporal ontology. I have tried to sketch – without intending seriously to defend – a proposal aimed at achieving this. The proposal comes with a serious metaphysical cost: the idea that the full and complete physical state of the world is *indeterminate* and subject to continual *refinement* as time passes. If these ideas had some independent support from another direction – either a utility for resolving the ongoing problems of quantum physics, or a philosophical utility of some sort – they might be worth developing further.

References

Albert, D. Z. (2001). *Time and Chance*. Cambridge, MA: Harvard University Press.

Albert, D. Z. (2015). *After Physics*. Cambridge, MA: Harvard University Press.

Bell J. S. (1984). "Beables for quantum field theory." CERN-TH.4035/84.

Frigg, R., and Hoefer, C. (2013). "The best Humean system for statistical mechanics." *Erkenntnis*, 80: 551–574.

Hájek, A. (2014). "Most counterfactuals are false." Unpublished manuscript, https://philarchive.org/rec/HJEMCA.

Hoefer, C. (2002). "Freedom from the Inside Out." In C. Callender (ed.), *Time, Reality and Experience (Royal Institute of Philosophy Supplement, Vol. 50)*, Cambridge: Cambridge University Press, pp. 201–222.

Hoefer, C. (2019). *Chance in the World: A Humean Guide to Objective Chance*. Oxford: Oxford University Press.

Ismael, J. (2016). *How Physics Makes Us Free*. Oxford: Oxford University Press.

Lower, B. (2022). "A new probabilistic account of counterfactuals." Unpublished manuscript, http://philsci-archive.pitt.edu/22971/

[13] For some criticism of the Mentaculus account of objective chances, see Frigg and Hoefer (2013) or Hoefer (2019: chapter 7).

Loewer, B. (2025), "The consequence argument meets the mentaculus." In C. Lopez and O. Lombardi (eds.), *The Arrow of Time: From Local Systems to the Universe*, Cambridge: Cambridge University Press, chapter 5.

Wallace, D. (2023). "The sky is blue, and other reasons quantum mechanics is not underdetermined by evidence." *European Journal for Philosophy of Science*, 13 (4): 1–29.

7

The Arrow of Time Is Alive and Well but Forbidden under the Received View in Physics

RUTH E. KASTNER

7.1 Introduction

Scientific research always takes place against a backdrop of particular ground-level assumptions. Such assumptions can be roughly categorized into those that are primarily (i) ontological, (ii) epistemological, and (iii) methodological. There is interaction among these categories; for example, particular ontological assumptions will dictate what is to be viewed as appropriate methodological approaches. (An illustration: it is now considered an established ontological fact that the Earth is round, so a theoretical model involving a flat Earth is viewed as methodologically inappropriate and not considered.) The set of such assumptions, along with the class of theories that is seen as comporting appropriately with them, can be described as the Received View of science at any particular epoch.

The current Received View of physics comprises (at least) four belief-assumptions.[1] I suggest categorizations for these foundational beliefs, but these are not necessarily definitive. They are:

1. The measurement problem of quantum theory is either in-principle unsolvable, or is not really a problem (i.e., not a disqualifying anomaly of standard quantum theory itself). Thus, models that claim to solve the measurement problem should be treated with a high degree of suspicion. (epistemological/methodological).
2. Field propagation is always unitary and causal (mechanistic metaphysics; so-called action at a distance is forbidden). (Standard quantum field theory picture.) (ontological/methodological)

[1] There are at least two others, which are not directly relevant to the current topic but are addressed in a separate work (Kastner and Schlatter 2023). These are:
  5. Thermodynamic entropy is equivalent to epistemic uncertainty. (epistemological/methodological)
  6. For thermodynamic purposes, measurement is copying. (methodological)

3. Physical theory, being time-reversal invariant, conflicts with the human experience of an "arrow of time" ("Two Times Problem"). (epistemological/methodological)
4. All physical systems exist in a spacetime container or against a background spacetime; acceptable theories must presuppose this background as fundamental. (A form of *actualism*) (ontological/methodological)

7.2 Critique of the Core Beliefs of the Received View

In this section, I subject the preceding Beliefs to a critical evaluation. I do not pretend that the evaluation is in any way exhaustive, since the topic under study is obviously of significant scope. However, the following considerations are the ones most relevant to the current issue, namely the alleged discrepancy between physical theory and the "human experience of time." In what follows, I abbreviate each of the Beliefs as B1, B2, and so on.

7.2.1 Belief 1: The measurement problem of quantum theory is either in-principle unsolvable or not to be considered a disqualifying anomaly for the standard theory.

The Measurement Problem (MP) can be defined as the inability of the standard formulation of quantum theory to specify what physical interaction counts as a "measurement" warranting attribution of an outcome-eigenstate to a measured system S, such that S can be accurately represented by a proper mixed state (as opposed to an improper mixed state resulting from tracing over other degrees of freedom). The MP is a direct consequence of B2, which assumes that the only *physically quantifiable* field propagation is unitary (and thus deterministic). While we will address B2 in more detail later, it plays a crucial role in the persistence of B1 and therefore the two unavoidably overlap and must be discussed together.

The most famous illustration of the MP is the Schrödinger Cat Paradox, which demonstrates the consequences of presumed-unbroken unitary evolution. The Cat Paradox was presented as a reductio ad absurdum of the theory by Schrodinger, although its original function – to demonstrate an anomaly of the standard theory – has largely been obscured. The assumed persistence of unitary evolution (B2) is a bedrock premise of the standard formulation of quantum theory (QT), whether or not QT is taken as involving a "collapse postulate" (CP). This is because a CP is not supported by any quantitative physical account under B2. Therefore, when the CP is invoked, it is an ad hoc instrumental device for consistency with the empirical data of determinate outcomes, which instantiate the Born Rule.[2] However, the Born Rule itself does not arise from the presumed ongoing physical unitarity and thus is also an ad hoc device in the standard theory.[3]

This situation leads to well-known inconsistencies (e.g., Frauchiger and Renner 2018) constituting problematic anomalies for the standard theory (Kastner 2020a, 2021, 2023). While B1 generally isn't stated explicitly, it is implied by a number of well-established behaviors among researchers. Among these are:

(a) extant attempts to portray standard quantum theory's inconsistency anomalies arising from the measurement problem as counterintuitive truisms about the world (e.g., Proietti et al. 2019, Bong et al. 2020, Brukner 2015, Baumann and Wolf 2018). This arguably constitutes anomaly neglect; see Kastner (2023), Gradowski (2022: iv, 154).
(b) ongoing neglect of an alternative formulation of quantum theory that offers a specific solution to the MP, along with a physical derivation of the Born Rule, but which denies aspects of the Received View.

Behavior (a) evades the theoretical inconsistencies arising from the MP by reinterpreting them not as shortcomings of the theory but as putatively valid predictions of incommensurable differences among different observers. Instances of (a) are critically discussed, for example, in Kastner (2021) and (2023a). These works point out that it is in fact untenable to use the inconsistencies as arguments for a conclusion that QT predicts incommensurable differences among different observers, because contrary to the usual assumption in discussions of the Frauchiger–Renner (FR) inconsistency, such observers could in fact communicate their experimental results, making these and their associated probabilities empirically comparable. In other words, the inconsistent results cannot be considered private and incommensurable in support of claims such as "different observers see irreconcilable facts"; instead, the standard theory itself (i.e., the theory under B2) generates empirically consequential inconsistencies. In view of this situation, the most charitable conclusion for the standard theory is that one of the observers must find that he is mistaken about his state assignment even though he has applied the theory in the standard way (i.e., upon obtaining an outcome, applying the relevant eigenstate to the system). The less charitable conclusion is that the theory is simply incoherent and unable to yield a consistent prediction at all. Even under the

[2] This ad hoc character is reflected in the so-called "Shifty Split" in which there is no principled physical account of which systems are to be taken as described by quantum theory under B2. On one side of the "split" the theory is simply suspended without theoretical justification. The resulting lacuna is often covered by recourse to "consciousness," but this is unsatisfactory since it leads to Wigner's Friend-type inconsistencies.

[3] The Born Rule was famously arrived at as an educated guess by Max Born (1926).

charitable conclusion, the theory is rendered a flawed instrument that may require updating of incorrect state assignments.

Regarding behavior (b), which evinces a methodological choice on the part of the mainstream physics community: a solution to the measurement problem is generally eschewed if it does not subscribe to the Received View, and as noted earlier, there cannot be an internally consistent solution to the measurement problem (at least not one fully consistent with the Born Rule of standard quantum theory)[4] under B2 of Received View. However, there does exist a specific solution which rejects B2, namely the Relativistic Transactional Interpretation (RTI; e.g., Kastner 2022). RTI has been available for over a decade (with a precursor form available since the 1980s),[5] but it has thus far not been admitted into mainstream discussion as an eligible approach. Gradowski (2022: 6, 154) discusses this phenomenon of epistemic insensitivity of the mainstream community to competitor theories.[6] Of course, the MP is not the focus of the current work, which concerns the arrow of time; nevertheless, the latter is not really separable from the topic of the MP, insofar as the transactional (RTI) solution to the MP yields not only the Born Rule but also a physically real arrow of time, consistent with our empirical experience of change (this point is reviewed in Section 7.3). Thus, entrenchment of the Received View leads to multiple obstacles and problems that are indeed readily resolved upon relinquishment of inappropriate metaphysical beliefs and their attending methodological impediments.

Before turning to the remainder of the Received View beliefs, we should acknowledge here that of course there are many who would claim that their preferred interpretation of the standard formulation of quantum theory (which supervenes on B2) does solve the MP. For example, it is often claimed that the Bohmian hidden variable theory (BHV) or Everettian ("Many Worlds") approaches solve the MP.[7] However, these approaches remain subject to the empirically fatal inconsistencies arising from the Wigner's Friend scenarios such as Frauchiger–Renner (2018) (again, see Kastner 2020a and 2023a for why these inconsistencies are empirically consequential). Since these inconsistencies arise directly from the standard theory's inability to define the conditions for the occurrence of an outcome *accurately meriting an eigenstate description* – which is the MP – these approaches

[4] E.g., so-called spontaneous collapse theories such as the Ghirardi–Rimini–Weber (1986) theory, which add an ad hoc term to the Schrodinger evolution, deviate from the Born Rule.

[5] E.g., Cramer (1986).

[6] Gradowski's observation (2022: 154) is especially relevant in this regard: "The inability or unwillingness of the mainstream to entertain promising alternative theories can lead to both theory-ladenness and anomaly neglect or even blindness, as when continental fixists called the jigsaw-fit of the continents an illusion."

[7] An argument by Norton shows that decision-theoretic derivations of the Born Rule, such as those appealed to under the Everettian approach, are circular since they presuppose the universal necessity of probabilities at the outset. (Norton 2021, Chapter 10). See also Kastner (2016) for a circularity problem in the Everettian approach.

fail to solve the MP despite their provision of narratives that seem to support the occurrence of an outcome.

7.2.2 Belief 2: Field propagation is always unitary; "action at a distance" is forbidden.

We now turn to further specifics of Belief 2, which stems from the metaphysical commitment to a mechanistic notion of causality inherited from classical physics. This is what we might call the "bucket brigade" (BB) picture, in which it is assumed that any influence among systems must always be communicated locally via a mediating field at sub-light or light speed in a future-directed process. This belief is implemented in current practice by way of quantum field theory (QFT), in which it is postulated that a quantum field operator be associated with every spacetime point, acting as an element of a "bucket brigade" that carries conserved quantities "through spacetime" from a source to a sink in a time-asymmetric way via the Feynman "causal" propagator exclusively. The mediating field is viewed as a necessary physical structure for the propagation of any physical influence. No other sort of propagation is permitted under this belief. In particular, so-called action at a distance is viewed as particularly offensive and scientifically unacceptable. However, this belief turns out to be inconsistent with B3. (We will see why when we deal with the specifics of B3 in Section 7.2.3.)

Underpinning B2 is the traditional classical adherence to a Democritan mechanistic metaphysics. Examples of the commitment of prominent physicists to this metaphysical picture are the following statements by Christian Huygens and William Thomson:

> In the true philosophy we can see the causes of all natural effects in terms of mechanical motions. And, in my opinion, we must admit this, or else give up all hope of even understanding anything in physics. *(Huygens,* Traité de Lumière, *1690)*

> It seems to me that the test of 'Do we or do we not understand a particular topic in physics?' is 'Can we make a mechanical model of it?'.
> *(Thompson,* Notes of Lectures on Molecular Dynamics, *1885)*

However, it should be noted that Heisenberg's breakthrough in QT began only when he gave up on precisely such mechanical models and instead allowed the empirical data to direct him to the correct form of the theory, which decidedly conflicts with the local realism implied by classical mechanistic metaphysics (cf. Kastner 2022, chapter 2). This historical fact should provide at least some modest support for questioning the preceding mindset, however esteemed were the classical physicists who subscribed to it. Yet it appears clear from current practice of insisting on a local, mechanical model of quantum fields (and more generally any physical process taken as legitimately explicated) that B2 is still in full force.

Einstein's abhorrence of "spooky action at a distance" remains in play to this day, as we see in ongoing attempts to "save locality" in QT (see Henson 2017, Kastner 2017b, and Kastner 2011 for a critical discussion of such attempts).

7.2.3 Belief 3: Physical theory, being time-reversal invariant, conflicts with the human experience of an "arrow of time."

Turning now to B3: time-reversal invariance is the property in which a theoretical quantity does not change upon reversal of the sign of the time index. Of course, this belief directly gives rise to the "Arrow of Time" problem. The view that physical law is time-reversal invariant is often stated as if it were established fact, for example in Buonamano and Rovelli (2023) and by Carroll in a 2010 interview. Carroll says:

> The weird thing about the arrow of time is that it's not to be found in the underlying laws of physics. It's not there. So it's a feature of the universe we see, but not a feature of the laws of the individual particles. So the arrow of time is built on top of whatever local laws of physics apply.[8]

However, the preceding statement, however categorically stated and consensus-endorsed, turns out to be false: the theoretical requirements for the propagation of conserved currents (such as energy) as well as the standard approach to QFT, both of which apply at the level of individual particles, *are time-asymmetric*. We explicate this point further in Section 7.2.3.2. For now, we note that the propagators of relativistic quantum theory are all explicitly temporally directed and decidedly not time-reversal invariant. Examples are the retarded, advanced, and Feynman propagators. Thus QFT reflects an arrow of time, as it must in order to propagate real conserved currents (see Section 7.2.3.2).

Belief 3 thus appears mistakenly to take time-reversible Newton mechanics (i.e., neglecting any dissipative forces) as representative of all physical law applicable to individual systems. This situation naturally gives rise to consternation about how to reconcile physical theory with the "human experience of a flow of time"; that is, to the so-called Two Times Problem (TTP). Since the Received View precludes a resolution of this discrepancy in terms of putative physical law, proposed resolutions attribute it to illusory or otherwise nonveridical features of human experience (i.e., aspects of psychology or neurology) or to ad hoc perspectival assumptions (e.g., we are "moving through" a putative block world).[9] The latter begs questions such as "Why and how are life forms 'moving through' the purported block?", and "Why and how do some systems count as 'moving through' (e.g., humans) and not other systems (e.g., rocks)?"

[8] Transcript of the interview is available at www.wired.com/2010/02/what-is-time/.

[9] See Kastner (2023b), footnote 2 for why purported empirical demonstrations of the truth of the perspectival account are unsound since they amount to affirming the consequent.

7.2.3.1 Belief 3 Involves a Double Standard Concerning "Empirical" Data

This situation raises a concern regarding what sorts of phenomena count as "veridical," and at a more basic level, what counts as empirical data serving as the standard for such a criterion. Physics is first and foremost an empirical science, in that its theories must be supported by empirical data and, at the very least, must be consistent with – as in, not contradict – empirical data. In this regard, it should be noted that considering the human dynamical experience of time as illusory in order to support beliefs such as B3 effectively contradicts the empirical data, which is, for example: "the system was at x_1 at time t_1 and at x_2 at time t_1," where those events were experienced as temporally separate (i.e., not simultaneously). Let us call those sorts of empirical observations *change-data*, where the term is defined by different states instantiated by a system in an experienced temporal sequence. That is what was seen and what can be corroborated among different observers; even though they may use different coordinate systems, no observer can dispute the change *empirically*. Thus, taking change-data as nonveridical constitutes a redefinition of what counts as "empirical data" in order to uphold beliefs such as B3, which is not actually based on empirical data but on certain theoretical or metaphysical models. Yet, alas, the theoretical models (such as wave equations) propping up B3 cannot actually support real energy propagation (this point is discussed later in Section 7.2.3.2).

Furthermore, this tacit redefinition of what counts as "empirical" constitutes a double standard. As is well known, in the late nineteenth century it was believed that classical physics was nearing a final triumph in successfully accounting for all the empirical data available. However, the two "small clouds" on the horizon were empirical data that classical physics failed to predict: (1) the blackbody frequency spectrum and (2) discrete atomic spectral lines. The physics community correctly concluded that classical physics was inadequate, since it was inconsistent with the empirical data in these cases. (As noted earlier, Heisenberg only made progress in developing QT when he abandoned classical mechanical model-making – that is, he rejected B2 as exemplified by the views of Huygens and Thompson). What we have today in the form of the so-called "Two Times Problem" is change-data that similarly refutes certain core beliefs. However, instead of discarding those inconsistent core beliefs, as was done for example in Heisenberg's successful development of QT, the prevailing approach is to disqualify change-data from being the empirical data that it in fact is, and instead to declare it illusory or otherwise ineligible to qualify as scientific data. That this constitutes a double standard is clear from the fact that at no time did physicists conclude that the phenomena of atomic spectral lines were illusory features of human subjective experience and attempt to look for psychological or neuroscientific explanations for these because classical physics –then the Received View – was inconsistent with the data. Since ultimately, all empirical data is "subjectively experienced," the double standard is to disqualify corroborable change-data while accepting other types of corroborable data, such as the blackbody spectrum and atomic spectral lines.

We can entertain a possible objection to the assertion that change-data is empirical data, as follows. The objector could attempt to claim that certain kinds of "empirical data" have been shown to be illusory, as for example in the case of the appearance of the Sun going around the Earth, and that change-data can therefore be illusory in the same manner. However, this would be a false equivalence. The idea that the Sun goes around the Earth is not the empirical data itself, but rather what seems to be an obvious *inference* from the data. The data itself is nothing more than: at a time t_1 the Sun was at location x_1 (in some local coordinate system) and at time t_2 the Sun was at location x_2. That data remains valid regardless of our theory about which body is going around which, and indeed the corrected theory – that it is the Earth that is going around the Sun – *is consistent with the same data* while better accommodating additional data (observed planetary motions). Change-data is precisely the same sort of basic empirical data as above. It is neither a theory nor a metaphysical assumption.[10] Thus, this objection fails.

In view of the above considerations, the Two Times Problem can be understood as a pseudo-problem that arises only from critical defects in the Received View, and that in fact physical theory is perfectly consonant with our empirical experience of change and becoming, namely, an Arrow of Time. This is a case that has been made in the literature before, in various forms (e.g., Reichenbach (1953, 1956)[11], Sorkin (2003), Kastner (2012, 2017a, 2022, 2023a), Schlatter and Kastner (2023), Schlatter (2021); but the relevant points have been disregarded or overlooked (again, apparently due to entrenchment of the Received View; see Gradowski 2022: iv, 154). They are reviewed again in Section 7.3, but the interested Reader is invited to peruse the relevant references for further details.

7.2.3.2 Belief 3 Is Contradicted by Current Theory

While Carroll's remark at least acknowledges that change-data are empirical data, his assertion that temporally oriented behavior is "not a feature of the laws of the individual particles" (i.e., B3) is falsified by the following facts:

(a) B3 does not actually describe current practice in the theoretical description of individual parties; and

[10] Of course, data can in principle be theory-laden. However, the only theory-ladenness in this sort of data is the invoking of some coordinate system, and we can recognize that the coordinate system is not absolute without discounting the data.

[11] In fact, Reichenbach argued that physical causation, in the sense of coming into being, was relevant to the Lorentz invariance of time order for timelike or null-separated events (Reichenbach 1956: 25).

(b) B3 actually precludes any unitarily propagating field as assumed in B2.

Specifically: the Feynman propagator D_F of standard QFT is not time-reversal invariant, as it describes positive energy propagating in the positive temporal direction and negative energy propagating in the negative time direction. It explicitly breaks time symmetry. Thus, standard QFT is, by construction, *not time-reversal invariant*. The only time-reversal invariant propagator is the time-symmetric propagator $\overline{D}$, which is the most general form of field propagation (both positive and negative energies in both temporal directions). However, $\overline{D}$ is the principal part of a complex function (e.g. Davies 1971).[12] This means that it cannot propagate any real (on-shell) energy, and thus cannot support unitary evolution understood as the propagation of any on-shell field excitation: $\overline{D}$ corresponds only to virtual (off-shell) fields. *The only way that propagation of real energy can occur is if time-symmetry is broken.* I italicize the previous point since it is generally not recognized, although the breaking of time symmetry is nevertheless imposed "by hand" in the service of upholding other foundational beliefs such as B2. In standard QFT this consists of imposing a "causal" propagator, such as the Feynman propagator, as an ad hoc choice driven by B2.

Thus, standard practice is to break time symmetry "by hand," at the level of individual particles; and thus B3 – the belief that the applicable physical laws are time-symmetric at the level of individual particles – is unambiguously false. The foregoing demonstrates that the Received View has self-contradictory aspects, such that subscribing to it necessitates overlooking these inconsistencies. In particular, one cannot claim that the laws of physics are time-reversal invariant (B3) while also insisting on a mechanistic model that propagates energy in a preferred temporal direction (B2). Of course, as noted earlier, we must have temporal symmetry breaking in order to have any propagation of real, on-shell energy, and that is why we can indeed have a real Arrow of Time consistent with physical theory. But this does not require a mechanistic metaphysics, and it can indeed allow for a form of time-symmetry to play a supporting, if not universal, dynamical role. We deal with this issue further in Section 7.3.

7.2.4 Belief 4: All physical systems exist in a spacetime container or against a background spacetime.

We turn now to Belief (4), the assumption that to be physically real, an entity must be an element of the spacetime manifold. This assumption is reflected in the

[12] Specifically, the Green's function $f(x)$ for a quantum field has the general form $f(x) = 1/x \mp i\pi\delta(x)$, and the time-symmetric propagator $\overline{D}$ is the real part of this function, which does not include on-shell energies. The latter appear in the delta function, that is, in the imaginary part.

reference literature, for example in this excerpt from the Stanford Encyclopedia entry on Quantum Field Theory:

> But is a systematic association of certain mathematical terms with all points in space-time really enough to establish a field theory in a proper physical sense? *Is it not essential for a physical field theory that some kind of real physical* properties *are allocated to space-time points*? This requirement seems not fulfilled in QFT, however. Teller (1995: ch. 5) argues that the expression *quantum field* is only justified on a "perverse reading" of the notion of a field, since no definite physical values whatsoever are assigned to space-time points. Instead, quantum field operators represent the whole spectrum of possible values so that they rather have the status of observables (Teller: "determinables") or general solutions. *Only a specific* configuration, *i.e., an ascription of definite values to the field observables at all points in space, can count as a proper physical field.* *(Meinard 2023)*

The latter excerpt features two key assertions of Belief 4, which I have italicized above: (i) the tacit ontological/theoretical requirement of a spacetime background or container, along with (ii) the declaration that anything physically real be an element of that manifold. Specifically: in (ii), the author declares that the requirement for a "real physical" entity is that it have definite values against this assumed spacetime background.[13] The preceding excerpt serves to illustrate the prevailing established view that the term "physically real" is equivalent to "is an element of the spacetime manifold," where the latter is understood as the delimiter of physical existence. This can be understood as a form of *actualism*. However, my use of the term is not an endorsement of the conventional definition of that term as the opposite of *possibilism*, where the latter is defined in terms of possible world semantics. In my usage, an *actual* is simply an element of the spacetime manifold, i.e., an invariant *event*. Such entities are to be distinguished from other real physical entities and processes – such as quantum systems and their interactions – that are not elements of the spacetime manifold. The latter can be understood as *physical possibilities* or a form of *res potentia* (Kastner, Kauffman, Epperson 2018) that are the ontological referents of QT. Thus, QT can be understood as describing a substratum for spacetime. While this is, of course, a startling idea given our phenomenal impression that we live in a spacetime container, there are compelling theoretical reasons to accept it; we briefly touch on these in what follows. (And it is worth mentioning in this regard that it was equally startling to consider the idea that the Earth was actually going around the Sun.)

[13] Technical point: "Having the status of observables" in the excerpt glosses over the fact that field creation and annihilation operators are not actually observables, since they are not operators on Hilbert or Fock space, which is the domain of the field excitation states constituting quanta. The latter are characterized by number operators, the legitimate observables. Thus, the Received View (which includes standard quantum field theory) has no theoretical place for the quantum field operators; they are not quanta in Hilbert space nor are they elements of spacetime. Yet standard QFT defines its basic theoretical entity in terms of such operators. This inconsistency is remedied in the heterodox direct-action picture of fields (Kastner 2015).

While in recent years some researchers have been entertaining the possibility that spacetime as a continuous entity or delimiter is not fundamental (as discussed for example in Becker 2022), the theories under consideration still presuppose a form of spacetime actualism. In particular, theories such as Loop Quantum Gravity tacitly assume that QT involves a spacetime background. However, the latter is a wholly optional metaphysical premise that is actually in tension with QT, as follows: relativistic quantum theory demotes spacetime indices to mere parameters, and there is no relativistic position operator nor time operator at any level of quantum theory. These are sound reasons to question the conventional assumption that QT implies or requires the ontological existence of a spacetime background. In contrast, conserved currents such as energy and momentum retain their observable status at all levels. This means that quantum systems retain the corresponding eigenstate descriptions independently of any spacetime properties, and in that sense the conserved currents are more physically fundamental than the spacetime construct.[14] In any case, at the relativistic level, the conserved currents are unambiguously "preferred observables" since the spacetime quantities are not observables at all.[15] In this picture, the conserved currents physically give rise to emergent spacetime symmetries, not the other way around as traditionally assumed in discussions of Noether's theorem (see Schlatter and Kastner 2023).

The consequences of B4, which basically consists in denying ontological existence to anything that does not constitute an element of spacetime (another consequence of Democritan metaphysics), are far-reaching. In particular, they severely restrict the debate around discussions of temporality, and have done so for centuries, even before the advent of relativity unified metrical space and time.[16] Relativity theory, which simply provides a covariant map of the metrical features of the spacetime manifold, is now uncritically taken as delimiting the entire territory of physical existence in keeping with the long-standing entrenchment of B4. This belief system leads to conclusions of an

[14] Of course, the uncertainty principle still applies, but this means that one can have an exact momentum eigenstate along with completely indeterminate position, while the inverse – an exact position eigenstate and completely indeterminate momentum – is not really physically possible, since there are no real position eigenstates. Traditional references to position "values" are limited to a nonrelativistic approximation and cannot legitimately be used to support ontological claims regarding spacetime.

[15] This shows how neglect of the relativistic level in discussions of the MP impedes its solution, since one alleged feature of the MP is basis arbitrariness. But such arbitrariness is restricted to the nonrelativistic theory, which is only an approximate limit. It also reveals a weakness in the usual decoherence arguments which, in seeking to explain the appearance of locality, help themselves to the position basis when there is actually no position observable. In contrast, the appearance of locality is fully accounted for in RTI through the localizing effect of emission and absorption of conserved currents (Kastner 2020b).

[16] Prior to relativity, the delimiting container or background was the Democritan "void" in an actualist metaphysics.

incompatibility of temporal passage or "becoming" with physical theory, as in the following:

> If on the other hand the world *is* fundamentally temporal in the way that Sellars insists it must be, then (at least as far a special relativity as a representation of that world is concerned), Sellars' famous scientific realism is compromised because what he deems for metaphysical reasons to be an 'essential feature' of a temporal picture of the world does not appear in a fundamental theory of spacetime. *(Savitt 2021)*

In the preceding, "the world" (presumably denoting the domain of all that exists) is tacitly taken as equivalent to (or at least delimited by) spacetime. Thus, the excerpt illustrates the usual ground-floor assumption of B4 – that there can be nothing ontologically real outside the spacetime manifold. However, one can reject B4 and instead consider what we call "spacetime" (really, a set of invariant events) to be a subset of the world's ontology, namely the realm of actuals, while allowing in addition the existence of ontological potentialities described by QT. Then, as Reichenbach suggested, "*The flow of time is a real becoming in which potentiality is transformed into actuality*," (Reichenbach 1953). (This transformation of quantum potentiality into spacetime actuality is quantitatively described in Schlatter and Kastner [2023] and yields a novel theory of quantum gravity in which QT and relativity are harmoniously reconciled.)

Thus, unburdened by B4, one can indeed be fully scientifically realist and still assert that the world has temporal features. In this regard, it is instructive to recall Ernan McMullin's observation that the realist is not restricted to conventional categories: "[I]maginability must not be made the test for ontology. The realist claim is that the scientist is discovering the structures of the world; it is not required in addition that these structures be imaginable in the categories of the macroworld"(McMullin 1984: 14).

The well-known ostensible tension of QT with relativity is a manifestation of B4, which effectively implies that quantum theory must not be referring to something real if that to which it refers seems to "violate" relativity theory. Once B4 is laid down, relativity theory and QT are not in conflict, since neither is presumed to exclusively describe "the world." And indeed their domains of applicability, as well as the relationship between the two theories, can be quantitatively and consistently delineated. This has already been demonstrated in the literature, for example, Kastner (2022, chapter 8); and, as noted earlier, Schlatter and Kastner (2023).

Moreover, even if one does not wish to consider any sort of "becoming" picture, B4 cannot legitimately be maintained even under standard general relativity (GR). This is because GR implies that matter sources of the spacetime manifold – which is the metric field of GR – are outside the field. Any curvature resulting from the existence of a mass M does not vanish anywhere in the vicinity of M, and therefore M cannot be part of the field – that is, cannot be part of spacetime. As alluded to earlier, material quantum systems – which are the sources of the spacetime metric field – can be understood as elements of a *quantum substratum* (Kastner 2022, chapter 8). This again is fully consistent with the fact that field sources are not elements of the fields they generate – they are always outside the field. For example, electric charge, as a source of the electromagnetic field, is not part of the electromagnetic field. We have no difficulty conceiving of the latter because we don't think of the electromagnetic field as the delimiter of all that physically exists. Spacetime, as a metric field generated by matter sources, is no more a delimiter of the physically real than is any other field. Thus the "masses in the void" Democritan metaphysical presumption of B4 now essentially has the status of a folk belief, since it is refuted by GR, even though that has generally not been noticed. This sort of blindness to the implications of theories, whether they be anomalies or unexpected ontological/structural aspects, is a drawback of belief entrenchment of the kind discussed in Gradowski (2022).

7.3 The Arrow of Time Can Be Found in Physical Law Unburdened by Inappropriate Metaphysical Beliefs

We have seen in the foregoing that the standard assertion that the arrow of time is "not to be found in the underlying laws of physics" is simply untenable, both in view of standard theoretical practice and in the requirement for the propagation of real (on-shell) conserved currents. The latter cannot occur without temporal symmetry breaking, namely, irreversibility. This instructs us that there must be such a breaking of time symmetry in Nature in keeping with the empirical facts observed, and in particular with the apparent predictive success of the "causal" Feynman propagator in standard QFT, which is required for the propagation of real energy, momentum, and angular momentum.

In the context of nonrelativistic QT, the claim that the basic laws of physics are time-reversible comes from the tradition of eschewing real reduction, that is, irreversible "collapse" in QT (i.e., from adherence to B2). However, if there is such collapse in Nature, that provides the missing link between an underlying time-symmetric physical theory and the empirically established fact of an arrow of time. In the formulation mentioned previously, the Relativistic Transactional Interpretation (RTI), one has exactly this. While quantitative details of RTI can be found in the references cited in the foregoing, we briefly review the basics here as a qualitative overview. Relativistic Transactional Interpretation is the relativistic elaboration of the Transactional Interpretation (TI) originally proposed by Cramer (1986). TI is based on the Wheeler–Feynman "absorber theory" or "direct action theory" (Wheeler and Feynman 1945, 1949). Let us denote the Wheeler Feynman Classical Absorber Theory as WFCAT to emphasize that *it is a classical theory only*, and as such does not fully describe the quantum level. (The quantum form of the direct-action theory is described in quantitative detail in Kastner (2022), chapter 5.) WFCAT involves a basic time-symmetric interaction between charged currents, which at the relativistic level can be described by the time-symmetric propagator $\overline{D}$. An additional process is that of "absorber response," in which absorbers generate their own $\overline{D}$ in response to the emitter. The sum of these fields is a causal field in the form of the retarded propagator in the classical version and the original TI. The WFCAT theory treated absorber response as universal, but it is crucial to recognize that this is not the case at the quantum level.

The quantum relativistic development, RTI, improves upon this basis by providing a quantitative account of the advent of "absorber response," which is taken as primitive in the original 1986 version. Absorber response is a nonunitary process which at the relativistic level transforms the time-symmetric propagator $\overline{D}$ into the Feynman propagator D_F. The interaction referred to as "absorber response" in the original TI is actually a relativistic interaction in which both emitter and absorber(s) (electromagnetic field sources) mutually participate. This process is fully quantified in RTI and essentially corresponds to decay rates in the standard theory (Kastner 2022, chapter 5; Kastner and Cramer 2018). Under RTI, the time-symmetric propagator $\overline{D}$ is identified as the Coulomb force involved in scattering (nonradiative) processes, while the imaginary part of the Feynman propagator D_F arising from the mutual nonunitary interaction is identified as a real photon, namely, a radiative process.

Thus, under RTI we have a time-symmetric theoretical foundation (in the form of the time-symmetric propagator $\overline{D}$) along with a specific mechanism for temporal symmetry breaking, which in addition to establishing a temporal arrow via the Feynman propagator D_F, yields objective reduction. At the relativistic level, the objective reduction corresponds to the transformation of $\overline{D}$ into D_F, so that the latter need not be imposed in an ad hoc manner as in standard QFT. The latter point was first observed by Davies (1970, 1971, 1972) and is further elaborated in Kastner (2022, chapter 5). It is also noted by Breuer and Petruccione (2000), who point out that the imaginary part of the temporally directed Feynman propagator D_F can be understood as a decohering factor arising from absorber response in the direct-action theory fields.[17] Since absorber response is a nonunitary process, it violates B2. However, the action corresponding to the Feynman propagator is already nonunitary (Breuer and Petruccione 2000; Kastner 2022, chapter 5). Thus, non-unitarity supporting a criterion for "measurement" is ironically already present in the standard theory despite the mainstream adherence to B1 and B2. This

[17] See Kastner (2020b) for a quantitative discussion of decoherence under RTI corresponding to objective reduction.

inconsistency illustrates, yet again, the unsupportable status of the Beliefs of the Received View despite their entrenchment.

An additional feature of RTI is that it accepts that quantum entities – systems described by Hilbert or Fock space vectors – are not elements of spacetime. This is a requirement for a realist approach to QT, since if quantum states are taken as referring to something physically real, that to which they refer cannot be a component of 3+1 spacetime, since Hilbert or Fock space elements are multidimensional and complex. Thus, RTI implicitly violates B4 (but so, also, does standard GR, as noted previously). An important dividend of RTI is a quantitatively well-defined account of "measurement," namely, a solution to the MP (e.g., Kastner 2017c, 2020b, 2023a). But since the latter violates B1, it has been seen not as a dividend but instead as a methodological violation; thus, the formulation has not been admitted into mainstream discussion.[18]

In any event, the RTI formulation straightforwardly yields an arrow of time and predicts precisely the same phenomena (e.g., propagation of conserved currents, outcomes in accordance with the Born Rule) as the standard theory. It improves on the standard theory by deriving the Born Rule from primary physical principles rather than postulating it, as a dividend of its solution to the MP. The existence of this formulation (as well as the fact that standard QFT possesses an arrow of time, however ad hoc the latter might be) falsifies the conventional claim that physical law is at odds with the directionality of time and the associated change-data.

7.4 Conclusion

This chapter has argued that the so-called Arrow of Time Problem or Two Times Problem (TTP) is essentially a myth. It is an article of faith that is contradicted by actual theoretical practice, in which the applicable physical theory does possess an Arrow of Time and must do so in order to account for the propagation of real energy and other conserved currents. Belief in the TTP is upheld only through entrenched adherence to a set of primarily metaphysical beliefs of a predominant Received View of physics that themselves are contradicted by the empirical facts, by current theory, and by inconsistencies among the beliefs themselves.

Moreover, the inadequacies in current standard theory – such as the MP, the ad hoc nature of the temporal symmetry breaking in standard quantum field theory, and

[18] The reader might object that other "spontaneous collapse" interpretations, such as the Ghirardi–Rimini–Weber theory (1986), have become part of mainstream discussion. However, these are frankly ad hoc approaches that change the Schrödinger equation and deviate from the Born Rule. In this respect they conform to B1, in that B1 holds that "real" quantum theory cannot have nonunitary collapse. RTI challenges that belief while GRW-type approaches do not. When objective reduction approaches are considered at all in mainstream discussion (e.g., Nurgalieva and del Rio [2018]), GRW-type approaches are taken as representative of all such approaches, which effectively serves both to misrepresent and suppress RTI.

logical inconsistencies pointed to by Haag's theorem (such as the lack of a basis for interacting quantum fields) – are all remedied by an extant formulation, the Relativistic Transactional Interpretation (RTI). Relativistic Transactional Interpretation is empirically equivalent to the standard approach (unlike the ad hoc "spontaneous collapse" theories currently admitted to mainstream discussion) and provides a more quantitative and precise foundation for the empirically observed Arrow of Time. However, due consideration of this remedy by the mainstream community, which has not occurred to date, apparently requires a willingness to recognize and to question the firmly held Beliefs of the Received View.

References

Baumann, V., and Wolf, S. (2018). "On formalisms and interpretations," *Quantum* 2, 99.

Becker, A. (2022). "What is spacetime really made of?" *Scientific American*, Feb. 1, 2022.

Bong, K.W. et al. (2020). "A strong no-go theorem on the Wigner's friend paradox," *Nature Physics*, Vol. 16: 1199–1205.

Born, M. (1926). "I.2." In J. A. Wheeler and W. H. Zurek (eds.). *Zur Quantenmechanik der Stoßvorgänge* [*On the quantum mechanics of collisions*]. pp. 863–867.

Breuer, H. P., and Petruccione, F. (2000). "Radiation damping and decoherence in quantum electrodynamics." In Breuer and Petruccione (eds.), *Relativistic Quantum Measurement and Decoherence*. Workshop, Naples, Italy. Berlin: Springer.

Brukner, C. (2015). "On the quantum measurement problem." In *Proceedings of the Conference "Quantum UnSpeakables II: 50 Years of Bell/s Theorem."* arXiv:1507.05255.

Buonamano, D., and C. Rovelli (2023). "Bridging the Neuroscience and Physics of Time." In *Time and Science*, edited by Lestienne & Harris. Singapore: World Scientific. 267–282.

Cramer J. G. (1986). "The transactional interpretation of quantum mechanics." *Reviews of Modern Physics*, 58, 647–688.

Davies, P. C. W. (1970). "A quantum theory of Wheeler–Feynman electrodynamics," *Mathematical Proceedings of the Cambridge Philosophical Society*, 68:3: 751–764.

Davies, P. C. W. (1971). "Extension of Wheeler–Feynman quantum theory to the relativistic domain I. Scattering processes." *Journal of Physics A: General Physics*, 6, 836.

Davies, P. W. (1972). "Extension of Wheeler–Feynman quantum theory to the relativistic domain. II. Emission processes." *Journal of Physics A: General Physics*, 5, 1025–1036.

Frauchiger, D., and Renner, R. (2018). "Quantum theory cannot consistently describe the use of itself," *Nature Communications* 9, Article number: 3711.

Ghirardi, G. C., Rimini, A., and Weber, T. (1986). "Unified dynamics for microscopic and macroscopic systems." *Physical Review D* 34 (*2*): 470–491.

Gradowski, Laura (2022). Facing the Fringe. New York: CUNY Academic Works. https://academicworks.cuny.ed

Henson, J. (2017). "How causal is quantum theory?" Presentation at New Directions in Physics workshop, University of Maryland, 2015. Powerpoints available at http://carnap.umd.edu/philphysics/hensonslides.pptx

Kastner, R. E. (2011). "Quantum nonlocality: Not eliminated by the Heisenberg Picture." *Foundations of Physics* **41**: 1137–1142.

Kastner, R. E. (2012). *The Transactional Interpretation of Quantum Mechanics: The Reality of Possibility*. Cambridge: Cambridge University Press.

Kastner, R. E. (2015). "Haag's theorem as a reason to reconsider direct-action theories." *International Journal of Quantum Foundations*, 1, 2: 56–64, https://ijqf.org/archives/2004.

Kastner, R. E. (2016). "The illusory appeal of decoherence in the Everettian picture: Affirming the consequent," excerpt from introductory chapter in J. Jeknic-Dugic, and G. Jaroskiewicz, eds., *Quantum Structural Studies*. Singapore: World Scientific. (2017). https://arxiv.org/abs/1603.04845

Kastner, R. E. (2017a). "On quantum non-unitarity as a basis for the second law of thermodynamics," *Entropy* 19(3), arXiv:1612.08734

Kastner, R. E. (2017b). "Why 'making sense of quantum theory' by denying that it describes the world does not really work," blog post, https://transactionalinterpretation.org/2018/09/22/why-making-sense-of-quantum-theory-by-denying-that-theories-are-about-the-world-does-not-really-work/ (accessed May 27, 2024).

Kastner, R. E. (2017c). "The status of the measurement problem: Recalling the relativistic transactional interpretation," *International Journal of Quantum Foundations*, 4: 128–141.

Kastner, R. E. (2020a). "Unitary-only quantum theory cannot consistently describe the use of itself: On the Frauchiger–Renner paradox," *Foundations of Physics* 50: 441–456.

Kastner, R. E. (2020b). "Decoherence and the transactional interpretation." *International Journal of Quantum Foundations*, 6: 24–39.

Kastner, R. E. (2021). "Unitary interactions do not yield outcomes: Attempting to model 'Wigner's Friend'," *Foundations of Physics* 51, 89. DOI: https://doi.org/10.1007.s10701-021-00492-3.

Kastner, R. E. (2022). *The Transactional Interpretation of Quantum Mechanics: A Relativistic Treatment*. Cambridge: Cambridge University Press

Kastner, R. E. (2023a). "Quantum theory needs (and probably has) real reduction," https://arxiv.org/abs/2304.10649.

Kastner, R. E. (2023b) "Physical time as human time," *Timing and Time Perception*, DOI: https://doi.org/10.1163/22134468-bja10081.

Kastner, R. E., Kauffman, S., and Epperson, M. (2018). "Taking Heisenberg's potentia seriously," *International Journal of Quantum Foundations*, 4 (2): 158–172.

Kastner, R., and Cramer, J. G. (2018). "Quantifying absorption in the transactional interpretation," *International Journal of Quantum Foundations*, 4 (3): 210–222.

Kastner, R., and Schlatter, A. (2023). "Entropy cost of 'erasure' in physically irreversible processes." http://philsci-archive.pitt.edu/22511/, https://arxiv.org/abs/2307.02643

McMullin, E. (1984). "A case for scientific realism." In J. Leplin (ed.), *Scientific Realism*, Berkeley: University of California Press, pp. 8–40.

Meinard, Kuhlmann (2023). "Quantum field theory." *The Stanford Encyclopedia of Philosophy* (Summer 2023 Edition), Edward N. Zalta & Uri Nodelman (eds.), https://plato.stanford.edu/archives/sum2023/entries/quantum-field-theory/.

Norton, J. (2021). *The Material Theory of Induction*. Calgary: University of Calgary Press.

Nurgalieva, N., and del Rio, L. "Inadequacy of modal logic in quantum settings." *Electronic Proceedings in Theoretical Computer Science*, 287 (2). DOI: https://doi.org/10.4204/EPTCS.287.16.

Proietti, M. et al. (2019). "Experimental test of local observer independence," *Science Advances*, Vol. 5, no. 9. DOI: https://doi.org/10.1126/sciadv.aaw9832.

Reichenbach, H. (1953). "La Signification philosophique du dualism ondes-corpuscles," in Andre George (ed.), *Louis de Broglie, Physicien et Penseur*, Paris: Michel Albin, p. 133.

Reichenbach, H. (1956). *The Direction of Time*. Berkeley: University of California Press.

Savitt, S. (2021). "Being and becoming in modern physics," *The Stanford Encyclopedia of Philosophy* (Winter 2021 Edition), Edward N. Zalta (ed.), https://plato.stanford.edu/archives/win2021/entries/spacetime-bebecome/.

Schlatter, A. (2021). "On the foundation of space and time by quantum-events." *Foundations of Physics* 52 (1): 1–17.

Schlatter, A. and Kastner, R. E. (2023). "Gravity from transactions: Fulfilling the entropic gravity program," *Journal of Physics Communications* **7**: 065009.

Sorkin, R. D. (2003) "Causal sets: Discrete gravity (Notes for the Valdivia summer school)." In A. Gomberoff and D. Marolf (eds.), *Proceedings of the Valdivia Summer School*. DOI: https://doi.org/10.48550/arXiv.gr-qc/0309009.

Wheeler, J. A., and R. P. Feynman (1945). "Interaction with the absorber as the mechanism of radiation," *Reviews of Modern Physics*, 17, 157–161.

Wheeler, J. A., and R. P. Feynman (1949). "Classical electrodynamics in terms of direct interparticle action," *Reviews of Modern Physics*, 21, 425–433.

Part III

The Arrow of Time and Time-Reversal Invariance

Part III

8

Time for Pancakes: Time Reversal and Ontology

VALIA ALLORI

8.1 Introduction

In this chapter, I wish to further motivate, defend and improve the nonstandard, or heterodox, account to time-reversal symmetry as proposed by Albert (2000) and Callender (2000), which has been recently dubbed by Roberts 'the pancake account'. According to this approach a time-reversal transformation is implemented by an operator which simply inverts the order of the physical states. Instead, the standard or orthodox approach is one in which the relevant operator also suitably changes some properties of the entities in the physical state.

In Section 8.2, I discuss the motivations for the standard approach: it is a "symmetry first" approach, in the sense that time-reversal symmetry, as many others, are required or imposed on the theory rather than inferred from the dynamics. This leads directly to the problem for this approach, discussed in Section 8.3, namely, that it is "metaphysics last": Albert and Callender argue that a scientific realist should take the theory at face value, and therefore read the properties of the theory from its ontology, rather than the other way around. Accordingly, if time reversal is simply seen as reversing the time ordering of events, and if the ontology of the theory is what it is commonly accepted that it is, then there is no reason why it should transform in such a way as to make the theory time-reversal invariant.

The defenders of the standard account have tried to provide these reasons, as presented in Section 8.4, thereby responding to the challenge posed by Albert and Callender. Nonetheless, as I argue in Section 8.5, the standard account is far from being problem free. I discuss several of such difficulties, the most challenging of which are that there is no justification of why the state should change under time reversal, and that the invariance of theories under such a definition of time reversal are not worth having.

In Section 8.6, I defend the pancake account from recent criticism by Roberts (2021) and, in Section 8.7, I propose an alternative approach, in which I endorse the time-reversal operator as defined by the pancake account, but I also restore the time-reversal symmetry of the theory by eliminating much of the ontology from the state. I conclude in Section 8.8, in which I evaluate the cost and benefits of each approach.

8.2 The Standard Account of Time Reversal

Consider a classical system such as a billiard ball on a frictionless table. According to every physics textbook, the state S of the system at a given time, namely what provides its complete description at that time, is given by its position x and its velocity v: $S = (x, v)$. Now consider various transformations the system can go through. For instance, a system can be rotated, or translated in space. Usually, one obtains the transformed system by applying some suitable operator to the original system. This operator is supposed to suitably describe the physical transformation the system goes through, as we observe it. That is, one can actually perform the transformation and later see its result. This is what allows us to select which mathematical operator to use to represent such transformation. For spatial translations, such an operator is one that takes the original system and moves it by an arbitrary constant. Alternatively, one can move oneself by that constant in order to describe the system from a translated frame. So, if one describes an unmoving billiard ball at one instant by its position x at that instant, $S = (x, 0)$, then the coordinate of translated ball will be described by $x + a$, for an arbitrary constant a, such that the transformed state is $S' = (x + a, 0)$. This translated system describes the transformation both passively (the ball seen from a translated frame) and actively (the ball being translated itself). For a rotation of an angle ϑ, the operator implementing the transformation is a suitable matrix R_ϑ: if S is the original state, then the state rotated by ϑ is $R_\vartheta S$.

These are not the only transformations one can perform on a system. Consider now the time-reversal operation. A new difficulty arises: how are we supposed to understand this operation? We cannot understand it in the same way as we understood rotations or translations, as we cannot actually reverse time in a physical process. So, it is not entirely obvious what time reversal does to the state of a system: it certainly inverts the order of the instantaneous states, but does it do also something else? To find an answer to this question, and thus to characterize time reversal, one usually resorts to the following metaphor. Imagine filming a body in motion (say, a billiard ball in a frictionless table) and then imagine projecting the resulting movie backwards. The backward projection is the time-reversed process. How is this process obtained, starting from the original system? The answer is: not only does one reverse the order of the events, but also the velocity changes sign. That is, according to the "standard account," namely the one that is found in physics textbooks, the time-reversal operator T is given by the composition of two transformations: (1) T inverts the temporal order with which states happen, that is $t \to -t$; (2) T changes the state of the system as to invert the direction of the velocities, that is $v \to -v$. Watching the backward movie suggests T to behave like this, but also the mathematical definition of the velocity as rate of change of position.

Symmetries, or invariances, of a given theory are associated with the transformations of the theory's dynamical law. If one wishes to use a quick slogan, one could say that a theory is invariant under a given transformation if and only if "transformed solutions are still solutions," where "solution" means "solution of the fundamental dynamical law of a theory." A solution of the fundamental laws expresses a possible way the world can be according to the theory. So, a theory is invariant under a given transformation if and only if both the original description provided by the theory and the transformed descriptions are both possible descriptions of the world. In classical mechanics, under translation and rotation, the coordinates of the system suitably change but the velocity does not: if the ball was moving in the original frame, it was also moving in the translated or rotated frame (even if the characterization of the coordinates of the velocity might change, the vector is the same). So, the state does not change under these transformations. That is, rotations and translations are symmetries of the theory. Or, one could say that the theory is invariant under these transformations. What about time reversal? As in the case of rotation and translation, there is a symmetry connected to the time-reversal operator, namely, time-reversal invariance. A theory is said to be time-reversal invariant if and only if both the original description provided by the theory and the time-reversed one are possible descriptions of the world according to the theory. In Newtonian mechanics the instantaneous state of the system is given by $S = (x, v)$; if T is the time-reversal operator defined earlier, then the transformed state is $T(S) = (x, -v)$. Newtonian mechanics is time-reversal invariant because transformed solutions, namely $x(-t)$, are also solutions. A movie of a billiard ball on a frictionless table projected forward and backward both describe ways the world can be. If we *impose time-reversal symmetry* on all other fundamental physical theories, then it follows that T, in addition to reversing the order of the states, should also change some of the other properties of the theory. More in detail, given that a temporal sequence of states is a dynamical trajectory, the time-reversal operator T acts on trajectories so that:

(1) T inverts the order with which states happen, $t \to -t$;
(2) T changes the state so that some properties are preserved, and some instead need to change, as illustrated in Table 8.1:

Table 8.1 *Property change under time-reversal transformation in the standard account.*

Theory	Changed	Preserved
Hamiltonian Mechanics	Momentum $p \to -p$	Position $x \to x$
Classical Electromagnetism	Magnetic field $B \to -B$	Electric field $E \to E$
Quantum Mechanics	Spin $\sigma \to -\sigma$	Kinetic energy $p^2/2m \to p^2/2m$
	Position wavefunction $\psi(x) \to \psi^*(x)$	Transition probabilities $\lvert\psi\lvert\varphi\rvert^2 \to \lvert\varphi\lvert\psi\rvert^2$

In the case of Hamiltonian mechanics, the state is $S = (q, p)$, where q and p may have general meaning, not necessarily position and momentum; the time-reversed state is $T(S) = (q, -p)$. If q and p are effectively position and momentum, then Hamiltonian mechanics is time-reversal invariant. Transformed solutions $q(-t)$ are also solutions. Now consider relativistic field theories such as classical electrodynamics. The state this time is given by $S = (x, E, B)$; while the transformed state is $T(S) = (x, E, -B)$. If so, then classical electrodynamics is time-reversal invariant: with these transformations, the transformed solutions are also solutions. What about standard quantum mechanics? The state is given by the wavefunction, $S = \psi$; and the operator T is such that $T(S) = \psi^*$. It is antiunitary, and antilinear, and this transforms solutions into solutions.

8.3 Against the Standard Account: The Pancake Account

As we have seen earlier, the movie analogy and the intrinsic definition of velocity led us to define the time-reversal operator so that it would do more than just reordering the states: it also changes something in the state, namely velocity. Then by imposing time-reversal symmetry on other theories, one finds the appropriate definition for T. With this definition, T guarantees time-reversal invariance of these theories. However, while this seems straightforward for classical mechanics, the situation is not as clear in the other theories. That is, the last two examples require a little more discussion. What about classical electrodynamics and quantum theory? Why does T flip the magnetic field, $B \to -B$? Why do we have this: $\psi(x) \to \psi^*(x)$?

Roberts (2021 and references therein) writes: "we want an operator that preserves position, and reverses momentum" and the magnetic field. But why is that? The idea is that we want the symmetries for the theory, and we define the time-reversal operator as to preserve this symmetry for us. For instance, in quantum theory, it is antiunitary because only antiunitarity preserves the Hamiltonian, and the properties transform as they do because it is part of their definition.

Albert (2000) and Callender (2000), however, find this reasoning question begging: we need to discover whether a theory has a given symmetry, on the basis of the ontology of the theory. Instead, what Roberts proposes is to impose the symmetries to constrain the dynamics of the theory to start with. If instead we adopt an ontology-first approach, a theory will have the symmetries that it has, as a consequence of its ontology, rather than an a priori theoretical imposition. In other words, according to Albert and Callender, in the standard approach one postulates the time-reversal operator to be defined as such in order to make the theories time-reversal invariant. That is, one postulates that T is defined that way to make the theory invariant. But why is this the right definition of T? What is the physical justification for the changes one would have for the magnetic fields and the wavefunction?

According to Albert and Callender, the time-reversal operator should not do anything more than reversing the order of the states, because this is what physically makes sense. Then, once T is defined that way, one can determine whether the theory is time-reversal invariant or not.

With such a definition, namely T, that merely reorders the states, Albert (2000, ch. 1) reasons that it makes sense for the velocity to change sign, and thus we can conclude that classical mechanics is time-reversal invariant. However, the situation is different in the other theories. Here is why. As we mentioned, in classical mechanics, typically, the state of a given physical system is taken to be constituted by the couple of positions and velocities. Albert instead thinks that such a couple should be called the dynamical condition D at an instant. In fact, for Albert the state should be complete but also genuinely instantaneous: the description at different times should be independent. Since the velocity depends on the position, it should not be in the state but only in the dynamical condition, as it is needed "in order to bring the full predictive resources of the dynamical laws of physics to bear" (Albert 2000: 17). Rephrasing, the instantaneous state represents what exists in the world at one instant, while the dynamical condition specifies what is needed at one time to determine the state of the system at another time. According to Albert, thus, in classical mechanics, S is given by the particles' positions, while D by the positions and the velocities: $S = (x); D = (x, v)$. Having said that, therefore in classical mechanics, the instantaneous state S remains invariant under T, and therefore classical mechanics is time-reversal invariant. The velocity transforms under T according to its definition as a function of x and t: $T(v) = T(dx/dt) = dx/(-dt) = -dx/dt = -v$. However, this does not affect time-reversal invariance because the velocity is not in the state. This is no longer true in classical electrodynamics or quantum theory. In fact, when considering classical electrodynamics, instead, Albert believes we need to add to S

also "the magnitudes and directions of the electric (E) and magnetic (B) fields at every point in space" (Albert, 2000:14). That is, the instantaneous state is given by the triplet (x, E, B). The fields are, unlike velocities, logically independent of the particles' positions and therefore they should be added to the state in order to complete the picture of the world at one time. That is, electromagnetic fields are real just as much as the particles are. According to Albert, as we already saw, the instantaneous state should not change under T. The velocity was not in the state and changed under T according to its definition as the rate of change of position. In contrast, magnetic fields are not the rate of change of anything. Rather, they are fields, and as such, he argues, they should be mathematically represented simply by vector functions, whose intensities represent the various field values in space. As such, therefore, there is no reason why they should change at all under T: $T(x, E, B) = (x, E, B)$. As a matter of fact, though, in order for classical electrodynamics to be time-reversal invariant, we need the magnetic field to flip sign under T. But since there is no reason why this should happen, classical electrodynamics is not time-reversal invariant.

Similar reasoning can be made in the case of quantum theory, as proposed by Callender (2000). Why should the wavefunction transform into its complex conjugate under T? The wavefunction, being a function of the particles' coordinates, is naturally understood as a scalar field in configuration space. If so, the wavefunction is not the right kind of object to transform into its complex-conjugate under T: a scalar field should not change its amplitude at all under a time-reversal transformation. What the time-reversal operator in fact should do is only change the order of the states, and nothing more. Since in order to have time-reversal invariance the wavefunction would have to change under T then, the argument concludes, the Schrödinger equation is not time-reversal invariant.

The basic idea is therefore this: first, find what is in the instantaneous state, namely the ontology of the theory. Then figure out whether there is a reason for the entities in the state to change under T as needed in order to preserve time-reversal invariance. In classical mechanics, by definition, the velocity changes sign under T and in any case it does not matter because it is not even thought of as belonging to the state. But in classical electrodynamics and quantum theory the situation is different. The magnetic field and the wavefunction are physical fields, mathematically represented as suitable functions of coordinates, and as such there is little reason for them to change during time reversal, Albert and Callender maintain. But if they do not change, then transformed solutions are no longer solutions, and thus the theories are no longer time-reversal invariant.

Therefore, here is, schematically, my reconstruction of the argument put forward by Albert and Callender against the standard account, and its consequence for the time-reversal invariance of a given theory:

1. The complete instantaneous description of any system is given by its state *S*;
2. In a given theory, the object *X* provides the state [this changes from theory to theory, and from one interpretation of the theory to another];
3. We both agree that the time-reversal transformation *T* transforms a sequence of states into their temporally reversed sequence: $S_i, \ldots, S_f \to T(S_f), \ldots, T(S_i)$;
4. According to the standard account, we should also add that the time-reversal transformation *T* also may need to change the properties in *S* to map solutions into solutions (to make the theory time-reversal invariant);
5. However, there is no reason why *S* should change under *T*: (4) is false;
6. Therefore, whether a theory is time-reversal invariant depends on what *X* is: if *X* does not change under *T*, the theory is invariant; otherwise, it is not.

In other words, the standard account does not provide us with any reason why properties change due to a time-reversal transformation, so, aside any further justification, we should not assume that they do. Albert applies this argument to classical electrodynamics:

1. The complete instantaneous description of any system is given by its state *S*;
2. In classical electrodynamics the state is given by (E, B) (in addition to position);
3. The time-reversal transformation *T* needs to change *B* into –*B* to map solutions into solutions;
4. However, there is no reason why*B* should change under *T*: "*It's not the rate of change of anything*" (ibid.);
5. Therefore, classical electrodynamics is not time-reversal invariant.

Callender has the same argument for quantum theory:

1. The complete instantaneous description of any system is given by its state *S;*
2. In quantum mechanics the state is given by the wave function ψ;
3. The time-reversal transformation *T* needs to change ψ into ψ^* to map solutions into solutions;
4. However, there is no reason why ψ should change under *T*;
5. Therefore, quantum mechanics is not time-reversal invariant.

(A more accurate conclusion would be to say that the Schrödinger equation is not time-reversal invariant. In fact, to argue that the quantum theory, understood as both the Schrödinger equation and the von Neumann collapse rule, is not time-reversal invariant, one would also have to show that the von Neumann rule is not time-reversal invariant. This is indeed easier to argue for, as it is intrinsically time directed, see Allori 2019a).

At this point, there are at least two possible reactions. Albert and Callender argue that there is no sensible definition for these quantities that would make sense of their change under time-reversal transformation. Because of this, they think this argument shows that the proponents of the standard account need to re-think what the time-reversal operator does: it does not change the properties of the entities in the state; rather it simply changes the temporal ordering at which the states happen. That is, Albert and Callender identify the time-reversal operation as simply an order-reversal operator. If so, however, the theories considered here are no longer time-reversal invariant.

Some have argued that the two accounts arrive at opposite conclusions because they talk past one another. For instance, Lopez (2019) has suggested that the proponents of the standard account think of time reversal as motion reversal: that is, T is the transformation which reverses the changes which have happened to the object during their forward temporal evolution. If so, then the transformation is not 'time' reversal at all, as it has nothing to do with time as such. This would be another objection of the standard account.[1] Be that as it may, I am going to assume that the two accounts share the same notion of time reversal, and I aim to show that the standard account is not satisfactory.

Proponents of the standard account were obviously not so quick at abandoning it. An obvious way to go is to reject premise 5 of the main argument (or 4 of the theory-specific arguments): find nonobvious reasons why the properties would change under T the way required for time-reversal invariant. This is discussed in the next section.

8.4 Defending the Standard Account

People have proposed ways to explain why the state changes under T as to make the theory retain time-reversal symmetry. In the case of classical electromagnetism, this type of approaches includes Arntzenius (2000, 2004), Earman (2002), Malament (2004). In quantum mechanics this route has been taken by Earman (2002), Roberts (2017).

Setting aside subtle differences, the general idea is that given a mathematical object, there is a natural way for it to transform under a particular transformation, which depends on its intrinsic or geometrical definition. This would fill the gap between classical mechanics and electromagnetism and quantum

[1] In his (2021), Lopez goes a step forward and says that the orthodox and the heterodox views hold different views of what time is. If one is relationalist with respect to time, then the orthodox view is quite natural, since time is change. Instead, if one is a substantialist, then one would want the time reversal to only flip the time coordinate.

theory: for the velocity we had a reason to change sign under T, based on its definition; now similar definitions are proposed for B and ψ. It is argued that electromagnetic fields are intrinsically defined as to transform under T to make classical electrodynamics time-reversal invariant. In fact, it is not correct to think of B as generic function; rather, it is an axial vector, also known as pseudovector. It is because of its so defined geometrical nature that it makes sense that the magnetic field flips sign under T. In fact, by definition, a pseudovector behaves like a vector function in many situations, but it changes into its opposite under T. A similar strategy has been proposed for quantum mechanics. In quantum theory, the wavefunction is to be understood not as a field in configuration space but as a ray in Hilbert space, that is, an equivalent class of vectors related by a phase θ of the form $\psi = \{\psi e^{i\theta}\}$. It is because of this geometrical definition that it makes sense that the wavefunction transforms into its complex conjugate under T.

With these definitions for the relevant objects, it is argued, one can provide reasons why they change the way they do under T, and therefore one can restore time-reversal invariance in these theories.

8.5 More Objections for the Standard Account

I am not convinced. Indeed, I think that there are more problems with the standard account than the one granted by Albert and Callender. In this section, I discuss first why I think that the defense provided is insufficient, and then I argue that the overall approach is too costly to be worth it.

8.5.1 No Independent Reasons

I think that the response just provided by the defender of the standard account begs the question. What has been shown is that there is another way of thinking of the electromagnetic fields and the wavefunction that is compatible with the invariance of classical electrodynamics or quantum mechanics. Nonetheless, no reason has been provided just yet as to why this new definition should be considered the correct one. In other words, one needs to provide additional, *independent* reasons to believe that the true nature of the magnetic field (of the wavefunction, or whatever) is the one captured by the mathematics of an axial vector (of a ray, or whatever), so that the theory turns out invariant. And no such additional reason has been provided so far. In any case, I think that even if one can provide independent reasons, the standard account would still be extremely problematic, as discussed in the next subsection.

8.5.2 Unjustified Changes in the Ontology

In this subsection, I am going to argue that Albert and Callender have been too kind with the standard account. In fact, their objection to the standard account was that there is no (obvious) reason why *S* should change under *T*. If this were the real problem, then one could simply provide some nonobvious reasons, as discussed in the last section, and save the account. Now, instead, I wish to argue that the situation is much worse than has been suggested so far. I wish to argue that *S* cannot change under *T*, otherwise one would have to provide a different meaning of what the state represents. If I am correct, and it does not make much sense for the state to change under time reversal, then the standard account has no chance of being fixed, providing further motivation to the alternative account to time reversal proposed by Albert and Callender.

Here is the reason why I think that it makes no sense for the state to change under a time-reversal transformation. Assume that the standard account is correct and that *T* changes the content of the state adjusting the properties so that the theory will turn out to be time-reversal invariant. However, think of what the state is supposed to represent. By definition, the state represents an instantaneous description of the world, namely it represents a picture of the world at one time. In other words, *the state gives you the instantaneous ontology of the theory*. This is what Albert assumes and no one in the standard account seems not to be willing to deny. However, if that is the case, namely, if *T* changes the content of the state at one time, then this means that *T* changes the ontology of the system at that time. In other words, what the world looks like depends on whether the picture of it we are looking at is taken from the movie going backwards or going forward. By why should it be like that? It would be as if, when playing 'Back to the Future' forward in time, Marty McFly is interpreted by Michael J. Fox, but when we play it backwards it is interpreted by Keanu Reeves. Or in forward 'Star Wars' Yoda is small and green but in backward 'Star Wars' he is tall and blue. Nothing can possibly justify such a change under a transformation labeled time reversal. In a nutshell, if the state of a system at one time represents the ontology of that system at that time, then the state cannot change under *T* because that would amount to changing the ontology, which is absurd.

8.5.3 Unmotivated Definition of Time Reversal

The proponents of the standard account may want to deny this absurdity. However, doing that, the burden of proof would be on them: the need to explain why the ontology changes under time reversal. That is, they need to explain what makes time reversal, which is supposed to reverse time, do something else entirely. In other words, now the question is about the time reversal transformation itself, and the question is: *why should we think that T, as time-reversal operator, changes the ontology, while, say, the translation operator does not?* We would not accept a translation operator such that it moves the body on the left, say, of a given distance, but also colors it all red. So, why should we accept something similar for *T*? What is special about *T*? At best, there is a mystery to be solved. Notice that one cannot reply by saying that this is the way to make the theory time-reversal invariant, because we have not yet provided any reason why one would want symmetries to start with. Since there seems to be no reason to have a time-reversal operator defined as such other than to preserve time-reversal symmetry to this change in ontology, I think we should reject this definition of the time-reversal operator.

8.5.4 No Value in Time-Reversal Invariance within the Standard Account

In addition, I wish to argue that, even if with the standard account of time reversal theories turn out to be time-reversal invariant, there is nonetheless no value in the symmetry so defined. To explain this, let me elaborate on why we want symmetries.

Some people have argued that symmetries are worth having because they are often used in theory construction: especially in quantum field theories, symmetries are imposed on the formal structure to select among the various possibilities. Symmetries are also used in theory evaluation: theories with more symmetries are usually preferred, all things being equal, to theories with less symmetries. Lopez and Esfeld (2023) have maintained that Humeanism is committed to the idea that symmetries, and time-reversal symmetry in particular, play a rule-prescribing role in the best systems, so they somehow play a role in theory evaluation: the more symmetries, the better the system. Others have argued that symmetries are important because of their connection with conservation laws. However, I think that the most important reason why symmetries are worth having is that they reflect our desire of providing a perspective-free description of the phenomena. When we require that a theory is rotation invariant, we mean that the description the theory provides of the phenomenon does not depend on whether we are observing the phenomenon from this frame of reference, or from a frame rotated with respect to this. The principle of relativity, that the form of the law should be valid in all (inertial) reference frames, just reflects this idea: the frame we are in should not influence our description of the phenomenon. In other words, the more symmetries a theory has, the less dependent on our point of view it is, and the more objective its description of the world is.

If that's the case, however, and if a time-reversal transformation is defined as to change the ontology of the system, then requiring a theory to be invariant under such T does not help at all our search for the most perspective-free description, so we have no reason for requiring such a symmetry. In fact, this invariance does the *opposite* of what we want, namely, it changes the description of a phenomenon depending on how we look at it. Instead, what we should want is a theory to be time-reversal invariant because in such a theory it would not be important whether we are gathering information about the world from its forward or its backward movie. In fact, we would want the instantaneous description not to change depending on whether it is taken from the backward or the forward movie, because we want our theory to be as perspective-free as possible. In other words, requiring time-reversal invariance does not seem to make sense in the standard account, if the reason why we want symmetries is to make the theory as objective as possible. In fact, the time-reversal operator as defined in the standard account makes the description provided by the theory perspective dependent, which was exactly what we wanted to prevent by requiring symmetries.

To conclude this section: these arguments, in addition to the original argument by Albert and Callender, in my opinion show that the time-reversal operator should only reverse the order of the states rather than messing around with its content, like proposed in the standard account.

8.6 Criticisms to the Pancake Account and Replies

If it is the case that the state should not change, because it is an instantaneous picture of the world, then we need to postulate that the time-reversal operator simply flips the order of the states but leaves them unaffected. This approach has been dubbed the "pancake account" because the states are stacked onto one another, like a stack of pancakes, without any connection with one another (Roberts 2021). I think that the name is meant to suggest that the account is too naïve to work. Nonetheless, I think this is the case in which simplicity is mistaken for naivete, understood as oversimplification. Anyway, if this is being simple-minded, I like naivete (and pancakes too), so I happily endorse the name.

Nonetheless, Roberts (2021) has recently criticized the originally proposed pancake account. In this section, I discuss these criticisms and provide my replies.

8.6.1 Misguided Motivation

According to Roberts, the original motivation for the account was that it did not make any sense for the magnetic field to change under T because "it is not the rate of change of anything." However, Roberts points out that properties can flip in this way without being rates of change. In order to show this, he uses the following example. Consider a soldier going to war. Assume he has the property of being brave, as he runs toward the enemy. If we take a movie of this and we time-reverse it, what the soldier does, namely running away from the enemy, can be characterized as 'cowardice'. If the pancake account were correct, it is argued, then we could not characterize the soldier as coward in the backward movie, and this seems wrong. Roberts also says that the same thing can be said for other, more physically interesting properties such as spin. So, the pancake account is mistaken.

As a first reply to this objection, let me notice that it is unclear in what sense "cowardice" is the time-reversed counterpart of "bravery." Both properties are much more than simply "running toward" or "running away" from the enemy. After all, someone could run toward the enemy for a variety of reasons: they may because they are brave, but also because they are being fool-hardy, or reckless, or irresponsible. Or also they may run toward the enemy because they are forced to do so, or they do it without realizing where they are going, just to mention a few obvious possibilities.

Regardless, I think that the "backward" soldier cannot be described as a "coward" in any sensible sense: they run away from the enemy but probably the enemy is also running away from them in the other direction, bombs are un-exploding, bullets are exiting bodies to "run toward" guns and rifles. That is, nothing makes sense in the usual language in the time-reversed movie, and that is because the macroscopic world is irreversible: we can tell which, among the forward and the backward movie, is physically possible, and the backward one is not. That is, not only cannot one characterize the backward soldier as coward; rather, the backward soldier is not a soldier at all.

Moreover, for this argument to work we need "cowardice" and "bravery" to be like the magnetic field, the wavefunction, and spin. The natural way of thinking of cowardice and bravery is thinking of them as dispositional properties. If so, then, also the magnetic fields, the wavefunction, and the spin should be understood in the same way. That is, they are dispositional properties of matter rather than part of matter themselves. Indeed, one may think that they would be more like velocity than we originally realized: they are properties with the suitable definition which lets them transform as needed to preserve time-reversal invariance. And since velocity is a property of matter, the magnetic field and the wavefunction are properties of matter too. In this way, as we classified velocities in D, rather than in S, we should do the same for the fields and the wavefunction as well. However, this would be a very unorthodox view. For starters, if the wavefunction is a property, what is matter made of in quantum theory? One could say particles, within the pilot-wave theory, but such an answer is not available within the standard theory. In turn, within classical electromagnetism this view would also imply a radical way of thinking about the fields. In fact, the magnetic field is not understood, as usual, as part of the ontology of classical electrodynamics but rather as a dispositional property of the ontology which presumably is given by particles. I have nothing against such a view (indeed, I am going to discuss it favorably later). Nonetheless, it would be a peculiar view to hold by someone who wants to defend the orthodoxy of the standard account.

In any case, this is question begging: just noticing that some stuff can be seen as doing what we want does not explain much. The point is to explain why certain objects have the properties that they do: why is spin flipping sign? Why is the magnetic field? Why does the wavefunction turn into its complex conjugate? What is the physical significance of a pseudovector? What is the physical significance of a ray in Hilbert space? What do they tell us about the nature of the world? In other words, if one wishes to take the theory ontologically seriously, one should be able to justify why the fields and the wavefunction are defined as they are. Instead, the standard account does not explain these features.

8.6.2 Incoherence

Roberts also argues that the pancake account is incoherent because it does not make sense of the transformation of the momentum under time reversal in Hamiltonian mechanics. Remember that on the pancake account the content of the state is left unchanged, only the order of the states is reversed. In Hamiltonian mechanics the state is (q, p), so the time-reversed state is the same, hence the momentum p does not change under T. However, since the velocity v is a rate, then it changes sign under T, even in the pancake account. So, the momentum p, given that $p = mv$, should change too, by definition. However, this contradicts what we have just seen, hence the pancake account is incoherent.

In reply, one could insist that this is just a misunderstanding of the pancake account: as we have seen earlier, according to the pancake account velocities are not in the state, they are in the dynamical condition D, and T can change D as long as what is in D is defined in a way that makes sense for them to change, like velocities. In this case, also momentum is a rate, hence the incoherence charge dissolves. (A counter reply could be that this means that Hamiltonian mechanics isn't an effective description of classical mechanics. However, the defender of the pancake account could simply deny this, as also in classical mechanics velocity is not in the state. Alternatively, one could bite the bullet and say that the Hamiltonian formalism is not physically significant: it is just a useful mathematical tool.)

8.6.3 Underdetermination

A third objection is that in the pancake account it is no longer clear why time reversal is implemented by $t \to -t$ and not some other transformation, like for instance $t \to \sqrt[3]{a - t^3}$. In the standard account the first transformation is selected by the requirement that T is an involution, that is, if it is performed again, it brings back to where one started. Instead, in the pancake account one only requires that the transformation inverts the time ordering. In this case, both transformations invert the order of the states, so why the former instead of the latter?

As a reply, I would say that in the pancake account the justification of T being $t \to -t$ is not the one provided. Rather, this transformation has a physical significance, while the other one does not. That is, we already know what the variable t stands for, and what the minus sign does, so why are we asking for more? To me, to ask for further justification is just like asking why the transformation $x \to -x$ represents a mirror transformation on the x axis, or why the transformation $x \to x + a$, where a is a constant, represents a translation along the x axis of a quantity equal to a. In these cases, we do not ask why we have, say, $x \to x + a$ rather than some other transformation, and this is because we know what this transformation represents. Why should the case of time reversal be different? In other words, why should we even think that the transformation $t \to \sqrt[3]{a - t^3}$ represents something physical at all, given that we already have specified what the various variables and the operations stand for?

8.6.4 Unwelcome Results

Another charge to the pancake account is that it makes classical electrodynamics and quantum theory no longer invariant under time reversal. This is supposed to be problematic because, as anticipated already, symmetries have been considered important features for a theory.

As we have seen in the last section, however, if we want symmetries because they are an indication of how objective a description is, then the time-reversal operator as defined in the standard account does not qualify as a valuable symmetry in this respect: we should be looking for a theory with most symmetries because such a theory would be as perspective-free as possible, but T as defined in the standard account makes the ontology change. That is, the ontology would be perspective dependent, because what we would say exists and their features depend on whether we are looking at the backward or the forward series. This is the exact opposite of what we look for. So, going to the standard account to preserve symmetries would not be of any help in this at all. What one would need is a theory to be invariant under the pancake account of time reversal, and so far, this is not what happens.

The proponents of the pancake account can, of course, bite the bullet and argue that the importance we give to symmetries is exaggerated and without much of a foundation. Nonetheless, it is a very radical step to make.[2]

8.7 The Best of Both Worlds

I am convinced that the time-reversal operator should be defined as proposed by the pancake account, namely as an ordering operator. However, I also think that the symmetry objection poses a real problem for the pancake account, as I value symmetries for the reasons discussed earlier. The question now becomes the following: is there a way to recover symmetries but also maintain the pancake account? In other words, we have a dilemma. In the standard account of time reversal, theories are invariant under time reversal, but we have the problem that the state, namely the ontology, changes depending on whether we take the state from the forward and the backward movies of the world. That is, this account is true to the symmetries (with qualifications, as we have seen), but not to the ontology. On the other hand, however, in the pancake account the state does not change but the theories lose time-reversal symmetry. That is, this account is true to the ontology but not to the symmetries. The goal is now to find an account of time reversal and of the ontology that keeps the best of both worlds: changes the order of the states without changing the ontology but also maps solutions into solutions. I think that this goal can be achieved by keeping the definition of T of the pancake account (reverses the order but does not change the states) but *redefining what is in the state*. Just proceed as Albert did in classical mechanics: put everything but x in the dynamical condition, out of the state.

Here's the situation so far. Both the standard account and the pancake account agree that in classical electrodynamics the state is $S = (x, E, B)$, and in quantum theory one has $S = \psi$. What if we challenge this? Albert already did this in classical mechanics: the velocity is in the dynamical condition D, not in S. Let's see what happens. Set aside quantum theory for the moment and focus on classical electromagnetism. Assume that the fields are not in the state, they are in the dynamical condition. That is, assume that there are only particles, so that $S = x$. If we start from this hypothesis, then T reverses the order of states but does not change them (as in the pancake account, contra the standard account). Because now the fields are no longer in the state, then classical electrodynamics is time-reversal invariant (as in the standard account, contra the pancake account). In this way we get both time-reversal invariance and no odd transformation for the ontology. Given that the electromagnetic fields are no longer in

[2] For an argument along these lines, see Lopez (this volume).

the state, they are no longer part of the ontology, so the ontology does not change by time-reversing the history of the world.

There is an obvious problem with this way of seeing things: if the electromagnetic fields are not in the state, then what are they? Are they even real? Aren't we used to thinking that fields are real? Isn't there energy associated to fields that suggests that they are, indeed, real?

As a reply, one could simply accept that they are real without being in the ontology of matter. That is, one could maintain that the electromagnetic fields are objective features of the world without thinking that they constitute something material. Fields should be understood as we understand forces or potentials: they are objective without being what tables and chairs are made of. They represent therefore, some "nomological" rather than "material" facts: they "tell" matter how to move, but they are not matter. In this approach, the ontology comes first, and symmetries are imposed on the nomology. So, in virtue of the role fields have in the theory, one can explain why B flips (contra the standard account): it is in D, it is an axial vector, and the reason why it is an axial vector is that this definition makes it change as to preserve the time-reversal invariance of the theory. This is compatible with the 'geometrical' transformation account: physicists defined B as an axial vector because in this way time-reversal symmetry is preserved.

Arguably, this is similar to the way in which fields have been originally introduced during the nineteenth century. The idea was to use them as a fiction to make interaction among particles local, mediated by the fields rather than in terms of forces acting at a distance and not through any intermediary matter. Charged particles were considered as giving rise to an electromagnetic field, which had a value at every point in space. "Nobody, however, thought of fields, early on, as anything more than a matter of bookkeeping. They were not considered a part of the fundamental ontology of the world, not something you might, say, stub your toe on, not (most certainly) *a real thing*" (Albert and Galchen 2009). Later on, however, the situation changed into what now we know as Maxwell's classical electrodynamics. It was observed that the energies and momenta of systems of particles were sometimes not conserved unless one would add the energies and momenta of the fields. This suggested that fields are real, in contrast with what was previously thought. This suggestion was reinforced by the fact that one can write an evolution equation for the fields, which have solutions independently of the presence of any particle. Also, the fact that these free fields could interact with themselves and that they oscillate at a velocity which is identical to the speed of light strongly suggested that they should be identified as light themselves.

So, the approach that I am now proposing is going back to the original way of understanding electromagnetism, in which conservation laws are broken and light is no longer identified as the oscillation of these fields (because there aren't any).

This requires a lot of reassessing, and it is certainly not cost-free. Still, we know that classical electrodynamics is not the final theory, so one may hope that in future theories these difficulties might compensate one another.

Indeed, what about other theories, like quantum mechanics? If this approach is correct, then the wavefunction is not in the state. Rather, it is part of the dynamical condition. But how can we even begin to make sense of this approach if the wavefunction is all there is in quantum theory? What is left in the ontology? We need to move beyond standard quantum theory. Indeed, we already knew that, albeit our reasons were different, as they were connected to the measurement problem. The measurement problem is the problem of dealing with unobserved macroscopic superpositions without imprecise rules such as von Neumann's collapse rule. This leads to the pilot-wave theory, the spontaneous localization theory, and the many-worlds theory. Now instead the motivation comes from thinking about ontology alone, and the way in which it makes sense to transform it. We have already seen this, but this is a different way of putting it: if we think that the ontology should not change during a time-reversal transformation, and if we also think that the wavefunction transforms under time-reversal transformation into its complex conjugate (because this is the way in which we can ensure that the theory is time-reversal invariant), then it follows that the wavefunction cannot be in the state (as we have seen, one can get around this by denying that the wavefunction actually transforms the way books say, and this is what Albert and Callender propose, at the price of the theory being no longer time-reversal invariant). If so, however, what can be in the state in place of the wavefunction?

For the pilot-weave theory, the answer is clear: it is a theory of particles so that $S = x$. For the other two theories instead, it is not clear at all, as they are both theories about the behavior of the wavefunction. So, to make sense, this approach needs to postulate the existence of other things which should be thought to be the ontology of the theory, in place of the wavefunction which is no longer in the state. This is indeed what the proponents of the so-called primitive ontology approach do (Allori et al. 2008, Allori 2013 a,b): GRW and many-worlds are not theories of the wavefunction, but they need to be "supplemented" with some other ontology, such as particles, a matter field, or spatiotemporal events. In this way, the completed state, rather than just the wavefunction, would instantaneously describe them. There are other reasons to have for ontology something other than the wavefunction. One of them is provided by the primitive ontology approach, which was developed to preserve the classical explanatory schema based on composition and dynamical reduction (Allori 2015a). To guarantee that such a reductive explanation is possible, one needs the ontology to be (suitably) microscopic and in spacetime. Thus, one needs to go beyond the wavefunction, as it is neither (obviously) macroscopic nor in spacetime. Here instead the argument has to do with symmetries: if the wavefunction transforms the way the textbooks say, then the theory is time-reversal invariant in both accounts of the time-reversal operator. However, only if you think of the time-reversal operator as in the pancake account, then it makes sense why the wavefunction transforms that particular way: it is because it does not belong to the state. To be clear, in this approach the wavefunction has disappeared from the state, namely it is not part of the ontology of matter. However, it should still be regarded as encoding something objective about the feature of the world, just like velocities in classical mechanics. Also, like in the case of electromagnetic fields, the wavefunction is not material but is better understood as expressing some nomological fact (see Allori 2021 for an account of what the wavefunction could be in this framework).

8.8 Evaluation

To summarize, we have seen three approaches: the standard approach, the pancake account, and the primitive ontology approach. None of them is perfect, namely in each of them costs are not light, otherwise we would have no debate at all.

In the standard account, we have the advantage that theories are invariant under time reversal, but the cost is that we need to define the time-reversal operator in such a way that the ontology changes depending on whether the state of a system belongs to the forward or the backward history of the world. This, I think, is a serious objection because it undermines the whole symmetry-first strategy: the time-reversal invariance so obtained does not do what we would like symmetries to do, namely to provide us with a perspective-free description. In the pancake approach, one does not have this issue, but by keeping certain quantities in the ontology the theory loses time-reversal symmetry. Finally, in the modified pancake account, in which one pairs the pancake account of time reversal with a bare understanding of the ontology (which ultimately is the core of the primitive ontology account), one has both time-reversal symmetry and the ontology behaving sensibly and restoring the value of symmetries. However, we have the problem of understanding the status of the entities which are not in the state but they are in D, namely they still encode objective information about the world, such as the electromagnetic fields and the wavefunction.

People will disagree about which approach is to be preferred based on which desiderata they care for more and depending on the costs they are more likely to swallow. So, very generally, I think we can conclude that we cannot have everything we want. That is, the three following claims (which we want all to be true) are incompatible (for more on this paradox, see Allori 2015b):

1. time reversal is a symmetry of the theory;
2. the ontology does not change under time-reversal transformation;
3. fields and wavefunction are in the state.

We can select the features we care about more, but we cannot have them all: one of them will have to go. This is the cost that each approach has. In fact, if, as both in the original and the modified pancake account, someone cares about an ontology which does not change under transformations like T, then they have to reject either 2 or 3. So, either, as in the pancake account, one accepts that the theory is no longer time-reversal invariant (rejecting 1). Otherwise, as in the modified pancake account, one has to assume that the fields and wavefunction are no longer in the state (rejecting 3).

Instead, if, as in the standard and pancake accounts, someone wants to keep fields and wavefunction in the state (thereby holding on to 3), then they have two other options. Either, as in the pancake account, one accepts that the theory is no longer time-reversal invariant (rejects 1). Or, as in the standard account, one accepts that the ontology changes under T (rejecting 2).

Finally, if someone cares about symmetries, as in both the modified pancake and standard accounts, then the options are the following. Either, as in the primitive ontology account, one can reject 3, namely accept that the fields and wavefunction are no longer in the state. Or, as in the standard account, one can reject 2, namely accept that the ontology changes under T.

My personal evaluation is the following. I think that allowing the ontology to change under T is too high of a cost, threatening the coherence of the whole symmetry-first strategy, and therefore I rule out the standard account, also considering that the T operator as defined does not provide an invariance we can care for. Between the two alternative pancake accounts, I favor the modified one because I value symmetries, so allowing that the fields and the wavefunction are not in the state seems to me as an acceptable cost. This is true especially in the case of quantum mechanics, in which I think there is already evidence that the wavefunction is not in the state. In other words, I think that it is much less problematic to make sense of the wavefunction not being in the state instead of the electromagnetic fields. In fact, the wavefunction not being in spacetime, thereby disrupting the classical explanatory schema based on compositionality and dynamical reduction, provides already a reason to reject the wavefunction as being part of the ontology (see Allori 2019b). This instead cannot be said for the electromagnetic fields, which are in spacetime. Moreover, the reasons we had to think of the fields as part of the ontology do not hold for the wavefunction. That is, the reasons we had to think of fields as part of the ontology were that fields carry energy, and that fields affect, and they are affected by particles. This is not the case for the wavefunction: it does not carry its own energy, and even if it affects the particles, it is not affected by them. The second fact indeed is one of the proclaimed mysteries of the pilot-wave theory: it is said that the theory is problematic because it does not explain why the wave acts on the particles, but the particles do not act back. In this framework instead, this asymmetry makes perfect sense: particles do not act back because there is no physical entity to act back on.

As a final remark, let me notice that, as Struyve (2022) has also discussed, the ontology of a theory is underdetermined by the formalism and whether a theory is time-reversal invariant depends on the ontology one considers, even with the same understanding of what the time-reversal operator does. For instance, classical electrodynamics can be written in terms of the electric field alone. In this case, the theory is time-reversal invariant also for the pancake account, because the electric field does not need to change sign under T in order to map solutions into solutions. Similarly, quantum theory can be written as a theory of the real part of the wavefunction alone, and in this case, again, the theory would be invariant also within the pancake account (see also Callender 2020). Thus, if we had a justification for considering classical electrodynamics as a theory about the electric field only, or quantum theory as a theory with the real wavefunction only, we would bypass the objection that these theories are not time-reversal invariant, and the ontology would still not change under T. So, why would we even need to kick the fields and the wavefunction out of the state? In other words, if we can read these theories in this way, then the paradox presented earlier does not arise, as we can have the three of them: time-reversal symmetry, fields, and wavefunction in the state and ontology not changing arbitrarily. So, why don't we obviously choose this? In reply, let me grant that we may not need to adopt the modified pancake account, and that we can make amends to the pancake account without losing symmetries as long as we re-interpret the relevant theories with enough ingenuity. However, the whole point rests on the viability of considering electrodynamics as a theory of the electric field alone, or quantum theory as a theory of the real component of the wavefunction. As also granted by Struyve, it is far from clear whether these theories are merely possible constructions or have a physical significance. For starters, one might also want other symmetries, in addition to time reversal, such as Galilean invariance or, preferably, Lorentz invariance. The viability of this approach is therefore strictly dependent on whether the ontology which makes the theory time-reversal invariant under the pancake account is the same ontology which makes the theory also Lorentz invariant, because if the answer is negative, then the approach is hopeless. The same objection does not apply to the modified pancake account, as symmetries are constraints on the nomology, and not on the ontology, and the nomology is designed to change as needed in order to present the various symmetries. So, a time-reversal quantum theory is one in which a Schrödinger evolving wavefunction changes into its complex conjugate as to map solutions into solutions; a Lorentz-invariant quantum theory is one in which, for instance, a Dirac evolving wavefunction transforms as needed as to map solutions into solutions, and so on. Be that as it may, let me conclude as follows. If these theories, namely classical electromagnetism and quantum mechanics respectively interpreted as theories of electric fields and real wavefunction alone, lack physical significance, then the argument proposed so far is an additional argument to favor the modified pancake account over the original one. If instead these interpretations turn out to be viable physical interpretations of these theories, then there is no additional argument for my proposed view coming from symmetry considerations.

In any case, these observations do not change the fact that the pancake account, in one form or the other, should be favored over the standard account.

References

Albert, D. Z. (2000). *Time and Chance*. Cambridge, MA: Harvard University Press.

Albert, D. Z., and Galchen, R. (2009). "A quantum threat to special relativity." *Scientific American* 300: 32–39.

Allori, V. (2013a). "Primitive ontology and the structure of fundamental physical theories." In D. Z. Albert and A. Ney (eds.), *The Wave Function: Essays in the Metaphysics of Quantum Mechanics*: 58–75. Oxford University Press.

Allori, V. (2013b): "On the metaphysics of quantum mechanics." In S. Lebihan (ed.), *Precis de la Philosophie de la Physique*: 116–151. Paris: Vuibert.

Allori, V. (2015a). "Primitive ontology in a nutshell." *International Journal of Quantum Foundations*, 1 (3): 107–122.

Allori, V. (2015b). "Maxwell's paradox: Classical electrodynamics and its time reversal invariance." *Analytica*, 1: 1–19.

Allori, V. (2019a). "Scientific realism without the wave-function." In J. Saatsi and S. French (eds.) *Scientific Realism and the Quantum*: 212–228. Oxford: Oxford University Press.

Allori, V. (2019b). "Quantum mechanics, time and ontology." *Studies in History and Philosophy of Modern Physics*, 66: 145–154.

Allori, V. (2021). "Wave-Functionalism." *Synthese*, 199: 12271–12293.

Allori, V., Goldstein, S., Tumulka, R., and Zanghi, N. (2008). "On the common structure of Bohmian mechanics and the Ghirardi–Rimini–Weber theory." *The British Journal for the Philosophy of Science*, 59 (3): 353–389.

Arntzenius, F. (2000). "Are there instantaneous velocities?" *The Monist*, 83: 187–208.

Arntzenius, F. (2004). "Time-reversal operation, representation of the Lorentz group and the direction of time." *Studies in Histories and Philosophy of Modern Physics*, 35(1): 31–43.

Callender, C. (2000). "Is time 'handed' in a quantum world?" *Proceedings of the Aristotelian Society*, 100: 247–269.

Callender, C. (2020). "Quantum mechanics: Keeping it real?" *The British Journal for the Philosophy of Science*. DOI: https://doi.org/10.1086/715032.

Earman, J. (2002). "What time-reversal invariance is and why it matters." *International Studies in the Philosophy of Science*, 16: 245–264.

Lopez, C., and Esfeld, M. (2023). "Humean Time Reversal Symmetry." *Synthese*, 202: 31.

Lopez, C. (2019). "Roads to the past: How to go and *not* to go backward in time in quantum theories." *European Journal for Philosophy of Science*, 9: 27.

Lorez, C. (2021). "The physics and philosophy of time reversal in standard quantum mechanics." *Synthese*, 199: 14267–14292.

Lopez, C. A. (2024). "Saving the appearances: The direction of time and time-reversal invariance." In C. Lopez and Lombardi (eds.), *The Arrow of Time: From Local Systems to the Universe*, Cambridge University Press, chapter 9.

Malament, D. B. (2004). "On the time reversal invariance of classical electromagnetic theory." *Studies in History and Philosophy of Modern Physics*. 35: 295–315.

Roberts, B. W. (2017). "Three myths about time reversal invariance." *Philosophy of Science* 84 (2): 315–334.

Roberts, B. W. (2021). "Time reversal." In E. Knox and A. Wilson (eds.), *The Routledge Companion to Philosophy of Physics*: 605–619. Routledge.

Struyve, W. (2022). Time reversal invariance and ontology. *British Journal for the Philosophy of Science*. arXiv:2307.07856v1 [physics.hist-ph].

9

Insights into Quantum Time Reversal from the Classical Schrödinger Equation

CRAIG CALLENDER

9.1 Introduction

Standard nonrelativistic quantum mechanics is widely presumed to be time-reversal invariant. Yet the justification for this claim has always been a bit mysterious. The literature produces many distinct justifications. Most are not very convincing. And the symmetry often feels imposed rather than discovered, as when author Paul Roman writes that "we should like, even if only by some artifice, to achieve full covariance [time reversal invariance]" (Roman 1960: 266). Add to this impoverished motivation the fact that the time-reversal operator is the only anti-unitary operator we use and it becomes natural to find the conventional wisdom a bit suspicious – or at the least, worth greater elucidation.

In what follows I provide a new insight into this issue. First, I show that the *exact same puzzle* that arises in quantum mechanics arises also in classical physics. I do this by introducing the "quantum-looking" representation of classical theory developed by Schiller (1962) and Rosen (1964). That this puzzle can arise classically suggests that the problem is due to the representation and not the theory. And just as classical physics is time-reversal invariant despite this puzzle, this observation provides hope that the conventional wisdom about quantum mechanics is right and that it is genuinely and unambiguously time-reversal invariant. Second, and more controversially, I use the counterpart of the classical resolution of the puzzle to vindicate a definitional link between spatial variation of the phase and temporal variation. I show how Bohmian mechanics (and some other interpretations) provides this crucial step. With Bohmian mechanics playing the same role Newtonian mechanics does in the solution to the classical puzzle, I provide a satisfying way out of the quantum puzzle.

Put loosely, I argue that the puzzle of quantum time reversal arises because we have been given only "half" of the theory. By putting classical mechanics in a representation that reproduces the same questions, we can see how we got into this mess and also how to get out of it. Classically a solution requires appeal to Newtonian ontology and laws. Quantum mechanically a solution needs the ontology and laws found in some quantum interpretations.

9.2 The Puzzle of Time Reversal in Quantum Theory

The conventional wisdom is that quantum theory – if we focus only on unitary evolution and ignore controversial features such as collapses – is time-reversal invariant. In the Schrödinger representation, the quantum state is described by a wavefunction. If we focus on one particle $\psi = \psi(x, t)$, the evolution of the state ψ through time is given by the time-dependent Schrödinger equation:

$$i\hbar \frac{\partial \psi}{\partial t} = \hat{H}\psi \tag{9.1}$$

where we'll assume throughout that we're working with a Hamiltonian $H = p^2/2m + V(x)$ that is time-reversal invariant. (9.1) is a first-order parabolic partial differential equation. As such, it is of the same family as the heat equation

$$\frac{\partial u}{\partial t} = k\nabla^2 u \tag{9.2}$$

where u represent the temperature and κ is a real-valued coefficient. The heat equation, of course, is the poster child of temporally irreversible equations. Given their similar form, how can it be that they differ in this symmetry?

To be a bit more precise, normally one associates time reversal in classical spacetimes with the physical operation corresponding to a temporal reflection about a spatial axis: the past is swapped for the future. Then, in natural coordinates over the spacetime $\{x, y, z, t\}$, a passive interpretation of time reversal corresponds to flipping the coordinates $t' = -t$ and using $\{x', y', z', t'\}$. Put actively, one sticks with the same but changes the world –reverses the sequence of physical states – so that the description of the world exactly matches the description in the passive transformation. Call this "no frills" time reversal. Assuming time-translation invariance, one flips the temporal sequence of states and that's it. If any other variable changes under the transformation (e.g., classical velocity) it happens "by logic and definition alone" from that flip.

Applying this operation to the Schrödinger equation, one immediately sees that it is not time-reversal invariant in this sense. A minus sign pops out due to it being first-order in time. Typically, if $\psi(x, t)$ is a solution of (9.1), then $\psi(x, -t)$ is not. A symmetry should take solutions to solutions and nonsolutions to nonsolutions. Solutions to the Schrödinger equation are not solutions to the time-reversed

Schrödinger equation, so the symmetry seems broken. That is precisely what we say about the heat equation when we explain that it is not time-reversal invariant.

Readers familiar with the topic will demand that we stop right here. They will point out that Wigner (1931) taught us that time reversal in quantum theory is represented by *two* operations, a temporal reflection and complex conjugation $\psi \rightarrow \psi^*$. And it is certainly true (9.1) is invariant under these combined operations. That is, if $\psi(x, t)$ is a solution of (9.1), then so is $\psi^*(x, -t)$, and vice versa. Quantum mechanics does indeed have this symmetry.

But is that the end of the issue? No. One can *call* the Wigner transformation "time reversal" if one wants to say that it vindicates the conventional wisdom, yet that obscures the fact that the two symmetries are different. Dub them whatever you like, the operation of flipping temporal sequence is different than that plus implementation of complex conjugation. Or so it seems. If someone described the operation $L : \kappa \rightarrow -\kappa$ and then said that the heat equation was time-reversal invariant because its solutions are invariant under the combined operation of time reversal and L, one would feel that was a bit of a cheat – or at least a kind of pun on the phrase "time reversal." Why in the world should κ flip sign due to reversing the temporal sequence? It doesn't follow from logic and definition alone. Same in quantum theory. Why should $K : i \rightarrow -i$ when reversing the temporal order? i is a number without spatiotemporal dimensions, so why should it flip? It is no wonder then that there has been a fair number of people in the philosophical foundations of physics puzzling over time reversal in quantum theory (e.g., Albert 2000, Allori 2022, Callender 2000, Earman 2002, Gao 2022, Lopez 2021, Malament 2004, Roberts 2022, Struyve 2022). Different camps emerged differing over whether the Wigner symmetry should count as time-reversal symmetry. One group holds that quantum theory is not invariant under time reversal, strictly speaking, but it is invariant under what we might call "motion reversal" (implemented via Wigner's symmetry). Given the interpretive complications with quantum theory, it wouldn't be surprising if two similar symmetries that overlap classically were conflated quantum mechanically. The other camp holds that Wigner symmetry is or warrants the label time-reversal invariance once this symmetry is properly understood.

Lately, however, there have been efforts at a kind of rapprochement between the two camps. Callender (2023) and Roberts (2022) hope to show that Wigner symmetry does follow from no-frills time reversal. The present argument is in this spirit. Success in this endeavor would be the best possible resolution. After all, one has a strong suspicion that there is still more to the story, that disambiguation (or not) is too simple. As Schrödinger (1931) remarks when comparing (9.1) with the Fokker–Planck equation (a parabolic equation like (9.2)), "whilst in both cases the differential equation is of first order in time, the presence of a factor $\sqrt{-1}$ confers to the wave equation a hyperbolic or, physically stated, reversible character at variance with the parabolic-irreversible character of the Fokker[–Planck] equation" (Chetrite 2021: section 2.5). So, the similarity with the heat equation may be superficial. Due to the i, (9.1) "feels" more like a hyperbolic equation than a parabolic one, which, after all, is why it's called a "wave equation" despite its parabolic nature. A more satisfying explanation would offer a deeper insight into the connection between complex conjugation and time reversal.

In what follows I provide such an explanation in three steps. The first step is to see that this puzzle can arise even in purely classical physics. The second identifies what it takes to extricate ourselves from the classical puzzle. The third simply takes the classical solution's counterpart in the quantum context. No doubt this last step will be controversial because it involves an interpretation of quantum theory. But if I'm right, the puzzle over time reversal in quantum theory arises because standard quantum mechanics gives us only "half" a theory.

9.3 The Classical Schrödinger Equation

In this section we'll stay entirely within the classical mechanical regime but shift to a representation that looks very quantum. We'll derive the so-called classical Schrödinger equation of Schiller (1962), Rosen (1964), and others. This equation arises from treating classical mechanics as a kind of field theory. The field theory is itself motivated from classical statistical mechanics. We'll use the Hamilton–Jacobi formulation of classical mechanics. Although it can describe a single particle, the Hamilton–Jacobi picture is naturally suggestive of a continuous ensemble of particles evolving in configuration space beginning with different initial conditions. And from this theory a field picture is natural, one with wavefronts evolving in configuration space. This picture is the reason why Hamilton–Jacobi theory is usually taught as a precursor to quantum theory. To get the classical Schrödinger equation we'll simply bundle these waves into a Schrödinger-like form.

We will consider particles of mass m evolving in a potential V in a Cartesian system $x = (x; y; z)$. We'll denote with q_i and p_i the generalized coordinates, where $i = 1, \ldots, n$ and n is the number of degrees of freedom of the system. The Lagrangian is

$$L(x, \dot{x}, t) = \frac{1}{\sqrt{2}} m\dot{x}^2 - V(x, t), \tag{9.3}$$

and the action is

$$S = \int_{t_{IqI}}^{t_{FqF}} L(x, \dot{x}, t), \tag{9.4}$$

where I is the initial time and location and F is the final time and location. Then using identities found in any good analytical mechanics textbook (e.g., Johns 2016), one derives the Hamilton–Jacobi equation:

$$\frac{\partial S}{\partial t} + \frac{(\nabla S)^2}{2m} + V(x, t) = 0, \tag{9.5}$$

The surfaces of constant S are regarded as wavefronts. These are sets of points q at time t for which $S(q, t) = constant$. Equation (9.5) describes the evolution of these wavefronts with time. While Hamilton–Jacobi wavefronts are often derived from knowledge of particle trajectories, it's important to recognize that this is not necessary. We can simply begin with (9.5) and regard it as an independent equation governing these wavefronts.

To extract individual trajectories from (9.5), one can associate particle paths with the characteristics of (9.5). Jacobi's Theorem tells us that $p_i = \frac{dS}{dx}$, and we can use this relation and a separation of variables to deduce the equation of motion for particular initial positions $x = (x(t, x_0))$. In this sense Hamilton–Jacobi theory encodes a kind of "particle-wave" duality, which is what attracted Schrödinger to it when he discovered his famous equation. One can begin with waves and extract particle trajectories, or vice versa.

The Hamilton–Jacobi theory is naturally suggestive of an ensemble picture. The S-function gives us momenta via $p = \nabla S$. Yet via the classical Lagrangian and its conjugate momentum we know that $p = m\dot{x}$. So $m\dot{x} = \nabla S$. The S-function therefore describes a velocity field all over configuration space. In the single-particle case, one particle "surfed" the "wave" described by S. But we can easily imagine an infinite ensemble of particles differing in their initial positions $x = (x(t, x_0))$ surfing the same S-function but tracing out their respective trajectories.

Let this ensemble be distributed with density $R^2(x, t) \geq 0$. As the particles evolve according to (9.2), we can watch this distribution $R^2(x, t)$ evolve in time. Since the particles move without sources or sinks, they must obey a continuity equation:

$$\frac{\partial R(x, t)^2}{\partial t} + \frac{\partial}{\partial t}\left(\frac{1}{m}\frac{\partial S(x, t)}{\partial x} R^2(x, t)\right) = 0. \tag{9.6}$$

This is a conservation law for probability, one implied by the classical Liouville theorem. Writing the Hamilton–Jacobi equation for an ensemble as the combined Equations (9.5) and (9.6) describes a statistical mechanics of classical particles. Solutions are obtained by specifying the initial S and R, $S(x,0) = S_0(x)$, and $R(x,0) = R_0(x)$. $S_0(x)$ determines an initial momentum field that shapes the development of the initial distribution. We arrive at a kind of field theory describing two partially coupled fields, $S(x,t)$ and $R^2(x,t)$.

Readers familiar with the Madelung decomposition of the Schrödinger equation know that if we write the wavefunction in polar form and separate real and imaginary parts we obtain two real equations, a "quantum" Hamilton–Jacobi equation and a continuity equation (Holland 1993). Since we have a Hamilton–Jacobi equation and continuity equation in Equations (9.5) and (9.6), we can perform a "reverse Madelung decomposition" and combine these equations to form a "classical" Schrödinger equation.

To get the classical Schrödinger equation, we simply package this familiar physics into a single compact equation. Introduce complex classical wavefunctions by joining together the action S and the previously introduced R to form:

$$\psi_{CL} = R(x,t)\exp(iS(x,t)/\hbar \tag{9.7}$$

where $\hbar$ is added merely to mimic the appearance of the quantum.[1] Combined together into one complex-valued equation, (9.5) and (9.6) are equivalent to

$$i\hbar\frac{\partial\psi_{CL}(x,t)}{\partial t} = \left(-\frac{\hbar^2}{2m}\frac{\partial^2}{\partial x^2} + U(x,t) - Q(x,t)\right)\psi_{CL}(x,t), \tag{9.8}$$

which is the so-called *classical Schrödinger equation* (Schiller 1962, Rosen 1964). Here U is the classical potential and we assume it to be time-independent $U(x,t) = U(x)$. The variable Q is what is sometimes dubbed the quantum potential. It has the form

$$Q = \frac{\hbar^2}{2m}\frac{1}{R}\frac{\partial^2 R}{\partial x^2}. \tag{9.9}$$

As we know from Bohmian mechanics (Bohm 1952), Q carries all the "quantum" aspects of nature. Intuitively, (9.8) is the Schrödinger Equation (9.1) with its quantum aspects "subtracted" out.

This "classical Schrödinger equation" should not be confused with the "classical Schrödinger equation" of Koopman (1931), von Neumann (1932), and Sudarshan

[1] It will drop out when equations of motion are derived.

(1976). The present equation is nonlinear and its wavefunction's domain is configuration space, whereas the Koopman–von Neumann–Sudarshan equation is linear and its wavefunction's domain is phase space.[2] The nonlinearity means that superpositions of solutions are not generally also solutions. If ψ_1 is a solution and ψ_2 is a solution, then $\psi_1 + \psi_2$ is not generally a solution; the exceptions are when ψ_1 and ψ_2 have no common support or when one is a multiple of the other. Interference patterns also do not develop, as one would expect.[3]

9.4 Time Reversal of the Classical Schrödinger Equation

Consider time reversal in these dynamics. Because each trajectory in the ensemble is governed by Newtonian dynamics, we expect the evolution to be time-reversal invariant in a no-frills sense. And it is. We're just running a bunch of Newtonian particles forward and backward in time. Newton's second law is time-reversal invariant, so the ensemble should be too.

What's interesting, and a bit surprising, is that in this formalism one achieves this temporally reversed evolution by *complex conjugation of* (9.8). We're in exactly the same situation with the classical Schrödinger equation as we were with the quantum Schrödinger equation! Just as with the Schrödinger equation, temporal reflection will bring a minus sign out in front of (9.8). The equation is nonlinear but still a first-order parabolic equation. Solutions to (9.8) will not be solutions to this equation:

$$-i\hbar \frac{\partial \psi_{CL}(x,t)}{\partial t} = \left(-\frac{\hbar^2}{2m}\frac{\partial^2}{\partial x^2} + U(x,t) - Q(x,t) \right) \psi_{CL}(x,t) \tag{9.10}$$

However, if we take the complex conjugate when we time reverse, then solutions of (9.8) will be mapped one-to-one to solutions of

$$i\hbar \frac{\partial \psi^{*}{}_{CL}(x,t)}{\partial t} = \left(-\frac{\hbar^2}{2m}\frac{\partial^2}{\partial x^2} + U(x,t) - Q(x,t) \right) \psi^{*}{}_{CL}(x,t), \tag{9.11}$$

which is the complex conjugate of (9.8).

We're in precisely the same situation as we were quantum mechanically. But the beauty of this case is that we know exactly what is going on because we have a good theoretical understanding of classical mechanics, its ontology, laws, and the derivation of the wave Equation (9.8). This understanding will help us extricate ourselves from the quantum puzzle.

[2] For a philosophical lesson we learn from Koopman–von Neumann–Sudarshan, see Callender (2023).

[3] For further discussion of (9.5), see Benseny, Tena, and Oriols (2016), Holland (1993), Ghose (2018), Rosen (1964), Schiller (1962), Schleich *et al.* (2013).

To drive the point home, let's be clear. Forget the derivation of (9.8). Suppose you were just given (9.8). This is a perfectly conceivable world. You've got your waves and from these you obtain your probability distributions. To get the distributions, you supplement (9.8) with a classical version of Born's rule. Perhaps you attach a Copenhagen-like interpretation to this, or something else. It is a theory along the lines suggested by Prigogine (1996) and Krylov (1979) and developed philosophically by McCoy (2020). You have an empirically adequate (to the classical world) theory in a classical statistical mechanical universe. You get along perfectly fine. You don't have or believe that you need a trajectory-based classical physics.

Still, you're a bit puzzled by time reversal. (9.8) is a first-order in time equation like the quantum Schrödinger equation. On a "no-frills" sense of time reversal, it is simply not time-reversal invariant. But you notice that if you take the complex conjugate of the wavefunction when time reversing, then solutions in one temporal direction map to solutions in the other. What reason would you have to take the complex conjugate when time reversing? As before, complex conjugation doesn't follow from temporal reflection by logic or definition alone. We could readily envision classical mechanical counterparts of all the concerns raised in the introduction.

If we assume that really the theory is time-reversal invariant – and we should because we know what's going on, to which we will turn in a moment – the problem is that we need the classical momenta to reverse under time reversal. In the other classical representations, we know that (q,p) going to $(q,-p)$ will cause the forward in time trajectories to retrace their paths in the reverse time direction. But in this representation – where we only have (9.8) – we lack reason to flip the momenta when we implement a temporal reflection. From Jacobi's theorem we have

$$p = \frac{\partial S}{\partial q}, \tag{9.12}$$

but this is *spatial variation,* not temporal variation. If we're strict about only reversing what follows from a simple temporal reflection, then we have no reason to turn the momenta around. To get the classical waves to turn around we need the classical phase to flip. Complex conjugation achieves this as it reverses ${}^{iS}/_{h}$ in (9.7), which in turn flips the momenta. With what we have, however, this isn't motivated from a no-frills temporal reflection.

9.5 The Classical Solution

There is no mystery in this case about what's going on. We know more than (9.8). In a classical world there are more than just waves rippling through high-dimensional configuration space. The wave description was built up from a statistical mechanics of individual particles and these particles ultimately are governed by Lagrangian or Hamiltonian or Newtonian mechanics. These resources associated with the particle ontology give us the ability to link the preceding momenta with temporal variation.

Suppose we have a classical free particle. It is governed by (9.5). Its solutions will be spherical wave fronts, and we'll know that the momentum p is given by Equation (9.12). But we also know a whole lot more if we assume that we have particles with trajectories. Importantly, we also know that the Lagrangian is $L(x,\dot{x},t) = \left(\frac{1}{2}\right)m\dot{x}^2 - V(x,t)$. The particle's canonical momentum is $p_i = {}^{\partial L}/_{\partial \dot{q}}$. With our Lagrangian, that means $p = m\dot{x}$.

So $m\dot{x} = m^{dx}/dt = \nabla S$ (see Holland 1993: 51). Alternatively, with a three-dimensional delta function we can extract the motion of a single particle directly from a solution S of (9.5) (see Rosen 1986: 690). This point is absolutely crucial. The canonical momentum on its own in our "wave picture" isn't necessarily tied to temporal variation. But when we bring in the resources and assumptions of a particle ontology traversing well-defined trajectories, we connect this to velocities. That connects the *temporal variation* of the ontology with the *spatial variation* of the phase. The classical Schrödinger representation does not give us this (or if you prefer, my imagined world where we're simply given (9.8) doesn't). Only when we look "under the hood" of this representation do we see this connection. By "under the hood" I mean classical particle mechanics.[4]

And of course, we know that classical mechanics is time-reversal invariant in a no-frills sense. Again, sticking with our single-particle Lagrangian, we can use the second-order Euler–Lagrange equations to derive Newton's equation of motion:

$$mx = -\nabla \ddot{V}|_{x=x(t)}, \tag{9.13}$$

which is time-reversal invariant in a no-frills sense.

In the classical Schrodinger equation representation, i does not go to $-i$ under time reversal. Or it shouldn't if one uses the full resources attached to the ontology. Really i never flips. It only looks like it flips if one ignores some of classical mechanics. What really happens is that time "flips," which in turn "flips" velocity, which in turn "flips" the classical phase; when that reflected phase is then used to compose a classical wavefunction, that wavefunction is the complex conjugate of the original classical wavefunction. The complex conjugation is simply a downstream consequence of reflecting time.

[4] One could also point out that classical hydrodynamics is an isomorphic situation. Like our systems here, the fluid is described by a continuity equation and an Euler equation (a field-theoretic expression of Newton's second law). With the usual restrictions, classical hydrodynamics is considered time-reversal invariant. The reason is that the current (density times velocity) reverses sign due to the velocity flipping. Without "opening the hood" on the physics of the fluid, we might not see that the theory is time-reversal invariant.

Opening the hood shows us that complex conjugation does indeed follow by logic and definition alone from a simple no-frills temporal reflection. The "problem" of time reversal in the Schrödinger representation of classical physics is therefore solved.

9.6 Back to Quantum Time Reversal

In standard quantum mechanics we are in the situation where we only have the quantum counterpart of (9.8). It's therefore hard to see a good reason to take the complex conjugate when time-reversing a system. Complex conjugation does not follow from $t \rightarrow -t$ if we restrict ourselves only to the Schrödinger equation. Hence time reversal is very puzzling. What makes this hard to tolerate is that quantum theory *almost* has the resources to extract itself as we did earlier. After all, a Madelung decomposition of the Schrödinger equation will yield two real equations, a quantum Hamilton–Jacobi equation and a continuity equation. Using a hydrodynamic analogy, we want to say that the probability current j defines a velocity for us via

$$v = \frac{j}{|\psi(x,t)|^2}, \tag{9.14}$$

or using the phase of the quantum Hamilton–Jacobi equation via

$$v = \nabla S, \tag{9.15}$$

We're prohibited from doing so because – we're told – we all know that quantum mechanics is incompatible with always-determinate trajectories, and hence there can be no velocities. This conventional wisdom is repeated again and again throughout quantum theory. In the Sakurai (1994) textbook the author uses scare quotes around the equal sign in Equation (9.15), and in Shimbori and Kobayashi (2000) the authors use scare quotes around the word "velocity." Sakurai uses scare quotes because he thinks always-determinate trajectories and velocities will contradict the quantum uncertainty principle[5] (see Romano 2021 for a convincing reply). Shimbori and Kobayashi (2000) don't say why they use scare quotes but presumably it is due to this conventional wisdom. In the philosophy literature, Earman (2002) and Gao (2022) recognize that the quantum phase must turn around, but neither explicitly connect this via logic and definition alone

[5] Sakurai: "we would like to caution the reader against a too literal interpretation of j as ρ times the velocity defined at every point in space, because a simultaneous precision measurement of position and velocity would necessarily violate the uncertainty principle" (1994: 102–103).

to a temporal reflection. Gao (2022) suggests that assuming the continuity equation invariant under time reversal is a requirement of meaning, and hence due to definition. However, in the next paragraph we see that the concept of the best explanation is really doing the work. Physics in the time-reversed world would be unexplained, he says, if we do not reverse the momentum – but that is just what happens in time-reversed worlds in a non-time-reversal invariant theory. Earman (2002) writes:

> How can the information about the direction of motion of the wave packet be encoded in $(x, 0)$? Well (when you think about it) the information has to reside in the phase relations of the components of the superposition that make up the wave packet. And from this it follows that the time reversal operation must change the phase relations. *(Earman 2002: 248)*

As congenial as what Earman says is to the present perspective, "when you think about it" and "has to" are not derivations. The spatial variation in the phase is only definitionally linked to the temporal variation in a temporal reflection when one removes the above scare quotes.

Remove the scare quotes and we remove the problem. What helped us understand complex conjugation's connection to time reversal in the case of the classical Schrödinger equation was that we already knew or posited the ontology and individual particle dynamics. We opened the hood and looked at the engine. What hinders us in the quantum case is that we don't know what's going on. The infamous measurement problem leaves us not knowing what the ontology or extra dynamics (if any) are.

However, there are interpretations of quantum mechanics that are especially well-suited for tackling this question. One is Bohmian mechanics. Bohmian mechanics can be derived via a Madelung decomposition of the Schrödinger equation to a "quantum" Hamilton–Jacobi equation and continuity equation – essentially the reverse of how we derived the "classical" Schrödinger equation. Nonrelativistic Bohmian mechanics also posits particles guided by a dynamical law, like Newtonian mechanics, and in many ways quantum mechanics is to Bohmian mechanics as statistical mechanics is to classical Newtonian dynamics. This similarity raises the possibility that we can understand complex conjugation's connection to time reversal in the same way as we did classically. Perhaps we can open the hood and see a rationale for complex conjugation when time reversing by studying the Bohmian engine.

Indeed, we can. Bohmian mechanics describes particles with always-determinate positions, X. The dynamics for these particles come in many equivalent forms. For our needs the "Newtonian" second-order formulation may be the most transparent. This formulation was the original one presented by Bohm. In this representation, the Bohmian dynamics for the particles are given by

Newton's second law, except now the preceding quantum potential joins the classical potential:

$$m\frac{d^2X(t)}{dt^2} = -\nabla(Q+V), \tag{9.16}$$

where we assume that all the particles have the same mass and that $V(x,t) = V(x)$. Because the quantum potential Q does not flip sign under time reversal, we can see that the Bohmian dynamics is time-reversal invariant in a completely simple "no-frills" sense. This theory is now written in terms of two real-valued functions, R, and S, just as classical Hamilton–Jacobi theory is. The dynamics requires that particle velocities satisfy $p = mv = \nabla S$, as in classical Hamilton–Jacobi theory. Here I stress that there is an "equals sign" without scare quotes.

We can now transform into the complex-valued wave representation. When we do, we see that in order to get back the time-reversed (in the no-frills sense) Bohm trajectories, we need to take the complex conjugate of the wavefunction. Without the underlying ontology, that transformation seems to have no rationale. But when we look under the hood at the assumed Bohmian ontology and laws, we see that complex conjugation does follow from $t \to -t$. As in the classical setting, this no-frills temporal reflection alters the sign of the velocities of the particles. And via $p = mv = \nabla S$, that changes the sign of the phase. That then demands complex conjugation.[6]

Essentially the same reasoning holds also in the "Newtonian Quantum Mechanics" interpretation of Hall *et al.* (2014) and Sebens (2015). According to this theory, there are many classical worlds evolving (and interacting) via a Newtonian force law that is manifestly time-reversal invariant in a no-frills sense. The time-reversal invariance of the fundamental ontology forces the phase to change sign. And when we compose the wavefunction from the more basic entities in the theory, complex conjugation occurs when time reversing, just as it does here when composing wavefunctions in both the classical and quantum Hamilton–Jacobi representations. Sebens claims that explaining time reversal is an advantage of his interpretation. I agree that it is an advantage of any interpretation that it link the spatial variation of phase information with time.

[6] Note that I must assume the so-called "nomological" interpretation of Bohm's theory for this to work (see Dürr, Goldstein, and Zanghì 1997; Callender 2015). On this picture, the particles are understood as the primitive ontology and the wavefunction is understood as nomological, like the Hamiltonian. If the wavefunction is on par with the particles, we end up with the original problem restated in the Bohm formalism. That is, perhaps the velocities flip as suggested in the text, but the phase does not; that would indicate that the guidance equation (Equation (9.14) above) doesn't hold under time reversal. That possibility exists because the wavefunction is its own entity on non-nomological views. On nomological interpretations, by contrast, the particles (or fields, or whatever beables are posited) are the "boss," and the wavefunction is part of what governs or best describes the particle motions. If velocity flips, then so does the phase. This restriction to the nomological view may be seen as a limitation of my proposal; alternatively, to the extent that it nicely resolves the puzzle of time reversal, it may be received as an additional reason in favor of the nomological picture of the Bohmian wavefunction.

9.7 Conclusion

Time reversal in quantum theory has long been puzzling. The main point of this chapter is that precisely this same puzzle arises also in classical physics, and in particular, in the representation of classical physics via the classical Schrödinger equation of Schiller and Rosen. I think many readers will be surprised to learn this. What does it mean? To me, it teaches us that the puzzle of time reversal in quantum theory originates from using a theory stripped of its ontology. Once the ontology and its dynamics are added, all becomes clear. Presented in its real-valued form with a full Bohmian ontology and laws (or other interpretation with similar effect), the theory is time-reversal invariant in a no-frills sense. We don't need to take the complex conjugate of anything. Translated into a complex-valued representation, complex conjugation is simply a consequence of $t \rightarrow -t$. Normally, we cannot see this because we are working with only "half" the theory. But if we imagine that a good interpretation provides the other half of the story, we recognize the engine that derives complex conjugation from temporal reflections.[7]

References

Albert, D. Z. (2000). *Time and Chance*. Cambridge, MA: Harvard University Press.

Allori, V. (2022). "On the Galilean invariance of the pilot-wave theory." *Foundations of Physics*, 52 (5): 1–21.

Benseny, A., Tena, D., and Oriols, X. (2016). "On the classical Schrodinger equation." *Fluctuation and Noise Letters*, 15, 1640011.

Bohm, D. (1952). "A suggested interpretation of the Quantum Theory in terms of 'Hidden' Variables, I and II." *Physical Review*, 85(2): 166–193.

Callender, C. (2000). "Is time 'handed' in a quantum world?" *Proceedings of the Aristotelian Society*, 100(3): 247–269.

Callender, C. (2015). "One world, One beable." *Synthese* 192: 3153–3177.

Callender, C. (2023). "Quantum mechanics: Keeping it real?" *British Journal for the Philosophy of Science*, 74(4): 837–851.

Callender, C. (2023). "The prodigy that time forgot: The incredible and untold story of John von Newton." In A. Bassi, S. Goldstein, R. Tumulka, and N. Zanghì (eds.), *Physics and the Nature of Reality: Essays in Memory of Detlef Dürr*, Springer, pp. 51–61.

Chetrite, R., Muratore-Ginanneschi, P., and Schwieger, K. E. (2021). "Schrödinger's 1931 paper 'On the reversal of the laws of nature'" ["Über die Umkehrung der Naturgesetze", *Sitzungsberichte der preussischen Akademie der Wissenschaften, physikalisch-mathematische Klasse*, 8 N9 144–153]. *EPJ* H 46, 28.

Dürr, D., Goldstein, S., and Zanghì, N. (1997). "Bohmian mechanics and the meaning of the wave function." In R. S. Cohen, M. Horne, and J. Stachel (eds.), *Experimental Metaphysics – Quantum Mechanical Studies for Abner Shimony*, Volume One, (*Boston Studies in the Philosophy of Science* 193), Boston: Kluwer Academic Publishers.

[7] Many thanks to Eddy Chen, Xavier Oriols, Bryan Roberts, Davide Romano, Chip Sebens, and Ward Struyve for very helpful comments.

Earman, J. (2002). "What time reversal is and why it matters." *International Studies in Philosophy of Science*, 16(3): 245–264.

Gao, S. (2022). "Understanding time reversal in quantum mechanics: A new derivation." *Foundation of Physics*, 52, 114. DOI: https://doi.org/10.1007/s10701-022-00634-1.

Ghose, P. (2018). "The quantum-like face of classical mechanics." DOI: https://doi.org/10.48550/arXiv.1801.02499.

Hall, M. J. W., Deckert, D., & Wiseman, H. M. (2014). "Quantum phenomena modeled by interactions between many classical worlds." *Physical Review*, X, 4, 041013.

Holland, P. (1993). *The Quantum Theory of Motion*. Cambridge: Cambridge University Press.

Johns, O. (2016). *Analytical Mechanics for Relativity and Quantum Mechanics* (2nd ed.). Oxford: Oxford University Press.

Koopman, B. O. (1931). "Hamiltonian systems and transformations in Hilbert space." *Proceedings of the National Academy of Science USA*, 17: 315–318.

Krylov, N. (1979). *Work on the Foundations of Statistical Physics*. Princeton: Princeton University Press.

López, C. (2021). "The physics and the philosophy of time reversal in standard quantum mechanics." *Synthese*, 199 (5–6): 14267–14292.

Malament, D. B. (2004). "On the time reversal invariance of classical electromagnetic theory." *Studies in History and Philosophy of Modern Physics*, 35: 295–315.

McCoy, C. (2020). "An alternative interpretation of statistical mechanics." *Erkenntnis*, 85, 1: 1–21.

Prigogine, I. (1996). *The End of Certainty*. New York: The Free Press.

Roberts, B. (2022). *Reversing the Arrow of Time*. Cambridge: Cambridge University Press.

Roman, P. (1960). *Theory of Elementary Particles*. New York: Interscience.

Romano, D. (2021). "On the alleged extra-structures of quantum mechanics." *Foundation of Physics*, 51, 29. DOI: https://doi.org/10.1007/s10701-021-00426-z.

Rosen, N. (1964). "The relation between classical and quantum mechanics." *American Journal of Physics*: 597–600.

Rosen, N. (1986). "Quantum particles and classical particles." *Foundations of Physics* 16(8): 687–700.

Sakurai, J. J. (1994). *Modern Quantum Mechanics*. Boston, MA: Addison-Wesley.

Schiller, R. (1962). "Quasi-classical theory of the non-spinning electron." *Physical Review*, 125: 1100–1108.

Schleich, W. P., Greenberger, D. M, Kobe, D. H., and Scully, M. O. (2013). "Schrödinger equation revisited." *Proceedings of the National Academy of Science USA*, 110 (14): 5374–5379.

Sebens, C. (2015). "Quantum mechanics as classical physics." *Philosophy of Science* 82(2): 266–291.

Shimbori, T., and Kobayashi, T. (2000). "'Velocities' in quantum mechanics," Appendix A of *Journal of Physics* A33 7637 DOI: https://doi.org/10.48550/arXiv.quant-ph/0004086.

Struyve, W. (2022). "Time-reversal invariance and ontology." *British Journal for the Philosophy of Science*. DOI: https://doi.org/10.1086/721089.

Sudarshan, E. C. G. (1976). "Interaction between classical and quantum systems and the measurement of quantum observables." *Pramana* 6(3), 117–126.

Von Neumann, J. (1932). "Zur Operatorenmethode in der klassischen Mechanik." *Annals of Mathematics*, 33, 587–642.

Wigner, E. P. (1931). Group Theory and *Its* Application to the Quantum Mechanics of Atomic Spectra. New York: Academic Press.

10

Saving the Appearances: The Direction of Time and Time-Reversal Invariance

CRISTIAN LÓPEZ

10.1 Introduction

My view is that the direction of time is so important for even a minimal understanding of our daily experience and of scientific phenomena that it must be regarded as primitive in any empirically well-based ontology. Isaac Newton and Gottfried Leibniz championed this view in the eighteenth century, *despite* their disagreement about the nature of time. In the contemporary literature, primitivism has rather been scarcely defended: Tim Maudlin, Michael Esfeld, and I have also defended primitivism recently, though on different grounds (see Maudlin 2002, 2007, Lopez and Esfeld 2025, Lopez 2024a). Typical reasons to uphold some form of primitivism come from explanatory necessity: a direction of time is postulated as a primitive structure of space-time because it is needed for explanation. In Maudlin's view, for instance, it follows from the necessary structure that space-time must have in order to get good working physical theories. Esfeld and Lopez run a more metaphysically oriented argument, defending primitivism as the best (if not the only) way to distinguish spatial and temporal relations in any ontology that involves, at least, change.

In this chapter, I defend primitivism about the direction of time on different grounds and in a nondirect, perhaps more winding way (readers are referred to Lopez and Esfeld 2025 and Lopez 2024a for a more direct and positive defense). I do this by, first, suggesting an alternative way to understand the problem at an epistemological level and, second, by arguing that one of the strongest and most persuasive arguments against primitivism (what I call the 'Time-Reversal Argument', **TRA** henceforth) can be rendered harmless. Let me briefly depict the overall argumentation.

To begin, quarrels about the direction of time in physics and philosophy of physics have generally assumed some form of *reliabilism* as a theory of epistemic justification[1] (see Goldman 1979, 1986): to generate justification for beliefs about the direction of time, a reliable, truth-conducive process must be adopted and followed. Without such a process, any positive belief about the direction of time is deemed unjustified. In a naturalistic framework, such a justification has relied on what physics can tell us about the direction of time, following in general a reductionist strategy. This epistemological assumption puts the burden of proof on the defender of a primitive direction of time: she is forced to deliver justification for the belief in a reliable way (i.e., in relying on physics). I do not want to argue that reliabilism as epistemological principle fails in general, but I want to suggest that, by shifting such an epistemological background, the burden of proof can be swapped. Instead of looking for epistemic justification in a reliable method, I adopt the view that all species of appearances are a prima facie source of justification for beliefs. In the epistemological literature this view is known as *phenomenal conservatism* (see Huemer 2007, Tucker 2013) and gives rise to what I call the 'Argument from Appearances': in the absence of defeaters, if the direction of time seems to be primitive to an agent, then there is some degree of justification for such an agent to believe that the direction of time is primitive. That the direction of time seems to be primitive (at least in the sense that it seems to be fundamental for the natural world) follows rather easily from the overwhelming evidence that time does seem to be directed. Note that this epistemological view has swapped the burden of proof: It is now on those who want to show that such a belief is unjustified because there are solid defeaters.

This brings us to the second part of the general argumentation. The fact that most laws of physics are time-reversal invariant has been an all-too-common argument in the literature against a primitive direction of time. This is the **TRA**, and it works as a defeater within phenomenal conservatism: despite the fact that time seems to be directed, any belief in a primitive direction of time is left unjustified because there is a solid defeater[2]. Yet, I argue that the **TRA** is actually harmless and thereby fails to be a solid defeater. In particular, I cast a series of doubts on the **TRA** that puts into question its strength and persuasive force, at least as it stands now. I divide my objections into three groups:

- Those that relate to which are the laws of interest
- Those that relate to what is the right concept of time reversal
- Those that relate to what is the role of symmetries in physics.

[1] Reliabilism is an externalist theory of justification since the factors that justify a belief are not necessarily conscious experiences or within one's own first-person perspective. It opposes to internalist theories of justification, according to which the factors that justify a belief must be within one's first-person sphere (e.g., conscious experience). *Evidentialism* as an internalist theory usually contraposes to reliabilism. Other views are the acquaintance theory, skepticism, phenomenal conservatism, among others.

[2] It is technically a *rebutting defeater*: time seems to be directed, but it is a false belief because there would be solid arguments in physics that teach us that it is fundamentally directionless.

All these points against the **TRA** play the role of showing that it is not really a solid defeater. Therefore, since phenomenal conservatism simply accepts that assuming things are the way they appear to us is the rational default position, the absence of defeaters is enough to uphold that the belief in the direction of time as primitive is justified.

The chapter is structured as following. In Section 10.2, I introduce in a general fashion the problem of the direction of time along with two epistemic views: *reliabilism* and *phenomenal conservatism*. In Section 10.3, I provide a careful reconstruction of the **TRA** and its main assumptions. In Section 10.4, I argue against some of the premises of the **TRA**, showing that it needs to be, at least, supplemented by additional premises. I also argue that such supplementary premises can be put into question, rendering the argument harmless. In conclusion, in the absence of solid defeaters, that time seems to be primitively directed to us is enough rational ground to have justification for believing that time is primitively directed.

10.2 What's at Stake? The Problem of the Direction of Time and an Epistemological Shift

10.2.1 The General Problem of the Direction of Time

The problem of the direction of time has largely been associated with problems in the foundations of thermodynamics and statistical mechanics (see Boltzmann 1872, 1964, Reichenbach 1956, Price 1996, Callender 1997, Albert 2000, Loewer 2012). In a classical book, Hans Reichenbach has famously introduced the problem of the direction of time as follows:

> The elementary processes of statistical thermodynamics, the motions and collisions of molecules, are supposed to be controlled by the laws of classical mechanics and are therefore reversible. The macro-processes are irreversible, as we know. How can this irreversibility of macro-processes be reconciled with the reversibility of micro-processes? It is this paradox which the physicist has to solve when he wishes to account for the direction of thermodynamical processes and for the *direction of time*. *(Reichenbach 1956: 109. Italics mine)*

More contemporarily, Craig Callender has said:

> Concisely put, the problem is that given a non-equilibrium state at t_2, it is overwhelmingly likely that (a) the non-equilibrium state at t_2 will evolve to one closer to equilibrium at t_3, but that due to the reversibility of the dynamics it is also overwhelmingly likely that (b) the non-equilibrium state at t_2 has evolved from closer to equilibrium at t_1.*(Callender 1997: S225)*

There is, of course, a deep and philosophically rich problem in the foundations of thermodynamics about the seeming contradiction between temporally directed macroprocesses and temporally undirected microprocesses (see Price 1996, Goldstein 2001). I, however, think that the philosophical debate over the direction of time greatly exceeds the more circumscribed issues around the foundations of thermodynamics, or of any other physical theory. In other words, I think this is a too narrow formulation of the problem of the direction of time, as if it were exclusively exhausted by the problem of the thermodynamical arrow of time (the same goes for the cosmological arrow of time; see Hawking 1985, Penrose 1989, 2004; the electromagnetic arrow of time, see Frisch 2000, Rohrlich 2005, Earman 2011; and the quantum-mechanical arrow of time, see Penrose 1989, Callender 2000, Hartle 2013). In consequence, I believe that the problem should be introduced in a more general way.

Let me then pursue a broader formulation. To begin, it is worth distinguishing the problem of the direction of time from related, but different issues. That time instantiates the problem of having a direction does not per se imply that time is dynamics, or that there is an objective becoming (Markosian 2004, Bourne 2006; see Price 2011 for a similar clarification). In metaphysics, it is common to distinguish between dynamical and static theories of time (Baron 2015); or, in John McTaggart's vocabulary, between A- and B-theories (McTaggart 1908, Gale 1966). While A-theories hold that time objectively flows (it is dynamic), B-theories deny it. Strictly, time is directed in both structures, regardless of whether time is dynamic or not. In other words, if time is directed, then it is *at least* B-structured. Some philosophers have argued that time may also be C-structured, in the sense that time is ordered, but directionless (see mainly Price 1996; Farr 2012, 2020). According to them, time may exist without a direction.[3] So, whether time is directed or not is not a problem about whether time has a A- or B-structure (i.e., whether it is dynamics or not), but about whether it has a B- or C-structure.

The second point is that, from a broad perspective, the problem of the direction of time is not about whether the direction of time is real (or whether it exists). That is, it does not hinge upon an "existence" question. Most explanations of the direction of time (even the most reductionist ones) try to recover it at some level, which implicitly implies that it must be taken as part of the ontology. A more metaphysics-oriented formulation of the problem would focus on *how* the direction of time exists, that is, about whether it is *primitive* (fundamental) or *derivative* (reducible to something else). In accordance with this, philosophical attitudes toward the direction of time can then be divided into *primitivism* and *reductionism*. In turn, reductionism comprehends two alternative views: *conservative* and *eliminativist* reductionism (see Esfeld and Sachse 2011). While primitivism takes the direction of time to be an irreducible, fundamental property of time, reductionism takes it to be reducible. It is now clear why existence questions are not at stake: reductionism

[3] I disagree with the view since I think that C-theories of time are not actually theories of time at all, but theories of something else.

does not mean per se that the direction of time is unreal, but that it is not primitive. Conservative reductionism will, for instance, endow time with a non-fundamental reality, as part of the secondary (or derivative) ontology. Primitivists and conservative reductionists do not then debate over the reality of the direction of time, but over whether it is part of the basic (or fundamental) ontology. In this sense, they are *realist* views of the direction of time. Eliminativist reductionism does defend that the direction of time is unreal, an illusion, nor being even part of the secondary ontology. In this way, only eliminativist reductionism (a scarcely defended view) holds that the problem of the direction is about its existence, in opposition to both primitivism and conservative reductionism.

To sum up, my view is that the problem of the direction of time is, first and foremost, the problem of whether time instantiates the property of having a direction or not. This basically means that if one direction obtains (say, from "past-to-future"), then the other does not (say, from "future-to-past")[4]. Second, it is about whether such a property is primitive in one's ontology (because, e.g., it plays a paramount role in explaining temporal asymmetries) or derivative (i.e., reducible to some material – *de facto* or *de iure* – asymmetry). The problem of the thermodynamical arrow of time, the cosmological arrow of time, the quantum-mechanical arrow of time, and so on, are sub-problems (or more specific problems) that presuppose that the direction of time is real but not primitive, calling for an explanation in terms of a nontemporal asymmetry.

10.2.2 An Epistemological Shift

Epistemological aspects have been barely discussed in the literature on the direction of time. It seems to me that most approaches to the problem of the direction of time have not only taken a reductionist attitude, but also implicitly adopted as epistemological principle some form of *reliabilism* for epistemic justification. This is an externalist account of epistemic justification which emphasizes that the *type of methodology* employed to form beliefs is what confers justification (see Goldman 1979, 1986, 1994). From this perspective, the epistemic problem is about whether we have *justified beliefs* about the direction of time as primitive (or not) by following reliable belief-generating processes. More precisely, reliabilists state that the property of being the outcome of a reliable belief-forming method is what confers justification for a given belief: the belief that time has a primitive direction can only be justified if it is the outcome of a reliable belief-forming

4 The labels "future" and "past" are conventional, so the direction of time does not hinge upon which direction is really the future and which really the past. The point is that if it is possible to fix a direction (which can conventionally be baptized as the "past-to-future," or the "positive," direction), then the opposite direction is a *different* direction (it might even be metaphysically or physically impossible).

method (in this case, if it is an outcome of physics, see Goldman 1994). Otherwise, our belief is rendered unjustified. The introduction of epistemological considerations is not trivial since they command the necessary criteria to ultimately have knowledge of something (although these criteria might be not sufficient). Unjustified beliefs perform very poorly in the competence of having knowledge of the world since knowledge will at least be formed by justified beliefs.

I do not want to argue that reliabilism is false (see Huemer 2007). My aim here is just to shift the epistemological principle at work in the debate to swap the burden of proof. If reliabilism (or something close to it) is assumed as an overarching theory of epistemic justification in the discussion, then the burden of proof is on who must show how a given belief is the outcome of a reliable process (e.g., physics). But this could be different under an alternative theory of epistemic justification. To make my point, I adopt here what has been called *phenomenal conservatism* (Huemer 2001, 2007; Tucker 2013). It basically states that if it seems to an agent that *p*, then, in the absence of defeaters, such an agent thereby has at least some degree of justification for believing that *p*. In terms of the topic of this chapter, phenomenal conservatism would imply that if it seems to an agent that the direction of time is primitive (e.g., by observation, by sensorial experience, by introspection, by explanatory necessity, or by paying attention to common ordinary and scientific explanations, etc.) then, in the absence of defeaters, such an agent has at least some degree of justification for believing that the direction of time is primitive.

Of course, further metaphysical arguments can be given to support the belief (e.g., explanatory power, simplicity in the basic ontology, etc.; see Maudlin 2002, Lopez and Esfeld 2025, and Lopez 2024a), but phenomenal conservatism, as a theory of epistemic justification, already sets the default rational position differently, swapping the burden of proof. In order for us to have justified beliefs on the direction of time we should now keep an eye on rebutting the defeaters, rather than positively defending why we are entitled to have some beliefs rather than others. Appearances (understood in a broad sense) already confer some degree of justification for believing that the direction of time is primitive (or at least that it is central in the explanation of our daily experiences and scientific processes), so the burden of proof is now placed on the power of the defeaters. From phenomenal conservatism, an "Argument from Appearances" can be sketched as follows:

The Argument from Appearances

1. In the absence of defeaters, if time seems to us to be primitively directed, then there is a prima facie justification for believing that time is primitively directed.
2. It is true that time *seems* to us to be primitively directed.

3. There are no defeaters.
C. Therefore, there is a prima face justification for believing that time is primitively directed.

P1 just states phenomenal conservatism applied to the problem of the direction of time. I will not argue in favor of **P2**, so I will just take it for granted.[5] Provided that **P3** holds, the conclusion follows by Modus Ponens. So, everything in this argument gravitates around **P3**. So, in the next section I introduce one of the main reasons to object to this argument: the **TRA**. To be clear, if the **TRA** holds, then the Argument from Appearances does not. In Section 10.4, I argue that the **TRA** does not hold (as least as it stands), cleaning the way for the Argument from Appearances to go through.

10.3 The Defeater: The Time-Reversal Argument

In the last years, philosophers of physics and metaphysicians alike have been growingly drawing their attention to physical symmetries to address various philosophical issues, such as the nature of space-time, the nature of natural properties and of the fundamental interactions, etc. The basic idea is that at least some physical symmetries would allow us to infer features or structures of the world (see, for instance, Baker 2010, North 2008, 2021). Shamik Dasgupta (2016) refers to this inferential mechanism as "the symmetry-to-reality inference." The **TRA** as I interpret it, can be viewed of as an instance of a symmetry-to-reality inference, where the symmetry of interest is time reversal, and the aspect of reality is the direction of time. One of the questions is whether the inference can be justified: Why are we entitled to draw metaphysical conclusions from physics or mathematics? (See Lopez 2023b for a critical assessment.) More particularly, why are we entitled to draw metaphysical conclusions about the direction of time from time-reversal invariance? (See Lopez and Esfeld 2023 for a critical stance on this.)

Another question is if the inference is really about what is real. I am a bit hesitant to take the argument in terms of what's real and what's not: even if all laws of physics come out time-reversal invariant, the direction of time would still be real! So, I take the inference as referring to what's primitive (or fundamental). In this line, the **TRA** states that because laws of nature are time-reversal invariant, the direction of time is not primitive. This can very well be explained because physical symmetries would be symptomatic of redundant or superfluous structure, so it is

[5] It can be argued that time does not appear to us to be *primitively* directed, but just merely directed. This is true. But if we take this as the ending point of the discussion, we are not yet debating the problem of the direction of time as I have introduced it in Section 10.2. Once we enter the debate, we must endorse either reductionism or primitivism. In this context, I hold that primitivism is the standard rational position, given the paramount role that a directionality of time has in any explanation. In this sense, time appears to us as primitively directed.

epistemically advisable to go with the most parsimonious structure, "slicing away superfluous structure" (Earman 1989: 46). So, in a general form, the "Symmetry-to-Primitive Argument" can be posed as follows:

The Symmetry-to-Primitive Argument

1. Take the laws of physics L.
2. There is a property P in the laws whose value can vary freely while keeping the laws true.
3. Then, if **P2** is true, P is redundant/superfluous/nonobjective, and so on.
4. We shouldn't accept redundant/superfluous/nonobjective properties as part of our primitives (Ockham's Razor).

C. Therefore, P cannot be part of our primitives (i.e., it is part of the derivative ontology).

There would be various points to discuss here, but I do not have enough room for providing a detailed discussion. It is just noteworthy that **P3** is crucial, since it provides a more robust, narrow concept of what a physical symmetry is. In **P2**, the notion of symmetry is too abstract to be conceptually useful; thus, it requires some additional content to have physical and conceptual meaning (see, for instance, Belot 2013, Dasgupta 2016). This is what **P3** delivers. There have been many ways to characterize such an additional content, but some of these narrowed concepts of symmetries have been related to the concept of objectivity (Weyl 1952, Nozick 2001), to undetectable properties (Robert 2008, Dasgupta 2016), to unmeasurable quantities (Ismael and van Fraassen 2003), or to physical equivalence (Saunders 2003). Others have been skeptical that the enterprise can be successfully achieved at all (Belot 2013, Wallace 2022).

I say that the **TRA** is an instance of the Symmetry-to-Primitive Argument. It relies on the fact that most of our fundamental laws of nature have the property of being time-reversal invariant. What is meant by 'time reversal' has been extensively discussed in the literature (see Sachs 1987, Albert 2000, Callender 2000, Earman 2002, Malament 2004, North 2008, Peterson 2015, Lopez 2021a-b, Lopez 2023a, Roberts 2022, and Allori 2019, 2024, among others), but it is enough for my purposes here to say that time reversal is a formal operation acting upon states, magnitudes and parameters in a dynamical equation that transforms them in such a way that represents an inversion of motion plus a reparameterization of the time coordinate (see Wigner 1932: 325, Gibson and Pollard 1976: 177; Sachs 1987: 6). If time-reversal invariance holds, then the space of solutions of the dynamical equation is preserved (i.e., time reversal maps solution to solutions and nonsolution to nonsolutions). More informally, it is said that time-reversal invariance "preserves" the truth of the laws of nature; or that it in principle means that laws of nature are "blind to," or "do not care about," the direction of time.[6]

If time-reversal invariance is a pervasive feature of dynamical equations in physics, why do we need a direction of time after all? At least at first glance, *if* our metaphysical beliefs ought to be tailored by our best physics, then it seems that the **TRA** suggests otherwise. In one way or another, a primitive direction of time seems unnecessary, superfluous. Why postulate it then?

The Time-Reversal Argument

P1 If the dynamical laws are time-reversal invariant, then a primitive direction of time is metaphysically unnecessary. [assumption]
P2 We shouldn't posit unnecessary structures. [parsimony]
P3. It happens that (most) dynamical laws are time-reversal invariant. [empirical premise]
P4. A primitive direction of time is then unnecessary. [MP with P1]
C. We shouldn't posit a primitive direction of time. [from P4 and P2].

The argument is of course valid. Whether it is true depends on the premises. Many in the literature on the direction of time have taken the argument as containing true premises, which undermines primitivism about the direction of time. According to the epistemological principle I have adopted in Section 10.2, if the **TRA** is true, then it is a solid defeater in the Argument from Appearances. If this is so, then any belief that time is primitively directed because it seems to be so is just unjustified. In the following section, I examine in detail the **TRA** and argue that it is not a solid defeater.

10.4 Defeating the Defeater

The **TRA** (or something very close to it) has had a huge influence in the philosophy of time and of physics. It puts into question a deep intuition about what the natural world is like. Yet, I believe the argument resorts to some premises that need thorough examination. I do not claim that the **TRA** is incorrect or incoherent; nor that it cannot be accepted under any circumstances. What I claim is that the truth of some of its premises depends on additional premises for them to preserve its persuasive force. Otherwise, the argument is much less convincing than it seems

[6] This is just an instrumental and very general definition of time-reversal invariance. The details vary across theories and contexts, and many subtleties should be mentioned. However, for my purposes here, this definition is good enough to introduce **TRA** as it is frequently introduced. It can be also argued that this definition of time-reversal invariance is too liberal and must be constrained. This is an important point to which I will come back in Section 10.4.

at first glance. I focus on two premises that I think are particularly troublesome – **P1** and **P3**. In a nutshell, my caveats, which I unfold in the next subsections, can be grouped into three questions:

- Which types of *laws* are time-reversal invariant?
- What is *time reversal*?
- What is the role of *symmetries* in physical theories?

None of these questions can get definite answers. They can rather be replied to in different ways. But more crucially, the replies require supplementary premises. The problem is that there are good reasons to reject these supplementary premises. If they are rejected, then the **TRA** is rendered harmless or, at least, much less persuasive than originally thought. This would show that one of the strongest defeaters of the Argument from Appearances is ineffective.

10.4.1 Which Types of Laws are Time-Reversal Invariant?

To begin, it is not strictly true that *all* laws of physics are time-reversal invariant. This has been already argued for in different ways, but it is worth emphasizing its importance for the debate. If it is not the case that *all* laws are time-reversal invariant, then **P3** in the time-reversal argument is just false. For instance, some have argued that the decay of neutral kaons or B-mesons in weak interactions are solid enough examples of time-reversal violation. In quantum field theories, the **CPT** theorem states that for any local relativistic quantum theory in Minkowski space-time, the combination of charge conjugation (**C**), parity transformation (**P**), and time reversal (**T**) is a symmetry. (For more details, see Greaves and Thomas 2014.) In 1964, Christenson, Cronin, Fitch, and Turlay carried out an experiment at the Brookhaven Alternating Gradient Synchrotron (**AGS**) in which they discovered that neutral kaons violated **CP**, thereby violating **T**.[7] In particular, they found that long-life neutral kaons (K_L^0) decay asymmetrically, violating **CP**, and thereby **T** (for further details, see Sachs 1987, Sozzi 2008, Bigi and Sanda 2009). Therefore, there is a fundamental violation of time reversal.[8]

Another way to argue against **P3** along the same line is by relying on quantum theories that explicitly introduce a time-asymmetric evolution for quantum systems. Theories of objective collapses (e.g., GRW, CSL) modify the Schrödinger equation in such a way that quantum states spontaneously and stochastically

[7] First experiments shown an *indirect* violation of **T**. But new experiments (see Angelopoulos et al. 1998) shown *direct* violations of **T**.

[8] There has been some discussion about whether this experimental result is relevant enough to conclude that there is a direction of time (see, for instance, Horwich 1987, Penrose 2004; also, Maudlin 2007, North 2021). Nonetheless, I am not interested here in showing that the experimental violation of **CP** is relevant for the direction of time debate, but that *within* the framework of the **TRA** it is evidence against **P3.**

undergo collapses that localize them in space-time (see Ghirardi, Rimini and Weber 1985, 1986; Ghirardi, Pearle and Rimini 1990). Due to the stochastic nature of the spontaneous collapses, these theories are not time-reversal invariant (see Arntzenius 1997, Esfeld and Sachse 2011, North 2021). So, if these theories are accepted as good ways to solve the measurement problem, then **P3** no longer holds. Of course, the downside of this argument is that it heavily depends on endorsing some version of the theories of objective collapse as an attractive way to solve the measurement problem, which can be contested. Be that as it may, it is enough to introduce some qualification in stating **P3**.

There is a third, much more interesting way to show that **P3** could be untrue. Premise **P3** implicitly assumes that only *the simplest and fundamental expressions of a dynamical law* are the relevant ones to assess time-reversal invariance. Philosophers of science have referred to them as "the general laws" or "covering laws" (Dray 1957, Hempel 1965, Cartwright 1983), arguing that such laws capture the general, highly idealized behavior of physical systems under unrealistic circumstances (see Lopez 2025 and Lopez 2024b). Yet, a vast number of instantiations of the covering laws (what is usually called *phenomenological laws*) will *not* be time-reversal invariant. Why is it then insisted that laws of physics are time-reversal invariant? The underlying assumption (which I call the *Fundamentality* assumption) is that only general and simple expressions of the dynamical laws are fundamental, meaning that they should be taken as ontologically privileged (latching on to reality at bottom, so to speak), while phenomenological laws (their more specific instantiations) are rather derivative. Craig Callender (1995) says:

> [in classical mechanics] we disregard 'phenomenal' forces and focus on the 'fundamental ones when we decide whether a theory and the systems it describes are TRI. When asking if the universe is TRI [time-reversal invariant], we desire to know whether it is *at bottom* time-reversal invariant. We make an ontological assumption. *(Callender 1995: 333)*

And then he continues: "classically, there are really only particles in motion and interparticulate (distance-dependent) forces" (Callender 1995: 333). All this means that **P3** actually requires an additional premise to do the job it intends to do: Only fundamental laws (in the sense of general or covering laws) are taken as relevant to assess time-reversal invariance.

There could be good motivations to adopt this premise, but I believe that it is a matter of one's underlying ontology after all. For instance, whoever endorses an ontology of capacities in science or a dispositional ontology as fundamental (see, for instance, Cartwright 1989, Mumford and Anjam 2011) would be more reluctant to accept that covering laws are fundamental in any sense. After all, won't laws of nature be redundant? Won't laws of nature be instrumental? If this is so, which are the reasons to take covering laws as ontologically fundamental under an alternative ontological framework? Under such a dispositional ontology, the situation seems to be in reality the other way around – more realistic, less idealized models (e.g., those that contain nonconservative forces) are likely to be ontologically more fundamental than the abstract covering laws! An analogous argument can be run from what has recently been called 'the open systems view' (Cuffaro and Hartmann 2024). According to the traditional view, closed systems are conceived of as fundamental. When assessing whether a dynamical law (or a model) is time-reversal invariant, it is also customary to focus on isolated systems, which goes in the same line as the **TRA**. Cuffaro and Hartmann, however, propose an alternative view in which systems interacting with their environment (i.e., open systems) are considered of as fundamental. Remarkably, if an open-system view is taken to assess time-reversal invariance, then *most* fundamental, now open-system models will be non–time-reversal invariant (for instance, many of them will involve nonconservative forces, see Lopez 2024a for an argument on this line).

I do not mean that dispositionalism or the open-system view is correct. What I want to point to is that the Fundamentality Assumption is crucial to run the **TRA** and alternative views may have good reasons to not accept it. Without something close to the Fundamentality Assumption, **P3** is left unwarranted.

10.4.2 What Is Time Reversal?

Putting aside the question about which types of laws are to be considered as relevant to assess time-reversal invariance, another pressing question is about the concept of time reversal itself. There has been some debate about what time reversal is in different theoretical frameworks in the last years. I will not get into much detail here (see Sachs 1987, Albert 2000, Callender 2000, Earman 2002, Malament 2004, North 2008, Peterson 2015, Lopez 2021a-b, Lopez 2023a, and Roberts 2022, among others), but I will circumscribe myself to show how this debate might affect the **TRA**, casting doubts on the scope of **P3**.

When it is claimed that time-reversal invariance holds, it is claimed that there is a formal operation that preserves some structure. It must also be assumed that there are good reasons to call such a formal operation 'time reversal'. A symmetry transformation is just a series of operations acting upon observables, states, and parameters in a specific way. The symmetry transformation *is* such operations. Of course, such operations are meant to represent something physical, as reflecting states in space, rotating systems by a certain angle, or reversing the motion of physical systems. It can be argued that the series of operations that are usually taken as formally implementing time reversal are actually representing something different (see Callender 2000, Lopez 2019). Let us call the formal operations '**T**'. Prima facie, **T** could stand for anything – for example, it could be a trivial transformation without physical meaning. Premise **P3** must not only state that **T** is a symmetry of most physical theories, but also that **T** fairly represents the *concept* of time reversal. This observation does not merely show that there might be a failure in the formal implementation of the symmetry transformation (i.e., that **T** should be replaced by another set of operations, say **M**), but that there might also be a failure in the association of a symmetry transformation with the *concept* it intends to formally implement. So, the argument against **P3** is that there exists a symmetry under **T**, but that it fails to fairly capture the notion of time reversal. In other words, the symmetry transformation should be rather associated with a different concept since it is not actually reversing time (see Callender 2000, Lopez 2019).

I know that this is a nonstandard argument, but I think it has some grip. In physics textbooks, the presentation of time-reversal invariance is often accompanied by the following clarification:

> In this approach we see that no metaphysical notion of reversal of the direction of the flow of time is involved. We are led to consider time reversed processes but not reversal of time itself. Although motion reversal and motion reversal invariance would be better names, we shall adhere to the accepted, if imprecise, usage. *(Gibson and Pollard 1976: 177)*

Eugene Wigner (1932: 325) and Leslie Ballentine (1998: 377) also qualify time reversal along the same line – the physical meaning of the time-reversal transformation is completely exhausted by an inversion of motion plus a reparameterization of the time coordinate. Although there can be very good physical reasons to adopt this definition (see Roberts 2017, Lopez 2021b for arguments in favor of it in nonrelativistic quantum mechanics), it is not free of assumptions. To begin, the association of time reversal with motion reversal at least rejects that motion (or change) can be independent from time. Temporal substantivalism, for instance, would defend that time is independent from matter in motion. More technically, that genuine temporal relations (e.g., instant-to-instant relations) are independent from temporal relations between matter (e.g., event-to-event relations). So, in the framework of temporal substantivalism, it would not be natural to suppose that the inversion of motion *means* the inversion of time (for an argument in this line, see Lopez 2019). Or to put it differently, it is not obvious that the inversion of event-to-event relations (motion) is equivalent to the inversion of instant-to-instant relations (time). A reliable time-reversal transformation should then be one in which **T** transforms instant-to-instant relations, upon which the inversion of event-to-event relation might supervene. The canonical implementation of time reversal, the argument goes, fails to implement time reversal properly. If the **TRA** is about the direction of time, **P3** is irrelevant (as space rotation would also be irrelevant).[9]

[9] Jill North (2008) says: "what is a time-reversal transformation? Just a flipping of the direction of time! That is all there is to a transformation that changes how things are with respect to time: change the direction of time itself" (North 2009: 212).

Besides whether this argument succeeds or not (it depends on whether temporal substantivalism is tenable or not; whether it can come up with a formal implementation that captures physicists' practice, etc.), it tells us that time reversal (and thereby, time-reversal invariance) can be conceived of differently. This gives rise to an "Argument from Underdetermination" (see Lopez 2023a), according to which the implementation of time reversal is not unique, but it depends on various metaphysical and epistemic theses about time. Once again, this forces us to put the **TRA** into perspective: **P3** turns out to be true *if* time reversal is implemented in a specific way, but it might turn out to be false *if* time reversal is implemented differently. One's underlying metaphysics of time can thus influence our conceptualization and implementation of time reversal. And it is clear that there is no guarantee that some dynamical equation will turn out to be time-reversal invariant if the formal implementation of time reversal is to be different (see Albert 2000, Callender 2000, and Lopez 2019). So, in the end, this Argument from Underdetermination recommends another qualification in the **TRA**, now concerning the meaning of time reversal and the right way to implement it.

The Argument from Underdetermination concludes that canonical implementations of time reversal (**P3**) might fail to capture the concept of time reversal. But here is another argument that rather states that canonical implementations already assume that the direction of time cannot be primitive (or fundamental). Hence, this new argument runs, **P3** should be rejected in the **TRA** because it begs the question against primitivism, and it cannot then be used nontrivially against it.

There is some consensus that time reversal reverses not only the order of a temporal series:

$$\boldsymbol{T}[A \to B \to C \to D] = D \to C \to B \to A,$$

but also the states themselves:

$$\boldsymbol{T} : X \to TX = X';$$

what gives

$$T[A \to B \to C \to D] = TD \to TC \to TB \to TA = D' \to C' \to B' \to A'$$

For instance, in Hamiltonian classical mechanics, time reversal is expected to transform the states by inverting the sign of conjugated momentum (T: $p \to -p$) but keeping the generalized positions unaltered (T: $q \to q$). In classical electromagnetism, time reversal transforms the states by changing the sign of the magnetic field (T: $\mathbf{B} \to -\mathbf{B}$) and the current ($T$: $\mathbf{j} \to -\mathbf{j}$) but keeping the electric field

(T: $\mathbf{E} \rightarrow \mathbf{E}$) and the electromagnetic force (T: $\mathbf{F} \rightarrow \mathbf{F}$) (see Albert 2000, Malament 2004, and Arntzenius and Greaves 2009, for discussion).

But what are the reasons to suppose that the time-reversed states are *physically possible* states? It can be argued that the direction of time is so crucial for the states to be even possible that the inversion of the direction of time will make them impossible. In other words, the assumption that time-reversed states are *physically possible* states is equivalent to assuming that the inversion of the direction of time is not important enough to affect the possibility of such states. But this indeed looks like an assumption that many primitivists about the direction of time can very well reject. If this is so, it follows that the canonical implementation of time reversal, in assuming that time-reversed states must be possible states, begs the question against primitivism: to properly assess whether time-reversal invariance holds, the possibility of physical states must not be presupposed. Therefore, **P3** should not be endorsed as it stands. Let me give an example.

Imagine a photon-emitting device (the emitter) and a photocell (the absorber). Now consider the following series in one temporal direction (say, '+'). First, the emitter radiates a photon at t_1 (event E) that a photocell absorbs at t_2 (event A). In the middle, the photon travels through space a distance $\Delta x = x_n - x_m$ for $\Delta t = t_2 - t_1$ (event T). It yields the following sequence of events:

$$+\boldsymbol{SEQ} \qquad E \rightarrow T \rightarrow A$$

The **TRA** requires us to time reverse the temporal series, that is, to transform $+\boldsymbol{SEQ}$ into its time-reversed $-\boldsymbol{SEQ}$. In accordance with the canonical implementation of time reversal, the order of the series must be transformed along with its elements (in this case, the events themselves). The time-reversed series would then look like the following:

$$-\boldsymbol{SEQT} \qquad \mathbf{T}(E \rightarrow T \rightarrow A) = \mathbf{T}A \rightarrow \mathbf{T}T \rightarrow \mathbf{T}E = A' \rightarrow T' \rightarrow E'$$

If **T** is a symmetry, then there is some relation of equivalence between $+\boldsymbol{SEQ}$ and $-\boldsymbol{SEQ}$ plus some preservation of structure or observational content. Therefore, the **TRA** concludes that positing a primitive direction of time that makes both sequences distinct is positing unnecessary structure or unobservable content. Since this should be avoided, a primitive direction of time is unwarranted.

Yet, one can wonder, what are A' and E'? It is clear what A and E are in the ordinary ('+') direction of time, but what is the time-reversed version of an emitter *emitting* a photon and a photocell *absorbing* a photon? Are they even physically possible events? Take the event E'. A plausible assumption is that E' is a time-reversed emitter that behaves like an absorber. But is an emitter the *kind* of thing that may absorb anything? An emitter can be defined by the role it performs. But it seems meaningless to talk about emitters that absorb photons in the opposite direction of time since it would not be an emitter any longer! It can in consequence be argued that if E' means a time-reversed emitter that absorbs photons (it would allegedly be the time-reversed process of emitting), then the event is not only impossible, but also meaningless. Even stronger, the very role of the object was destroyed under time reversal. Therefore, it seems that we are not entitled to even conceive of a time-reversed version of E without begging the question against primitivism, which presumes that to fix a direction of time is essential to explain what an emitter is and what it does.

But this construal of E' can be wrong. It can alternatively be said that E' is not a time-reversed emitter that absorbs photons, but that a time-reversed emitter *is* a photocell in the reversed direction of time ('−'). Naturally, photocells are indeed the right kind of things that absorb photons, so the previous argument is blocked. Nevertheless, this is even worse because the time-reversal transformation becomes trivial. If E' is to be construed as a photocell that absorbs photons in the negative direction of time, then the time-reversal transformation becomes merely a relabeling: $+\boldsymbol{SEQ}$ and $-\boldsymbol{SEQ}$ are rendered symmetrical because time reversal *only* changes the labels of the events and the sign of the time coordinate. To say it differently, time reversal becomes the transformation that transforms, for instance, the ' $+t$' into ' $-t$' and that *names* ' $+\boldsymbol{SEQ}$' as ' $-\boldsymbol{SEQ}$'. But if this is so, then it is legitimate to not take the **TRA** metaphysically seriously: while the question about the direction of time was a metaphysical question, **P3** in the **TRA** is just merely stating a trivial linguistic feature, namely, that a change of labels should not change the physics.

To sum up, all these caveats call for a revision of the scope and limits of **P3**. As it stands, it is either underdetermined or unspecific. Any determination and specification require supplementary premises, which must be assessed separately. Be that as it may, the **TRA** is weakened.

10.4.3 Going Deflationary: The Role of Symmetries in Physics

Let us suppose that problems 4.1 and 4.2 can be overcome. And let us now focus on **P1**, rather than on **P3**. As was said previously, **P1**'s role is mainly to connect time-reversal invariance with the direction of time (that is, the formal apparatus of a physical theory with what the world is like fundamentally). But why, and how, is this inference justified? A way to resist the **TRA** could be *to deflate* the meaning of time-reversal symmetry in **P1**. That is, the overall argument erroneously supposes that time-reversal invariance as a formal property can shed light on the direction of time as a metaphysical issue. But a metaphysical conclusion requires metaphysical premises somewhere. It can hence be argued that **TRA** fails to properly bridge this gap between the theory and the world without additional assumptions. Since a complete exposition would lead us too far in the investigation of the role of symmetries in metaphysics, I circumscribe myself here to giving a hint of a response.

Some have held that symmetries are real or, even stronger, fundamental (see Weinberg 1987, 1993, French 2014, Schroeren 2020). If such a philosophical position is adopted, then it is straightforward to connect time-reversal symmetry with a metaphysical conclusion about the direction of time. After all, symmetries are located in the basic ontology to begin with. Others, less clearly, have had it that some symmetries entitle us to draw metaphysical conclusions (see North 2009, 2021; see Lopez 2023b for arguments against). Nevertheless, others have taken symmetries (or some of them) as "epistemic virtues" or, more radically, mere conventions (see Martin 2002, Guay 2004, Bird 2007, and Lopez 2024b). These epistemic readings of symmetries (or, at least, of some symmetries) would deflate the metaphysical status of time-reversal symmetry. After all, that time-reversal symmetry holds just exhibits an "epistemic" virtue of physical theories, not a feature of what the world is like. To say it differently, whether a physical theory is time-reversal invariant or not is a theoretical problem about the formulation of physical theories; whether the direction of time is primitive or not is a metaphysical issue (see Lopez and Esfeld 2023).

To briefly illustrate this point, let me show an epistemic version of time-reversal symmetry. Suppose the best system approach to laws of nature is a Humean framework. Laws of nature then supervene on the Human Mosaic, but they are axioms (or theorems) in the best system, striving for simplicity and informativeness (Lewis 1986, Loewer 1996). In some sense, the laws of nature are assessed epistemically at the same time that they objectively describe the regularities in the Mosaic, but without fixing its content. What the basic ontology (the Humean Mosaic) is like will be a metaphysical issue, on which the laws of nature do not play any role. What about time-reversal invariance? It can mean two different things. Time-reversal invariance as a property of the laws of nature is a formal property of the axioms of the best system. Time-reversal invariance is thus involved in the epistemic assessments of the axioms in terms of its simplicity and informativeness. It may happen that axioms that have the property of being invariant under time reversal are simpler than axioms that have not. Under this construal, the property of being invariant under time reversal has no take in assessing the content of the Humean Mosaic (for instance, whether it contains a primitive direction of time). This Humean interpretation of time reversal cannot accept **P1** as it stands, which undermines the **TRA** (for a more complete exposition of what Humean time-reversal symmetry would look like, see Lopez and Esfeld 2023).

It can be said that this reading of time-reversal invariance is too weak. According to a second reading, time-reversal invariance is actually a property of the Humean Mosaic; that is, it temporally reverses the patterns in the Humean Mosaic, transforming its temporal relations and states into their time-reversed counterparts. Then, time-reversal invariance holds when both Mosaics are equivalent in some respects (e.g., their supervenient observational content is preserved). This "metaphysical" reading of time reversal is stronger than the previous one, but it is not so clear to me that the canonical implementations of time reversal are compatible with it. I find this approach highly problematic for various reasons. First, since laws supervene on the Humean Mosaic, there is no reason to suppose that equivalent laws are to obtain in a time-reversed Mosaic; however, we have got to it. We should have independent means to time-reverse the Humean Mosaic, without going through the laws. Second, even if we can guarantee equivalent laws, there is no reason to assume that properties of the laws that supervene on certain regularities of a Mosaic are to be preserved in a hypothetical time-reversed mosaic. And third, which reasons independent from the laws of nature can we have to suppose that a time-reversed Mosaic exists to begin with? Those reasons may run into begging-question arguments, as pointed to in Section 10.4.2.

The main aim of this deflationary argument is to point out that the **TRA** must presume some way to bridge the gap between a symmetry (as a formal feature of a physical theory) and a metaphysical claim, which in turn seems to demand a metaphysically stronger view of time-reversal symmetry. It is this stronger view which would entitle us to take time-reversal symmetry metaphysically seriously. So, in order for the **TRA** to work as intended, **P1** should be complemented by another premise stating that **P1** is true only if a "realist" view of symmetries (and of time-reversal symmetry) is also true. As I argued, this supplementary premise can be rejected by endorsing a more deflationary view of time-reversal symmetry. But are there independent reasons to adopt a deflationary view? Let me just give one reason that could suggest that we should go deflationary when thinking of time-reversal invariance (and of symmetries in general).

In a 2009 paper, Frank Artzenius and Hillary Greaves say that the implementation of time-reversal invariance in classical electromagnetism follows what they call "the textbook account":

> Next let us consider the electric and magnetic fields. How do they transform under time reversal? Well, the standard procedure is simply *to assume* that classical electromagnetism is invariant under time reversal. From this assumption of time reversal invariance of the theory [. . .] it is inferred that the electric field E is invariant under time reversal [. . .].
>
> *(Arntzenius and Greaves 2009: 6. Italics mine)*

The procedure is also recommended in a more general fashion by Detlef Dürr and Stephan Teufel (2009) when introducing the fundamentals of Bohmian Mechanics:

> A symmetry can be a priori, i.e., the physical law is built in such a way that it respects that particular symmetry by construction. This is exemplified by spacetime symmetries, because spacetime is the theater in which the physical law acts (as long as spacetime is not subject to a law itself, as in general relativity, which we exclude from our considerations here), and must therefore respect the rules of the theater. *(Dürr and Teufel 2009: 43–44)*

In a nutshell, the idea is that dynamical equations of motion *must* respect some symmetries as constraints (in that sense, they can be regarded of as "a priori." See Redhead 1975.) They must thus be assumed in order to write down the dynamical equations in the right way. This suggests a 'by-stipulation' view of some symmetries at least: they seem to play the role of constraining the dynamics as they are part of the set of necessary assumptions that are required to formulate a physical theory (see Lopez 2024b for the difference between by-stipulation and by-discovery views). For instance, in nonrelativistic quantum mechanics, Robert Sachs (1987) says that "[i]n order to express explicitly the independence between the kinematics and the nature of the forces, we *require* that the transformations leave the equations of motion invariant when all forces or interactions vanish"(Sachs 1987: 7. Italics mine).

So, in order to set up the kinematics and dynamics of nonrelativistic quantum mechanics, it is stipulated that the time-reversal transformation must keep the equations of motion invariant in some circumstances, which amounts to stipulating that time-reversal symmetry holds under some circumstances. This is not per se an argument for a deflationary view of time-reversal symmetry (and other symmetries, more generally), but I do think it emphasizes the architectonic aspect that time-reversal symmetry plays in physical theories. In other words, it seems to be part of some of the necessary assumptions that help physicists build acceptable and good physical theories, as it were, part of the virtues that an acceptable physical theory must exhibit. The reasons can vary, but it is widely accepted that symmetries in general, and time-reversal symmetry is not an exception, play a role in simplifying the formal structure of a physical theory and in identifying superfluous structure (see Dewar 2019). This strong theoretical-epistemic role points toward the deflationary camp, rather than toward the realist one.

10.5 Final Remarks

The aim of this chapter was to defend that it is justified to believe that the direction of time is real and primitive because it *seems* so. I began by shifting the epistemological background of the discussion. By endorsing phenomenal conservatism, the epistemic starting point is to assume that if things appear to us in a specific way, then we have justified beliefs that they *are* in such a way, if there are no reasons to believe otherwise (i.e., if there are no defeaters). In terms of the problem of the direction of time, phenomenal conservatism implies that if the direction of time seems to us to be primitive (or fundamental), then we have justified beliefs that it *is* so, if there are no defeaters. The main argument was:

The Argument from Appearances

1. In the absence of defeaters, if time seems to us to be primitively directed, then there is a prima facie justification for believing that time is primitively directed.
2. It is true that time *seems* to us to be primitively directed.
3. There are no defeaters.

C. Therefore, there is a prima face justification for believing that time is primitively directed.

As I pointed out, phenomenal conservatism as epistemological background swaps the burden of proof: it is not necessary to justify a belief by going beyond how things appear to us (as a reliabilist epistemology would recommend to us), but to argue against the potential defeaters that can be brought up. In the debate about the direction of time, the *par excellence* defeater is what I have called the Time-Reversal Argument (TRA) as presented in Section 10.3. This argument directly attacks **P3** in the Argument from Appearances, making the conclusion untenable.

The Time-Reversal Argument

P1 If the dynamical laws are time-reversal invariant, then a primitive direction of time is metaphysically unnecessary. [assumption]
P2 We shouldn't posit unnecessary structures. [parsimony]
P3 It happens that (most) dynamical laws are time-reversal invariant. [empirical premise]
P4 A primitive direction of time is then unnecessary. [MP with **P1**]
C We shouldn't posit a primitive direction of time. [from **P4** and **P2**]

Notwithstanding its persuasive force, I have argued in Section 10.4 that the argument must be substantially qualified to work as intended, which amounts to adopting additional premises with respect to laws of nature, time reversal and the role of symmetries:

- To begin, I have focused on **P3**, arguing that it is not true that all laws of nature are time-reversal invariant and that the premise also presumes that only so-called covering (or general) laws are ontologically relevant to assess time-reversal invariance. This Fundamentality Assumption can be nonetheless challenged by adopting alternative ontological frameworks.

- Next, I have centered in the concept of time reversal, arguing that it suffers from underdetermination (the Argument from Underdetermination): Not only can time reversal be understood differently (bringing about different formal implementations), but its canonical implementations can also be challenged. To a large extent, **P3** either is underdetermined or begs the question.
- Finally, I have focused on **P1**, assessing the role of symmetries in physics. **P1**, I have argued, presupposes a realist interpretation of symmetries, which can be challenged. By adopting a deflationary view, it is harder to justify why we are entitled to move from a feature of physical theories (time-reversal invariance) to an aspect of the world (the direction of time). By challenging **P1**, the whole **TRA** is rendered harmless.

I haven't argued that phenomenal conservatism is true, and that it is enough to endorse primitivism with respect to the direction of time. Nor have I argued that the Argument from Appearances is necessary to uphold primitivism. My thesis is conditional: if an alternative epistemological background is adopted, then the Argument from the Appearances has some grip, and the burden of proof is rather on the defeaters. Since I find the main defeater unconvincing, phenomenal conservatism can be employed to claim that we are justified in believing that the world is fundamentally time directed.

Primitivists about the direction of time could well be in need of a more positive defense of primitivism. They could also reject phenomenal conservatism as an epistemic theory of justification. Nonetheless, by removing the main argument against primitivism, a prima facie epistemic justification for it can be offered. In the end, I fully endorse Charles Peirce's words: "let us not pretend to doubt in philosophy what we do not doubt in our hearts."

References

Albert, D. Z. (2000). *Time and Chance*. Cambridge, MA: Harvard University Press.

Allori, V. (2019). "Quantum mechanics, time and ontology." *Studies in the History and Philosophy of Science Part B: Studies in History and Philosophy of Modern Physics*, vol. 66: 145–154.

Allori, V. (2024). "Time for pancakes." In C. Lopez and Lombardi (eds.), *The Arrow of Time: From Local Systems to the Universe*, Cambridge University Press, chapter 7.

Angelopoulos, A. et al. (1998). "First direct observation of time-reversal non-invariance in the neutral kaon system," Phys. Lett. B **444**: 43–51.

Arntzenius, F. (1997). "Mirrors and the direction of time." *Philosophy of Science*, 64: 213–222.

Arntzenius, F. and Greaves, H. (2009). "Time reversal in classical electromagnetism." *The British Journal for the Philosophy of Science*, 60: 557–584.

Baker, D. (2010). "Symmetry and the metaphysics of physics." *Philosophy Compass* 5: 1157–1166.

Ballentine, L. (1998). *Quantum Mechanics. A Modern Development*. Singapore: World Scientific

Baron, S. (2015). "Feel the flow." *Synthese*, 194 (2): 609–630.

Belot, G. (2013). "Symmetry and equivalence," in R. Batterman (ed.), *The Oxford Handbook of Philosophy of Physics*, Oxford: Oxford University Press.

Bigi, I. I., and Sanda, A. I. (2009). *CP Violation*. Cambridge: Cambridge University Press.

Bird, A. (2007). *Nature's Metaphysics: Laws and Properties*. Oxford: Clarendon Press.

Boltzmann, L. (1872). "Weitere Studien über das Wärmegleichgewicht unter Gasmolekülen," in F. Hasenöhrl (ed.), *Wissenschaftliche Abhandlungen*, New York: Chelsea Publishing Company, 1968.

Boltzmann, L. (1964). *Lectures on Gas Theory* 1986–1898. Berkeley, CA: University of California Press.

Bourne, C. (2006). *A Future for Presentism*. Oxford: Oxford University Press.

Callender, C. (1995). "The metaphysics of time reversal: Hutchison on classical mechanics." *The British Journal for the Philosophy of Science*, 46: 331–340.

Callender, C. (1997). "What is 'the problem of the direction of time'?" *Philosophy of Science*, 64: S223–S234.

Callender, C. (2000). "Is time 'handed' in a quantum world?" *Proceedings of the Aristotelian Society*, 100: 247–269.

Cartwright, N. (1983). *How the Laws of Physics Lie*. Oxford: Clarendon.

Cartwright, N. (1989). *Nature's Capacities and Their Measurement*. Oxford: Clarendon.

Cuffaro, M., and Hartmann. S. (2024). "The open system view". *Philosophy of Physics*, 2(1): 6, 1–27.

Dasgupta, S. (2016). "Symmetry as an epistemic notion (twice over)." *British Journal for Philosophy of Science*, 67: 837–878.

Dewar, N. (2019). "Sophistication about symmetries." *British Journal for the Philosophy of Science*, 70: 485–521.

Dray, W (1957). *Laws and Explanation in History*. Oxford: Oxford University Press.

Dürr, D., and Teufel, S. (2009). *Bohmian Mechanics: The Physics and Mathematics of Quantum Theory*, Berlin: Springer.

Earman, J. (1989). *World Enough and Space-Time: Absolute versus Relational Theories of Space-Time*. Cambridge, MA: MIT Press,

Earman, J. (2002). "What time reversal is and why it matters." *International Studies in the Philosophy of Science*, 16: 245–264.

Earman, J. (2011). "Sharpening the electromagnetic arrow(s) of time." In C. Callender (ed.), *The Oxford Handbook of Philosophy of Time*. Oxford: Oxford University Press, pp. 353–381.

Esfeld, M., and Sachse, C. (2011). *Conservative Reductionism*. London: Routledge.

Farr, M. (2012). "Towards a C-theory of time: *An* appraisal of the physics and metaphysics of time direction." PhD Thesis, Bristol, UK: University of Bristol.

Farr, M. (2020). "C-theories of time: on the adirectionality of time." *Philosophy Compass*, e12714: 1–17.

French, S. (2014). *The Structure of the World*. Oxford: Oxford University Press.

Frisch, M. (2000). "(Dis)Solving the puzzle of the arrow of radiation." *British Journal for the Philosophy of Science*, 51: 381–410.

Gale, R. M. (1966). "McTaggart's analysis of time." *American Philosophical Quarterly*, 3(2): 145–152.

Ghirardi, G., Pearle, P., and Rimini, A. (1990). "Markov processes in Hilbert space and continuous spontaneous localization of systems of identical particles." *Physical Review A*, 42(1): 78–89.

Ghirardi, G. C., Rimini, A., and Weber, T. (1986). "Unified dynamics for microscopic and macroscopic systems." *Physical Review D*, 34: 470.

Gibson, W. M., and Pollard, B. R. (1976). *Symmetry Principles in Elementary Particle Physics*. Cambridge: Cambridge University Press.

Goldman, A. (1979). "What is justified belief?" in George S. Pappas (ed.), *Justification and Knowledge: New Studies in Epistemology*, Dordrecht: Reidel, pp. 1–25.

Goldman, A. (1986). *Epistemology and Cognition*, Cambridge, MA: Harvard University Press.

Goldman, A. (1994). "Naturalistic epistemology and reliabilism." *Midwest Studies in Philosophy*, 19(1): 301–320.

Goldstein, S. (2001). "Boltzmann's approach to statistical mechanics." In J. Bricmont, D. Dürr, M. C. Galavotti, G. Ghirardi, F. Petruccione, and N. Zanghi (eds.), *Chance in Physics: Foundations and Perspectives* (Lecture Notes in Physics 574), Berlin: Springer-Verlag.

Greaves, H., and Thomas, T. (2014). "On the CPT theorem." *Studies in History and Philosophy of Science Part B: Studies in History and Philosophy of Modern Physics*, Volume 45, February, pp. 46–65.

Guay, A. (2004). "The arbitrariness of local gauge symmetry." Manuscript unpublished, http://philsci-archive.pitt.edu/3361/

Hartle, J. (2013). "The quantum mechanical arrows of time." Talk at the conference to honor Yakir Aharonov's 80th birthday, *Fundamental Aspects of Quantum Theory: A Two-Time Winner*. Orange: Chapman University.

Hawking, S. (1985). "Arrow of time in cosmology." *Physical Review D*, 32: 2489.

Hempel, C. (1965). *Aspects of Scientific Explanations*. New York: The Free Press.

Huemer, M. (2001). *Skepticism and the Veil of Perception*. Lanham, MD: Rowman & Littlefield.

Huemer, M. (2007). "Compassionate phenomenal conservatism." *Philosophy and Phenomenological Research*, vol. LXXVI, 1: 30–55.

Ismael, J., and van Fraassen, B. (2003). "Symmetry as a guide to superfluous theoretical structure." In E. Brading and E. Castellani (eds.), *Symmetries in Physics: Philosophical Reflections*. Cambridge: Cambridge University Press, 371–392.

Lewis, D. (1986). Collected *Papers* (Vol. II). Oxford: Oxford University Press.

Loewer, B. (1996). "Humean supervenience." *Philosophical Topics*, 24(1): 101–127.

Loewer, B. (2012). "Two accounts of law and time." Philosophical Studies 160, pp. 115–137.

Lopez, C. (2019). "Roads to the past: How to go and not to go backward in time in quantum theories." *European Journal for Philosophy of Science*, 9: 27.

Lopez, C. (2021a). "Three facets of time-reversal symmetry." *European Journal for Philosophy of Science*, 11: 51.

Lopez, C. (2021b). "The physics and philosophy of time reversal in standard quantum mechanics." *Synthese*, 199: 14267–14292.

Lopez, C. (2023a). "The metaphysical under-determination of time reversal." *Synthese*, 201: 29.

Lopez, C. (2023b). "Should physical symmetries guide metaphysics? Two reasons why they should maybe not." *European Journal for the Philosophy of Science*, 13: 23.

Lopez, C. (2025). "Against symmetry fundamentalism." *Erkenntnis*. 90: 1821–1845.

Lopez, C. (2024a). "Time direction essentialism." Manuscript.

Lopez, C. (2024c). "Fundamentally epistemic symmetries". Manuscript.

Lopez, C., and Esfeld, M. (2023). "Humean time reversal." *Synthese*, 202: 31.

Lopez, C., and Esfeld, M. (2025). "Relational primitivism about the direction of time." *International Journal for the Philosophy of Sciences*, vol. 38 (2): 83–99.
Malament, D. B. (2004). "On the time reversal invariance of classical electromagnetic theory." *Studies in History and Philosophy of Modern Physics*, 35, 295–315.
Markosian, N. (2004). "A Defense of Presentism." In *Oxford Studies in Metaphysics*, Vol. 1, Dean W. Zimmerman (ed.), Oxford: Oxford University Press, pp. 47–82.
Martin, C. (2002). "Gauge principle, gauge arguments and the logic of nature." *Philosophy of Science*, 69: S221–S234.
Maudlin, T. (2002). "Remarks on the passing of time." *Proceedings of the Aristotelian Society*, 102: 237–252.
Maudlin, T. (2007). *The Metaphysics within Physics*. New York: Oxford University Press.
McTaggart, J. M. E. (1908). "The unreality of time." *Mind*, 17: 457–473.
Mumford, S., and Anjam, R. (2011). *Getting Causes from Powers*. Oxford: Oxford University Press.
North, J. (2008). "Two views on time reversal." *Philosophy of Science*, 75: 201–223.
North, J. (2009). "The structure of Physical theories." *Journal of Philosophy*, 106: 57–88.
North, J. (2021). *Physics, Structure, and Reality*. Oxford: Oxford University Press.
Nozick, R. (2001). *Invariances: The Structure of the Objective World*. Cambridge, MA: Harvard University Press.
Penrose, R. (1989). *The Emperor's New Mind*. New York: Oxford University Press.
Penrose, R. (2004). *Road to Reality*. London: Jonathan Cape.
Peterson, D. (2015). "Prospect for a new account of time reversal." *Studies in History and Philosophy of Modern Physics*, 49: 42–56.
Price, H. (1996). *Time's Arrow and Archimedes' Point: New Directions for the Physics of Time*. New York: Oxford University Press.
Price, H. (2011). "The flow of time," pp. 276–311, in C. Callender (ed.), *The Oxford Handbook of Philosophy of Time*. Oxford: Oxford University Press.
Redhead, M. (1975). "Symmetry in intertheory relations." *Synthese*, 32, v. 1/2: 77–112.
Reichenbach, H. (1956). *The Direction of Time*. Berkeley: University of California Press.
Roberts, B. (2017). "Three myths about time reversal invariance." *Philosophy of Science*, 84, 2: 315–334.
Roberts, B. (2022). *Reversing the Arrow of Time*. Cambridge: Cambridge University Press.
Roberts, J. (2008). "A puzzle about laws, symmetries and measurability." *The British Journal for Philosophy of Science*, 59: 143–168.
Rohrlich, F. (2005). "Time reversal invariance and the arrow of time in classical electrodynamics." *Physical Review E*, 72: 057601–1–3. S223–S234.
Sachs, R. (1987). *The Physics of Time Reversal*. London: University Chicago Press.
Saunders, S. (2003). "Physics and Leibniz's principles." In E. Brading and E. Castellani (eds.), *Symmetries in Physics: Philosophical Reflections*. Cambridge: Cambridge University Press, 2003, 21–22.
Schroeren, D. (2020). "Symmetry fundamentalism: A case study from classical physics." *Philosophical Quarterly*, 71 (2): 308–333.
Sozzi, M. S. (2008). *Discrete Symmetries and CP Violation*. Oxford: Oxford University Press.
Tucker, Chris. 2013. "Seemings and justification: An introduction." In Chris Tucker (ed.), *Seemings and Justification: New Essays on Dogmatism and Phenomenal Conservatism*, Oxford: Oxford University Press, pp. 1–29.
Wallace, D. (2022). "Observability, redundancy and modality for dynamical symmetry transformations." In James Read and Nicholas Teh (eds.), *The Philosophy and Physics of Noether's Theorem*, pp. 322–353, Cambridge: Cambridge University Press.

Weinberg, S. (1987). "Towards the final laws of physics." In R. Mackenzie and P. Durst (eds.), *Elementary Particles and the Laws of Physics: The 1986 Dirac Memorial Lectures*, Cambridge: Cambridge University Press, 61–110.

Weinberg, S. (1993). *Dreams of a Final Theory*. New York: Vintage.

Weyl, H. (1952) Symmetry. Princeton: Princeton University Press; extract in E. Brading and E. Castellani (eds.), *Symmetries in Physics: Philosophical Reflections*. Cambridge: Cambridge University Press, 2010: 21–22.

Wigner, E. (1932). *Group Theory and Its Application to the Quantum Mechanics of Atomic Spectra*. New York: Academic Press (1959).

Part IV

The Whole Universe

11

Following Earman's Time Direction Heresy: From the Global Arrow of Time to Local Irreversible Processes

OLIMPIA LOMBARDI

11.1 Introduction

The problem of the arrow of time has traditionally been analyzed in terms of entropy by relating the direction past-to-future to the direction of the entropy increase in isolated systems. In turn, many authors have considered that time-reversal invariance plays a relevant role in picking the arrow of time insofar as it links a property of the dynamics of physical theories to a topological feature of spacetime. In a series of papers published during the last decades, with Mario Castagnino we developed a global and nonentropic approach to the arrow of time that follows John Earman's "time direction heresy" (1974). According to that approach, the problem of the arrow of time can be addressed in terms of the geometry of spacetime, independently of entropic arguments and without appealing to non–time-reversal invariance. In this way, the global arrow of time emerges from the topological properties of spacetime as a whole (Castagnino, Lara, and Lombardi 2003a; Castagnino, Lombardi, and Lara 2003; Castagnino and Lombardi 2004), which can be manifested at a local level as a tetradimensional flow of energy (Castagnino, Lara, and Lombardi 2003b; Castagnino and Lombardi 2009) that breaks the symmetry of local theories (Aiello, Castagnino, and Lombardi 2008).

The aim of this chapter is to present a review of the global and nonentropic approach to the arrow of time, and to consider some aspects that were not discussed in detail in those original works. With this purpose, the chapter is organized as follows. In Section 11.2, the concepts of time-reversal invariance, reversibility, and arrow of time are clearly disentangled. This will allow us, in Section 11.3, to formulate the problem of the arrow of time in a precise way and from a perspective purged of temporal intuitions. Section 11.4 provides a justification for the adoption of a global, nonentropic approach. In Section 11.5, the conditions for the existence of a global nonentropic approach are introduced: time-orientability, cosmic time, and time-asymmetry. Section 11.6 explains why this approach is highly generic and does not require a "Past Hypothesis." Section 11.7 is devoted to show how the global arrow, defined in terms of the topological properties of the spacetime, can be transferred to local contexts as a local energy flow that points to the same direction of time all over the universe and, as a consequence, breaks the symmetry derived from local time-reversal invariant laws. Section 11.8 is a reflection on the extent to which the arrow of time can still be defined if the preceding conditions are not satisfied. In Section 11.9, the role of time-reversal invariance in the proposed global and nonentropic approach is discussed. Section 11.10 faces certain issues related to the arrow of time from the proposed approach, in particular, issues about contingency, fundamentality, reducibility, and objectivity. Finally, Section 11.11 introduces some concluding remarks, opening the way to analyze how the substantivalism–relationalism debate about time may have an impact on the global and nonentropic approach to the arrow of time.

11.2 Disentangling Concepts

The first step for an accurate presentation of this global and nonentropic approach to the arrow of time is to clearly elucidate the concepts involved in the debates. In particular, I will focus on the concepts of time-reversal invariance, reversibility, and arrow of time, and the associated problems.

11.2.1 Time-Reversal Invariance

From a general point of view, time-reversal invariance (hereafter, t-invariance) can be characterized as follows.

> A dynamical law is t-invariant when the equation that represents it is invariant to the application of the time-reversal operator **T**.

Of course, this characterization rests essentially on how the operator **T** is defined, and this has triggered a heated controversy in the context of the philosophy of physics.

From an openly heterodox position, David Albert (2000) argues that all the physical theories that are traditionally considered t-invariant are not really so. For example, electromagnetism would be non–t-invariant because the magnetic field should not invert its sign upon application of the operator **T** since no field represents a rate of change. John Earman (2002) severely criticizes Albert's position by pointing out that, if A produces B and A changes its sign due to time reversal, then B must also be affected by such a reversal. This is the case of the magnetic field: if the velocities of the charges passing through a wire are reversed, then the magnetic field produced by the electric current is also reversed. Castagnino,

Manuel Gadella, and Lombardi (2005) express the same idea in definitional terms: the time-reversal operator **T** is, by definition, an operator that performs the transformation $t \rightarrow -t$ and reverses the sign of all dynamic quantities whose definitions as a function of time are not invariant under the transformation. Thus, in electromagnetism, **T** leaves invariant the electric field, $\mathbf{T}E = E$, and inverts the velocity of the charges, $\mathbf{T}v = -v$, and the magnetic field, $\mathbf{T}B = -B$, since this field is defined as a function of such velocity.

The problem becomes more acute in the case of quantum mechanics. According to the orthodox view, the time-reversal operator **T** is antiunitary: it not only performs the transformation $t \rightarrow -t$ But also introduces the transformation $i \rightarrow -i$. The Schrödinger equation is t-invariant under its application. The reasons given for using the anti-unitary operator **T** instead of the unitary operator $\mathbf{T}^*$, which only reverses time, are diverse. In some cases, the justification is based on a classical analogy: the application of $\mathbf{T}^*$ does not lead to the transformation of momentum as $P \rightarrow -P$ (Earman 2002). In other cases, it is argued that $\mathbf{T}^*$ would invert the energy values, so that the energy spectrum would not be bounded to a minimum value, which is physically unacceptable (Sakurai 1994). In this sense, some authors directly require the Hamiltonian of the closed system to be invariant under time reversal (Gasiorowicz 1966). And this is sometimes justified by arguing that only in that way does the Schrödinger equation turn out to be t-invariant (Sachs 1987). In this line of argument, there are those who consider certain symmetries of physical laws to be a priori (Dürr and Teufel 2009). Challenging orthodoxy and following the work of Giulio Racah (1937) and Satosi Watanabe (1955), some authors advocate a unitary time-reversal operator for quantum theories (e.g. Costa de Beauregard 1980, North 2008). From this perspective, the Schrödinger equation turns out to be non–t-invariant (Albert 2000, Callender 2000, López 2019). It has also been claimed that the antiunitary operator T actually does not represent time reversal but motion reversal (Lopez and Lombardi 2019; López 2021a).

The purpose here is not to analyze in detail the debate about the definition of t-invariance and to weigh the arguments in favor of the different positions (see Roberts 2017, 2022; López 2021a,b). Regardless of the interpretation of the operator **T** adopted, a formal fact already noted in 1912 by Paul and Tatiana Ehrenfest (1959) must be taken into account: if $e(t)$ is a solution of an equation representing a t-invariant law, then $\mathbf{T}e(t)$ is also a solution. In general, any t-invariant law gives rise to "*time-symmetric twins*" (Castagnino, Lara, and Lombardi 2003a), that is, two mathematical structures symmetrically related by the time-reversal transformation: each "twin" is the temporal mirror image of the other. A traditional example of time-symmetric twins arises in electromagnetism, where the dynamical equations always have advanced and retarded solutions. From the point of view of the local theory that produces them, both twins are nomologically possible. Traditional arguments for discarding one of the twins and retaining the other invoke time-asymmetric notions that are not justified in the context of the local theory. For example, the retarded nature of radiation is usually explained by de facto arguments referring to initial conditions: the advanced solutions of the wave equations correspond to convergent waves that require a "miraculous" cooperative emitting behavior of spatially distant regions at the origin of the process (one of the first authors to appeal to this argument was Karl Popper, in a series of short untitled notes appearing in *Nature* between 1956 and 1967). As Huw Price (1996) has repeatedly warned, this type of argument is not legitimate in the context of the problem of the arrow of time, since it introduces the arrow "by hand" by presupposing the difference between the two directions of time from the beginning. I will return to this issue later.

11.2.2 Reversibility

Irreversibility has traditionally been an essential mark of thermodynamic processes, which spontaneously evolve toward equilibrium. During the early days of thermodynamics, reversibility was often conceived as recoverability: a process is irreversible if it is impossible to recover its initial state, leaving everything else unchanged (see Lombardi and López 2023). However, as Jos Uffink (2001) points out, the notion of reversibility as recoverability is rarely found in recent literature on the subject. In thermodynamics, a process is said to be reversible if it is so slow that the system always remains close to equilibrium. The seemingly paradoxical idea of an "infinitely slow" process has also sparked much debate (see, e.g., Rechel 1947; Norton 2016), but will not be discussed here. In the context of statistical mechanics, the notion of reversibility must be reconsidered. According to Earman (1986), a process is irreversible if it approaches equilibrium, represented by the microcanonical distribution. From a more generic point of view, Castagnino and Lombardi (2004) define reversibility in terms of the absence of attractors at finite or infinite points of the phase space.

Regardless of the precise definitions of t-invariance and reversibility, the point that deserves to be emphasized is that they are two completely different concepts, which even apply to different entities: whereas *t-invariance* is a property of a dynamical *law* (represented by a dynamical equation) and, a fortiori, of the set of the evolutions compatible to it (represented by the equation's solutions), *reversibility* is a property of a single *evolution* (solution). The two properties are not even correlated: it is possible to obtain reversible evolutions of non–t-invariant laws and irreversible evolutions of t-invariant laws (see examples in Castagnino, Lara and Lombardi 2003a,b).

On the basis of the preceding characterizations, the *problem of irreversibility* can be posed in a simple way: how to explain irreversible evolutions in terms of t-invariant laws. Once it is admitted that reversibility and t-invariance apply to different entities, it becomes clear that there is no conceptual obstacle to the resolution of the irreversibility problem: nothing prevents a t-invariant law from leading to irreversible evolutions. Huw Price (1996) illustrates this point with the analogy of a factory producing equal numbers of dextro-rotating and levo-rotating corkscrews: the production as a whole is completely symmetric, but each individual corkscrew is spatially asymmetric. Of course, while the conceptual answer is simple, a great deal of theoretical work is required to obtain irreversible evolutions from an underlying t-invariant dynamics (appeal to coarse-graining, partial traces, restricted algebras, Gamov vectors, etc.). This was the problem faced by the founders of statistical mechanics when they tried to describe the irreversible evolutions of thermodynamics by means of the t-invariant laws of classical mechanics. But the point to emphasize here is that the question of the arrow of time need not be invoked to address the problem of irreversibility. In fact, when one speaks of entropy-increasing or -decreasing processes, one presupposes an entropy increase or decrease into the future; when one considers a process that evolves from nonequilibrium to equilibrium, one implicitly locates equilibrium in the future. In general, any evolution tending to an attractor is conceived as approaching it in the future. This means that the distinction between past and future is usually presupposed in the traditional treatments of the problem of irreversibility. But then, what is the arrow of time, which introduces the distinction between the two temporal directions?

11.2.3 Arrow of Time

Let us begin by pointing out the difference between a conventional distinction and a substantial distinction in physics. Two objects are identical when, by permuting them in the description of a whole system, the resulting description is as correct as the original one. When different labels are used to designate identical objects, the distinction is *conventional*. If two objects are not identical, they are substantially different. In this case, even if the labels chosen are conventional, the difference is still *substantial*. For example, the labels "positive" and "negative" applied to electric charge or to spin directions are conventional. The same happens with time-asymmetric expressions, such as "initial/final conditions," "past/future light cone," or "increasing/decreasing time." In fact, nothing in theoretical physics distinguishes between past and future in a substantial way, as they are distinguished in everyday language. Consequently, the problem of the direction of time in physics cannot be formulated in terms of identifying the future direction of time, since 'future' is a conventional label in theoretical physics. On this basis, in the next section I will focus on the problem of the arrow of time.

11.3 The Problem of the Arrow of Time

As is well known, we can move in any direction in space. In contrast, to move in time we are essentially limited to a single direction, which we call "future." Indeed, we intuit the two directions of time in very different ways: we remember the past and we predict the future. The nature of time and its apparent asymmetry have been the subject of reflection since the very dawn of philosophy, and the discussions have acquired special relevance in twentieth-century metaphysics (see, e.g., Emery, Markosian, and Sullivan 2020). In this chapter, I will not analyze these discussions but will focus on the problem of the arrow of time in the context of physics, which arises when looking for a physical correlate of the intuitive asymmetry between past and future.

As stressed in the previous section, in physics the labels "past" and "future" are used in a conventional way. Then, the problem cannot be formulated in terms of identifying one direction of time as the future. When we address the problem of the arrow of time in physics, we cannot appeal to our assumptions about past and future without begging the question. In other words, we must take a perspective purged of temporal intuitions, adopting a "view from nowhen" (Price 1996): a point outside of time from which reality is analyzed in atemporal terms. Once this constraint is accepted, the traditional expression coined by Eddington can be understood by analogy. Since the head and the tail of an arrow can be distinguished on the basis of its geometrical properties, the two directions, head-to-tail and tail-to-head, can be defined independently of any particular perspective. Analogously, "the problem of the arrow of time becomes the question about the possibility of establishing a substantial difference between the two directions of time on the basis of exclusively physical arguments, with no appeal to our temporal intuitions" (Castagnino and Lombardi 2009: 6).

11.4 A Global and Nonentropic Approach

As advanced in the Introduction, the present proposal addresses the problem of the arrow of time from a global and nonentropic perspective. Here I will begin by justifying the decision to adopt such a perspective.

11.4.1 Why Global?

According to the traditional local approach to the problem of the arrow of time, the direction past-to-future is the direction of time in which the entropy of isolated systems increases. But this approach faces two obstacles, one related to the t-invariance of the fundamental laws, and the other to the local nature of the involved theories.

In the context of the attempts to reduce the second law of thermodynamics to statistical mechanics, Paul and Tatiana Eherenfest (1959) noted as early as 1912 that, if entropy is defined in statistical-mechanical terms, any increase of entropy into the future corresponds to an identical increase into the past: this is a direct consequence of the t-invariance of the underlying laws of classical mechanics. As Lawrence Sklar (1993) rightly points out, any answer that appeals to the assumption that the probabilities of later events are determined by the probabilities of earlier events is inadmissible, since probability is blind to temporal direction. In Huw Price's terms (1996), such a strategy violates the "nowhen" standpoint. As we will see later in Section 11.9, a global approach is immune to this problem, derived from the t-invariance of fundamental laws.

Let us suppose that we can describe all the irreversible processes of the universe, in particular, the decaying processes, by means of a certain local theory – a theory that only describes phenomena confined to small regions of spacetime. However, this does not yet tell us in which time direction each process decays. The arrow of time could be defined if it could be ensured that all processes decay in the same temporal direction. But this is precisely what cannot be guaranteed by local theories: only from the perspective of a global theory can one speak of the temporal direction toward which all or most processes decay.

11.4.2 Why Nonentropic?

The first global approach to the arrow of time was that presented by Boltzmann himself in his debate with Loschmidt and Zermelo (see Brush 1976). Since then, many authors have defined the past-to-future direction as the direction of the entropy increase of the universe (see, e.g., Reichenbach 1956; Feynman, Leighton, and Sands 1964; Davies 1974). However, this perspective faces two difficulties. First, it supposes that a global entropy can be defined for the whole universe, which is a very controversial issue (see, e.g., Lineweaver 2014). Second, it assumes that there is a single time that plays the role of the parameter for all processes in the universe, with respect to which the entropy gradient can be calculated; but, according to general relativity, the existence of a global time imposes geometric constraints on spacetime prior to any entropic consideration.

It is precisely for these reasons that the present proposal follows Earman's "Time Direction Heresy," according to which the arrow of time, if it exists, is an intrinsic feature of spacetime "which does not need and cannot be reduced to non-temporal features" (Earman 1974: 20). In fact, entropy is a phenomenological property, whereas the arrow of time should be a property of time itself. On this basis, the global and nonentropic approach defines the arrow of time in terms of the geometrical properties of spacetime.

11.5 Conditions for the Global Arrow of Time

Once it has been accepted that (i) the problem of the arrow of time consists in explaining the substantial difference between the two temporal directions, and (ii) such a difference must be found in the geometrical properties of spacetime, then the necessary conditions for a spacetime to have an arrow of time can be identified.

11.5.1 Time-Orientability

Let us recall that, in general relativity, a spacetime is an ordered pair (M, g), where M is a pseudo-Riemannian manifold, and g is a Lorentzian metric for M such that $\nabla g = 0$. A spacetime (M, g) is *time-orientable* if there exists a continuous non-vanishing vector field $\gamma^{\mu}(x)$ on M which is everywhere nonspacelike. The field $\gamma^{\mu}(x)$ splits the set of all light semicones of the spacetime into two subsets: C_{+}, the subset of the semicones that contain the vectors of the field, and C_{-}, the subset of the semicones that do not contain the vectors of the field.

If a spacetime is non–time-orientable, the distinction between "+" and "–" semicones cannot be globally defined. For this reason, there is a broad consensus that time-orientability is a requirement for the arrow of time (see, e.g., Grünbaum 1963; Earman 1974). Nevertheless, as we will see in Subsection 11.8.1, this assumption has been challenged in the literature on the subject.

11.5.2 Global Time and Cosmic Time

There is a hierarchy of conditions that, applied to a time-orientable spacetime, avoid "anomalous" temporal features such as closed timelike curves (see Hawking and Ellis 1973). In particular, a spacetime (M, g) possesses a *global time* τ if there exists a function $\tau : M \rightarrow \mathbb{R}$ whose gradient is everywhere timelike. The global time defines a foliation of the spacetime by splitting it into hypersurfaces of simultaneity (see Schutz 1980, Torretti 1983).

The mere existence of a global time does not univocally define a foliation of spacetime. In general, cosmology works with spacetimes with cosmic time, corresponding to a foliation that fulfills two additional conditions: (i) the global time τ corresponding to that foliation is such that there exists a continuous set of worldline curves which are orthogonal to all the hypersurfaces $\tau = const.$ (orthogonality recovers the notion of simultaneity of special relativity for tangent hyperplanes of those hypersurfaces) (see Misner, Thorne, and Wheeler 1973), and (ii) the proper time interval between two hypersurfaces $\tau = \tau_1$ and $\tau = \tau_2$ is the same when measured on any curve of that continuous set (for details, see Castagnino, Lombardi, and Lara 2003). In this case, the metric can be expressed as

$ds^2 = dt^2 - h_{ij}x^ix^j$, where h_{ij} is the three-dimensional metric of each hypersurface of simultaneity and t is the *cosmic time*.

Although in principle the arrow of time could be defined in a spacetime endowed with global time, only the cosmic time guarantees the possibility of coordinating all the process of the universe according to a single time and, therefore, makes it possible to speak of the arrow *of that time*. I will come back to this point in Subsection 11.8.2.

11.5.3 Time-Asymmetry

Some authors (e.g., Grünbaum 1963) correctly point out that mere time-orientability does not yet provide time-orientation. But what does it mean to define the time-orientation of spacetime? As already discussed, it is not a matter of identifying the future direction, since in physics the words "past" and "future" are used only conventionally. In other words, given the continuous nonvanishing non-spacelike vector field $\gamma^\mu(x)$ that can be defined all over the manifold if the spacetime is time-orientable, the choice between $\gamma^\mu(x)$ and $-\gamma^\mu(x)$ as the time-orientation is a purely conventional choice. Therefore, the problem of the arrow of time must be formulated not as the search for the time-orientation of spacetime but, as already stated, in terms of the substantial difference between the two directions of time. A strategy to establish such a difference would be to appeal to a local non–t-invariant law, such as the law governing the decay of the K meson. However, insofar as the law is local, there is strictly no global argument to prevent the decay from matching $\gamma^\mu(x)$ in one bounded region of spacetime and $-\gamma^\mu(x)$ in another region (I will come back to this point in Subsection 11.8.1).

An advantage of the present approach to the arrow of time is that, being global, it does not depend on the t-invariance of local laws. In fact, the global arrow can be defined on the basis of the time-asymmetry of spacetime. A time-orientable spacetime (M, g) with cosmic time t is *time-symmetric* with respect to some spacelike hypersurface $t = t_S$, where t_S is a constant, if (i) there is a diffeomorphism d of M onto itself which reverses the time orientations but preserves the metric g (the spacetime is *time-isotropic*) and (ii) that diffeomorphism d leaves fixed the hypersurface $t = t_S$. Intuitively this means that, from the hypersurface $t = t_S$, the spacetime looks identical in the two directions of time. Therefore, if the spacetime is time-asymmetric, any hypersurface $t = t_A$ splits the manifold into two sections, $t > t_A$ and $t < t_A$, which are *substantially different* from each other with respect to their geometrical properties, analogously to the substantial geometrical difference between the head-to-tail and tail-to-head directions in a material arrow.

At this point, it can be conventionally said that the vectors of $-\gamma^\mu(x)$ point toward $t < t_A$ and the vectors of $\gamma^\mu(x)$ point toward $t > t_A$, or vice versa, but in any case $-\gamma^{\mu}(x)$ and $\gamma^{\mu}(x)$ are substantially different. Such a difference can be used to ground the substantial difference between the subsets C_{-} and C_{+} of the light semicones of the spacetime, a difference that remains merely conventional without time-asymmetry.

11.6 A Generic Approach: No Need of "Past Hypothesis"

In a review article of the main debates on the arrow of time to date, Steven Savitt (1996) distinguishes between two issues: (i) the "how possible" question, that is, how is it possible to formulate a time-asymmetric model by means of t-invariant laws?, and (ii) the "why probable" question, that is, for what reason can time-asymmetry be assumed to be probable? As shown in the previous section, the global nonentropic approach offers a straightforward answer to the first question. Now, regarding the second question it is time to ask: how generic is the proposed approach?

One of the major difficulties that the global entropic approach has to face is the need to assume what David Albert (2000) called the "Past Hypothesis," that is, the hypothesis of very low entropy at the origin of the universe. As Paul Davies clearly explains, "the origin of the arrow of time always refers back to the cosmological initial conditions. There exists an arrow of time only because the universe originates in a less-than-maximum entropy state" (Davies 1994: 124). The status of the hypothesis is a highly controversial issue: for some authors it expresses a contingent fact (Callender 2004; Price 2004), for others it is a law (Penrose 1989, 2004), whereas for Earman (2006) it is not even false (for more recent discussions, see Wallace 2017; Roberts 2022). Therefore, the fact that an approach to the arrow of time does not require the Past Hypothesis counts in its favor.

In the context of the Arnowitt–Deser–Misner (ADM) formalism, in which the foliability of the spacetime is assumed, a spacetime is time-symmetric with respect to a certain hypersurface Σ with geometry g_{ij}, with $i,j = 1,2,3$, if the conjugate momentum π^{ij} is zero. This means that the space of the Cauchy data leading to time-symmetry, $\{g_{ij}(x^1,x^2,x^3), \pi^{ij}(x^1,x^2,x^3) = 0\}$ is a proper subspace of the space of all the possible Cauchy data $\{g_{ij}(x^1,x^2,x^3), \pi^{ij}(x^1,x^2,x^3)\}$ (for a full argument, see Castagnino and Lombardi 2004). This is a case of the intuitive idea that symmetry is overwhelmingly improbable: in general, if the set of all the models of a certain theory has a measure, the subset of the symmetric models has measure zero. Therefore, the arrow of time is highly *generic*: its existence does not need extremely improbable initial conditions.

11.7 From the Global Arrow to Local Irreversibility

Up to this point, the arrow of time in a time-asymmetric universe was characterized in mathematical terms. Now the question is whether it can be defined in terms of magnitudes with more physical content and with relevance in local contexts.

11.7.1 The Global Arrow as Energy Flow

The first step is to connect the vector field $\gamma^{\mu}(x)$, which expresses the arrow of time, with the energy-momentum tensor $\boldsymbol{T}_{\mu\nu}$. Such a connection can be established if two conditions are satisfied everywhere: (i) the density of matter-energy is nonzero, and (ii) the energy-momentum tensor satisfies the *dominant energy condition*. Condition (i) means that there is no perfect vacuum in any region of the spacetime. Condition (ii) means that the energy component dominates the other components of $T^{\alpha\beta}$ in any orthonormal basis: $T^{00} \geq |T^{\alpha\beta}|$ for each α, β (see, e.g., Hawking and Ellis 1973; Visser 1995); in other words, the matter–energy density is nonnegative and the energy flow is nonspacelike for any local observer.

Let us now consider a continuous orthonormal basis field $\{\boldsymbol{V}_{(\alpha)}(x)\}$, where $\boldsymbol{V}_{(\alpha)}(x) = \left(V^0_{(\alpha)}(x), V^1_{(\alpha)}(x), V^2_{(\alpha)}(x), V^3_{(\alpha)}(x)\right)$, with $\alpha = 0, 1, 2, 3$, is a tetrad of four unitary vectors. Then, the four-magnitude $T^{0\alpha}(x)\boldsymbol{V}_{(\alpha)}(x)$ can be conceived as representing the *four-dimensional energy flow* at each point x of the spacetime, whose coordinates in the basis $\{\boldsymbol{V}_{(\alpha)}(x)\}$ are the $T^{0\alpha}$. Under conditions (i) and (ii), it can be proved that such an energy flow is nonspacelike everywhere (see Castagnino and Lombardi 2004, 2009).

Although $T^{0\alpha}(x)\boldsymbol{V}_{(\alpha)}(x)$ is usually referred to as the "energy flow vector" at point x (e.g., Hawking and Ellis 1973), strictly speaking, it is not a vector because it does not transform in a covariant way; as a consequence, it cannot be directly identified to $\gamma^{\mu}(x)$. Nevertheless, this fact does not prevent the arrow of time from being associated with that energy flow, because the dominant energy condition is a *covariant condition*: if $T^{0\alpha}(x)\boldsymbol{V}_{(\alpha)}(x)$ is nonspacelike in a reference frame, it is nonspacelike in all reference frames. This means that, no matter which basis $\{V_{(\alpha)}(x)\}$ is selected, the energy flow in that basis, represented by $T^{0\alpha}(x)V_{(\alpha)}(x)$, can define the arrow of time.

It is worth noting once again that, for each $T^{0\alpha}(x)\boldsymbol{V}_{(\alpha)}(x)$, there is a $-T^{0\alpha}(x)\boldsymbol{V}_{(\alpha)}(x)$ that can also legitimately represent the energy flow, because the difference between them is, a priori, just conventional. It is the time-asymmetry of spacetime that makes them substantially different. The time directions of $T^{0\alpha}(x)\boldsymbol{V}_{(\alpha)}(x)$ and $-T^{0\alpha}(x)\boldsymbol{V}_{(\alpha)}(x)$ can be called "future" and "past" respectively, or vice versa, since names are always conventional. The usual convention in physics is to call the direction of the positive energy flow 'future', so that such a flow is contained in or belongs to the light semicone $C_+(x)$.

11.7.2 From the Global Arrow to Local Arrows

If the global arrow is to be transferred to local contexts, a well-known difficulty arises: $T^{0\alpha}(x)$ does not represent the energy flow as it is conceived in local physics. The conservation equation with ordinary derivatives is $\partial_\mu \tau_{\mu\nu} = 0$, where $\tau_{\mu\nu}$ – which is not a tensor – is defined as $\tau_{\mu\nu} = \sqrt{-g}(T_{\mu\nu} + t_{\mu\nu})$, and $t_{\mu\nu}$ is an homogeneous and quadratic function of the connection $\Gamma^\lambda_{\mu\nu}$ (see Misner, Thorne, and Wheeler 1973).

Analogously to the case of $T_{\mu\nu}$, in any orthonormal basis $\{V_{(\alpha)}(x)\}$, $\tau^{0\alpha}(x)V_{(\alpha)}(x)$ is not a vector. But, in this case, the dominant energy condition is of no help because it cannot be expressed in terms of $\tau_{\mu\nu}$. Nevertheless, in the particular case of *local inertial frames*, $\Gamma^\lambda_{\mu\nu} = 0$, which makes $t_{\mu\nu} = 0$ and then $\tau_{\mu\nu} = \sqrt{-g}T_{\mu\nu}$. As a consequence, the dominant energy condition leads to $\tau^{00} \geq |\tau^{\alpha\beta}|$ for each α, β. This means that, at any point x, if $\{W_{(\alpha)}(x)\}$ is the basis of the local inertial frame, with $W_{(\alpha)}(x) = \left(W^0_{(\alpha)}(x), W^1_{(\alpha)}(x), W^2_{(\alpha)}(x), W^3_{(\alpha)}(x)\right)$, then the *local energy flow*, represented by $\tau^{0\alpha}(x)W_{(\alpha)}(x)$ and satisfying the usual conservation equation, is nonspacelike (see Castagnino and Lombardi 2004, 2009).

In addition, at each point x, the local energy flow is contained in or belongs to the same light semicone in which the global energy flow $T^{0\alpha}(x)V_{(\alpha)}(x)$ is contained or to which it belongs. In other words, $\tau^{0\alpha}(x)$ inherits the time direction defined at the global level and so transfers the global arrow of time to local contexts, turning it into a usual magnitude of local physical theories.

11.7.3 Breaking Local Symmetries

As explained in Section 11.2.1, any t-invariant law gives rise to time-symmetric twins, each one of which is the temporal mirror image of the other. Since, from the viewpoint of the local theory that produces them, the twins are only conventionally different, retaining one of them and discarding the other amounts to introducing the arrow of time "by hand." From a "nowhen" standpoint, the challenge is to provide a nonconventional criterion, based on theoretical arguments, to choose one of the twins as the physically relevant one, thus establishing the substantial difference between the two.

The desired criterion, which cannot be legitimately provided either by local theory or by our pretheoretical intuitions, is supplied by the local energy flow expressed in terms of $\tau^{0\alpha}$. Once it has been conventionally decided that, at each point x of the spacetime, the local energy flow defines the "future" light semicone $C_+(x)$, then the energy emitted at x must be contained in or must belong to $C_+(x)$. Therefore, in any t-invariant theory, the twin corresponding to this "positive" energy flow is the member of the pair that must be retained as physically meaningful.

In the traditional case of electromagnetism, the criterion based on the local energy flow selects retarded solutions, since they describe waves which cannot propagate outside of the future light semicone. Besides this traditional case, the symmetry of the time-symmetric twins can be broken by this same strategy in any t-invariant theory.

For example, axiomatic quantum field theory includes a postulate according to which the spectrum of the energy momentum operator P^μ is confined to the future light semicone, that is, its eigenvalues p^μ satisfy $p^2 \geq 0$ and $p^0 \geq 0$ (see, e.g., Haag 1996). That condition selects "by hand" one of the time-symmetric twins, $p^0 \geq 0$ and $p^0 \leq 0$, which would arise if the postulate were not included. Appealing to the local energy flow, by contrast, provides a substantial criterion for the selection. In fact, as explained in the previous subsection, $\tau^{\mu\nu}$ can be conceived of as the local conservative version of $T^{\mu\nu}$ in the orthonormal coordinates of a local inertial frame; as a consequence, $\tau^{0\alpha} = \tau^{\alpha 0}$. Therefore, if the local energy flow, represented by $\tau^{0\alpha}$, points to the future, the local linear momentum density, represented by $\tau^{\alpha 0}$, also points to the future. Since $\tau^{\alpha 0}$ is precisely the relativistic magnitude corresponding to the classical p^μ in quantum field theory, the requirement that p^μ points to the future is justified in terms of a previously defined arrow of time rather than being imposed as a starting point of quantum field theory (for the full argument, see Castagnino and Lombardi 2004, where the argument is also developed for ordinary quantum field theory).

In irreversible quantum mechanics, the analytical continuation of the energy spectrum of the system's Hamiltonian in the complex plane leads to at least a pair of complex conjugate poles, one in the lower half-plane and the other in the upper half-plane. Such poles correspond to a pair of Gamov vectors, $|\Psi_G^+\rangle$ and $|\Psi_G^-\rangle$ in the rigged Hilbert space formalism, which represent the exponentially growing and decaying part of resonant unstable states, respectively (see Bohm and Gadella 1989). The two only conventionally different Gamov vectors are another example of time-symmetric twins. Arno Bohm breaks the symmetry "by hand" by introducing the so-called "preparation-registration arrow" as a postulate of the theory (Bohm, Antoniou, and Kielanowski 1994a, 1994b). By contrast, on the basis of the local energy flow pointing to the future, the decaying Gamov vector $|\Psi_G^-\rangle$ turns out to be the physically relevant object, since it represents the decaying process that emits energy toward the future (for a detailed discussion, see Castagnino, Gadella, and Lombardi 2005, 2006).

Many phenomenological laws, such as the Fourier law and the second law, are not only non–t-invariant, but also describe irreversible evolutions. In this case, only the processes conceived as future-directed are considered as physically relevant. However, in general those laws include some phenomenological coefficients that are assumed as positive. But when these coefficients are deduced from underlying fundamental t-invariant theories, a negative counterpart can always be discovered: the hidden t-invariance is revealed, and the corresponding pair of time-symmetric twins can be identified. Furthermore, many phenomenological coefficients that are normally considered positive, such as viscosity and thermal conductivity, have a negative counterpart that comes to light when they are derived from underlying t-invariant laws. All these cases are examples of time-symmetric twins whose symmetry can be broken by a future-directed energy flow (for many detailed examples of breaking symmetry in phenomenological laws and coefficients, see Aiello, Castagnino, and Lombardi 2008).

11.8 Relaxing Conditions

The aim of developing Earman's "Time Direction Heresy" led us to define a global and nonentropic arrow of time under the conditions of temporal orientability, existence of global/cosmic time, and temporal asymmetry. This global arrow can be conceived as an energy flow and can be transferred to local contexts if the dominant energy condition is met. At this point, it is legitimate to wonder whether those conditions can be relaxed and still have an arrow of time.

11.8.1 Time-Orientability

If the spacetime is not time-orientable, a timelike vector pointing to a certain time direction can be transformed into a timelike vector pointing to the opposite time direction by means of a continuous transportation that always keeps timelike vectors timelike. In this case, the distinction between the two temporal directions cannot be established globally. But is it still possible to define regional arrows of time when the global arrow is not obtainable?

In explicit controversy with Earman, Geoffrey Matthews (1979) argues that the importance of the global structure of the universe for the solution of the problem has been overestimated. According to the author, a spacetime may have regional arrows of time without a global arrow if the criterion to define the local arrows is given by local physical laws, for example, the phenomenological law of entropy increase or the fundamental law governing the decay of the K meson. However, things are not that simple. Let us suppose that the arrows of time are defined by a local non–t-invariant law L, and that the directions of the so-defined arrows in two regions of the spacetime disagree when compared by means of continuous timelike transport. In this case, the transport trajectory necessarily crosses the border between the two regions at a point where the arrow is not univocally defined, which amounts to a break in the validity of *L* at the border point. This fact violates a widely accepted methodological principle of current cosmology: the principle of universality, according to which the laws of physics are valid at all points of the spacetime. Therefore, the possibility of arrows of time pointing to opposite directions in different regions of the spacetime, although logically admissible, requires ignoring global consideration coming from cosmology.

11.8.2 Global Time and Cosmic Time

In the context of the proposed global and nonentropic approach, the arrow of time is defined in terms of the time-asymmetry of the spacetime with respect to the cosmic time, which implies that the spacetime is also endowed with a global time. The first question is to what extent the existence of global time is an indispensable condition to define an arrow of time. If the spacetime has no global time, then the set of all the events of the universe cannot be partitioned into equivalence classes such that (i) each class is a spacelike hypersurface, and (ii) the hypersurfaces can be ordered in time. This is the case when there are closed – or almost closed – timelike curves (CTCs) in the spacetime.

However, the existence of CTCs does not seem to be an insurmountable obstacle to having an arrow of time that distinguishes between the two directions of "rotation" on the curve. Indeed, in their discussion on time machines, Earman, Christian Wüthrich, and J. B. Manchak define a CTC as "a parameterized closed spacetime curve whose tangent is everywhere a future-pointing timelike vector. A CTC can be thought of as the world line of some possible observer whose life history is linearly ordered in the small but not in the large: the observer has a consistent experience of the 'next moment,' and the 'next,' and so on, but eventually the 'next moment' brings her back to whatever event she regards as the starting point" (Earman, Wüthrich, and Manchak 2021: 6). This means that, in this discussion, the arrow of time on the CTC is presupposed: regardless of the criterion to distinguish between the two directions of time, "CTCs are always future-directed curves" (Earman, Wüthrich, and Manchak 2021: 33). Of course, in a CTC the arrow of time cannot be defined in entropic terms: from any point in the CTC, if the entropy increases during an interval, it must decrease later to reach the original value at the starting point. The criterion to distinguish between the two directions of time can be given by a non–t-invariant law, such as that ruling the decay of the meson K. Nevertheless, the global character of the arrow must be preserved; otherwise, the methodological principle of universality would be violated, as pointed out in the previous subsection.

The next natural question is: can the arrow be defined in terms of the global time in the case that there is no cosmic time? If the spacetime has no cosmic time, it cannot be guaranteed that all the processes of the universe can be coordinated by a single time, which plays the role of the parameter of the evolution of the universe as a whole. For instance, this is the case in a "trousers spacetime," in which a single "trunk" bifurcates into two "legs," each with its own global time: the events of a spacelike hypersurface on one "leg" are disconnected from the events of a spacelike hypersurface on the other "leg." Given its time-asymmetry, this "trousers spacetime" admits the definition of an arrow of time for each global time, on each "leg," even though it has no cosmic time (for independent reasons for rejecting branching spacetimes, see Earman 2008).

However, it could be argued that, in a universe without cosmic time, there is no single time whose arrow must be defined. Furthermore, some authors consider that the time dimension of a time-branching spacetime should not be considered legitimately "time." For example, Sean Gryb and Karim Thébault (2016), in their criticism of Rovelli's relational view of time, claim that time must be "globally defined in terms of a linear sequence" (684) because "the basic one-dimensional ordering structure [. . .], in our view, is constitutive of time." (666). Regardless of these conceptual debates, it must be recognized that Friedmann–Lemaitre–Robertson–Walker models of current cosmology are endowed with cosmic time: since they are spatially homogeneous and isotropic on the large scale, a family of spacelike hypersurfaces can be found, which can be labeled by the proper time of the worldlines that orthogonally cross them. Those labels define the cosmic time, represented by the variable t of the Robertson–Walker metric, and the scale factor a is a scalar-only function of t.

11.8.3 Dominant Energy Condition

From this approach, the dominant energy condition allows the transference of the global and nonentropic arrow to local contexts, as a flow of energy pointing in the same time direction at all points of spacetime. The requirement that the energy-momentum tensor satisfies the dominant energy condition everywhere is not a strong constraint; as Hawking and Ellis point out, such a condition "holds for all known forms of matter and there is in fact good reason for believing that this should be the case in all situations" (Hawking and Ellis 1973: 91).

Nevertheless, certain strange cosmological "objects" lead to universes where the dominant energy condition does not hold in certain regions of the spacetime. For example, the condition is violated in the vicinity of a wormhole throat, since wormholes require "exotic" matter with negative density of mass-energy (see Visser 1995). The existence of wormholes can undermine the precondition of global time existence, generating closed timelike curves, and this would lead us to the discussion of the previous section. But if this is not the case, the fact that the dominant energy condition does not hold everywhere is not an obstacle to the global and nonentropic approach. Indeed, the dominant energy condition is a local condition that does not define the arrow of time but only allows it to be transferred to local contexts. Therefore, in the regions of the spacetime where the condition is not satisfied, the arrow of time can still be defined, but it cannot be translated as a local energy flow.

11.9 Arrow of Time and Time-Reversal Invariance

A traditional view in the philosophy of physics literature has been to closely relate the concept arrow of time to that of time-reversal invariance (here, t-invariance). What is the relationship between the two concepts in the present global and nonentropic approach?

11.9.1 The Role of Time-Reversal Invariance

According to Paul Horwich (1987), the symmetries of the dynamics of a theory make it possible to infer certain properties of the underlying spacetime. Huw Price (1996) takes a similar position when he concludes that, given that the laws of physics are t-invariant, there is no objective direction of time. Though from a different perspective, Tim Maudlin shares a similar diagnosis: "The usual approach sets the problem as follows: the fundamental physical laws have a feature called "Time Reversal Invariance." If the laws are time-reversal invariant, then it is supposed to follow that physics itself recognizes no directionality of time" (Maudlin 2002: 266). The underlying assumption of this so-called symmetry-to-reality inference (Dasgupta 2016) is that dynamical symmetries shed some light on the structure of spacetime (for a criticism, see López 2023). In Lawrence Sklar's words:

> Some of the important symmetries of physical laws are the invariance of laws under translation in space, invariance under translation in time, invariance under spatial rotation, invariance under change of inertial frame, invariance under spatial reflection, and so-called time-reversal invariance. All of these invariance principles are intimately related to the structure of the spacetime in which the material happenings of the world occur. (Sklar 1974: 359)

However, the symmetry-to-reality inference only provides a sufficient condition for the arrow of time: it does not prevent the arrow from existing even under t-invariant laws. By quoting Sklar again: "Do the proponents of this position really wish to allege that, if the laws of nature all do turn out to be time-reversal invariant, our whole impression that the world of events is a world in which an asymmetric temporal priority relates event to event is an illusion?" (Sklar 1974: 399). From the global and nonentropic approach proposed here, the answer to this question is clearly negative: since t-invariance does not prevent the construction of time-asymmetric models, the arrow of time can be defined in terms of the time-asymmetry of a spacetime described by the t-invariant laws of general relativity.

11.9.2 Does Symmetry Threaten Again?

Let us recall that, if the spacetime is endowed with cosmic time t, then the metric can be expressed as $ds^2 = dt^2 - h_{ij}x^i x^j$, where h_{ij} is the three-dimensional metric of each hypersurface of simultaneity. Since the Einstein field equations are t-invariant, if (h_{ij}, t) is a solution, then $(h_{ij}, -t)$ is also a solution (for the case of a superspace including a scalar matter field, see Halliwell 1994). In other words, a pair of time-symmetric twins is obtained, which are both possible according to the laws of general relativity. With this result, the ghost of symmetry seems to threaten again. Admitting the need of selecting one between the two solutions implicitly rests on the assumption that they describe different possible universes, one the temporal mirror image of the other. However, this assumption loses its validity in the global context of general relativity.

As is well known, symmetry transformations admit, in principle, two interpretations: the active one, which consists in changing one system for another, and the passive one, which consists in changing the descriptive perspective on the system. According to the usual view, only the active interpretation makes sense in the case of discrete transformations:

> The passive interpretation of continuous symmetries, like spatial rotation, is meaningful since one can suppose at least in principle that an idealized observer can rotate himself in space in correspondence with the given spatial rotation [. . .]. But how is an observer, even an idealized one, supposed to "rotate himself in time?" (Earman 1974: 26–27)

This claim is correct when the observer is embedded in the same spacetime as the observed system. However, when the system is the universe as a whole, how is the observer supposed to rotate in space in order to describe the system from a different spatial perspective? Since there is no space or time outside spacetime, rotating in space is as impossible as rotating in time. But this does not imply that the active interpretation is the correct one for discrete transformations: what is the conceptual meaning of the idea of two identical universes, one rotated in space or time with respect to the other? Moreover, the two interpretations are also meaningless when applied to continuous transformations over the entire universe: how would an observer be spatially displaced with respect to the universe, or how to conceive of two identical universes spatially displaced with respect to each other? In brief, both the active and the passive interpretations of symmetry transformations collapse into conceptual nonsense when applied to the universe as a whole.

In current cosmology, symmetry transformations are neither actively nor passively interpreted. Two mathematical models of the universe, (M, g) and (M', g'), are considered equivalent if they are *isometric*, that is, if there is a diffeomorphism $\theta : M \rightarrow M'$ that carries the metric g into the metric g'. Since symmetry transformations are isometries, two models m and $S(m)$, where S is a symmetry transformation, are considered equivalent descriptions of one and the same universe. This assumption is what Earman and John Norton call "Leibniz equivalence" in the framework of the "hole argument"; denying this principle is "at odds with standard modern texts in general relativity, in which this equivalence is accepted unquestioningly in the specific case of manifolds with metrics" (Earman and Norton 1987: 22; for a detailed discussion about the equivalent thesis, see Castagnino, Lombardi, and Lara 2003). Therefore, in the global context of general relativity, the symmetry threat is only apparent: the problem of selecting one of the two time-symmetric twins vanishes once it is accepted that both twins represent the same universe.

11.10 The Nature of the Arrow of Time

A good argument to accept that we live in a universe with an arrow of time is that current cosmological models, which we accept as describing, even approximately, the actual universe, meet the conditions of time-orientability and cosmic time, and are also time-asymmetric. More specifically, Big Bang–Big Chill and Big Bang–Big Rip universes are manifestly time-asymmetric. In the case of Big Bang–Big Crunch universes, the scale factor $a(t)$ has maximum value for a definite value of the global time t and, then, the spacetime might be time-symmetric with respect to the time of maximum expansion; nevertheless, as explained in Section 11.6, time-symmetry is an exceptionally extraordinary condition that we have no reason to suppose corresponds to our actual universe.

However, even if we accept that we live in a time-asymmetric universe, this does not cancel out the fact that the global arrow of time is not an inevitable result of the laws of general relativity. By contrast, it is a feature of certain models that satisfy specific topological properties and, as a consequence, it is contingent. Somebody might consider that, metaphysically, the fact that time has a privileged direction is part of the "essence" of time itself, that is, a property that time cannot lack without ceasing to be time (Maudlin 2007). But here it is worth recalling something already emphasized in Section 11.3: the problem of the arrow of time in the context of physics is not the metaphysical problem of the ontological nature of time but arises from the search for a physical correlate of the intuitive asymmetry between past and future. In particular, the challenge is to establish a substantial difference between the two directions of time on the basis of exclusively physical arguments, with no appeal to our temporal intuitions. In the context of this problem, the arrow of time is given by that substantial difference which, according to the global and nonentropic approach, is not necessary but contingent. In a certain sense, the arrow, when it exists, emerges from the topological properties of spacetime. But it is not emergent in the sense of contingency used by Barry Loewer (2012a), who considers that the direction of time is not a fundamental property of the world, but a secondary, derivative, or emergent one because the t-invariance of the fundamental laws of physics generally holds.

A usual criterion for classifying the positions on the arrow of time is based on whether it is considered fundamental and irreducible or not. Traditionally, those who adopt the first option have tried to base the difference between the two directions of time on non-t-invariant fundamental laws (e.g., Horwich 1987; Maudlin 2002; Roberts 2022). Those inclined to the second option consider that the fundamental laws of physics are t-invariant and, therefore, the arrow of time must be reduced to nontemporal relations, such as those defined by the entropy gradient of closed systems (e.g., Loewer 2012b). On the basis of the preceding presentation, it is clear that the approach proposed here follows Earman's heresy and, therefore, the global arrow is an intrinsic property of spacetime because it does not reduce to nontemporal features. However, it is not defined by non–t-invariant laws and does arise from intrinsic properties of spacetime itself. On the other hand, the global arrow is fundamental in the sense that it can be defined entirely in the context of a fundamental theory such as general relativity. However, in such a context it can be considered as emerging from certain topological properties that spacetime may or may not have. In summary, a global approach that defines the arrow of time in a general relativistic framework does not fit comfortably into the traditional classification in terms of the fundamentality and irreducibility of the direction of time.

An issue related to the preceding remarks is whether the arrow of time refers to a property of time itself *or* to a property of the disposition of things in time. The nonentropic approach proposed here agrees with Earman's heresy in regarding the arrow of time as a property of time itself. However, the disjunction in the first sentence should be taken with caution. First, it should be noted that this is an exclusive disjunction only from a substantivalist conception of time; for the relationalist, time itself is nothing beyond the structure of relations between events (e.g., Barbour 2001). Second, in the framework of general relativity, the two conceptually different terms of the disjunction are closely related: the Einstein field equations express the lawlike connection between the geometrical properties of spacetime and the matter–energy distribution on it.

Finally, it is worth saying a few words about objectivity. Authors who consider that the only means of defining the arrow of time is from non–t-invariant fundamental laws, under the assumption that physics does not include such laws, conclude that the difference between the two directions of time is not objective (e.g., Weyl 1952; Nozick 2001) or is superfluous (e.g., North 2009; Belot 2013). Since the proposed global nonentropic approach does not rely on the non–t-invariance of physical laws, the fact that the arrow of time is contingent and can be conceived of as emerging from certain topological properties of spacetime does not undermine its objectivity: this fact does not imply that the difference between the two directions of time is a mere illusion. Although not an inevitable consequence of the laws of physics, the global arrow of time is a result of the objective properties of our universe, that is, properties that do not depend on our means of empirical access to reality or on our particular perspective as observers.

11.11 Concluding Remarks

After unraveling the concepts of time-reversal invariance, irreversibility, and arrow of time, I have introduced a global and nonentropic approach that, following Earman's "Time Direction Heresy," defines the arrow of time as the time-asymmetry of a time-orientable spacetime with cosmic time. I have also shown how, under the dominant energy condition, the global arrow is locally expressed as a local energy flow pointing in the same direction of time all over the spacetime. It is such an energy flow that breaks the symmetry of all the pairs of time-symmetric twins resulting from the t-invariant fundamental local laws of physics.

In principle, the present approach is independent of the conception of time and spacetime that is adopted (for a detailed analysis of the substantivalism versus relationalism debate, see Earman 1989). However, the relevance of the position taken on the substantival or relational nature of spacetime to the understanding of t-invariance has been recently pointed out (López 2021a). Therefore, it is to be expected that the substantivalism–relationalism debate will also bring to light new questions about the nonentropic global arrow of time, especially in view of the still ongoing discussion about the nature of the spacetime of general relativity.

Acknowledgments

I am grateful to the participants of the workshop *The Arrow of Time: From Local Systems to the Whole Universe*, Buenos Aires, February 2023, for their enlightening contributions. This article was made possible through the support of Grant 61785 from the John Templeton Foundation and Grant PICT-2018–04519 from the National Agency of Scientific and Technological Promotion of Argentina.

References

Aiello, M., Castagnino, M., and Lombardi, O. (2008). "The arrow of time: From universe time-asymmetry to local irreversible processes." *Foundations of Physics*, 38: 257–292.

Albert, D. (2000). *Time and Chance*. Cambridge, MA: Harvard University Press.

Barbour, J. (2001). *The End of Time: The Next Revolution in Physics*. Oxford: Oxford University Press.

Belot, G. (2013). "Symmetry and equivalence." In R. Batterman (ed.), *The Oxford Handbook of Philosophy of Physics*. Oxford: Oxford University Press, pp. 318–339.

Bohm, A. and Gadella, M. (1989). *Dirac Kets, Gamow Vectors, and Gel'fand Triplets*. Berlin: Springer.

Bohm, A., Antoniou, I., and Kielanowski, P. (1994a). "The preparation/registration arrow of time in quantum mechanics." *Physics Letters A*, 189: 442–448.

Bohm, A., Antoniou, I., and Kielanowski, P. (1994b). "A quantum mechanical arrow of time and the semigroup time evolution of Gamow vectors." *Journal of Mathematical Physics*, 36: 2593–2604.

Brush, S. (1976). *The Kind of Motion We Call Heat*. Amsterdam: North Holland.

Callender, C. (2000). "Is time 'handed' in a quantum world?" *Proceedings of the Aristotelian Society*, 100: 247–269.

Callender, C. (2004). "There is no puzzle about the low-entropy past." In C. Hitchcock (ed.), *Contemporary Debates in Philosophy of Science*. London: Blackwell, pp. 240–255.

Castagnino, M., Gadella, M., and Lombardi, O. (2005). "Time's arrow and irreversibility in time-asymmetric quantum mechanics." *International Studies in the Philosophy of Science*, 19: 223–243.

Castagnino, M., Gadella, M., and Lombardi, O. (2006). "Time-reversal, irreversibility and arrow of time in quantum mechanics." *Foundations of Physics*, 36: 407–426.

Castagnino, M., Lara, L., and Lombardi, O. (2003a). "The cosmological origin of time-asymmetry." *Classical and Quantum Gravity*, 20: 369–391.

Castagnino, M., Lara, L., and Lombardi, O. (2003b). "The direction of time: From the global arrow to the local arrow." *International Journal of Theoretical Physics*, 42: 2487–2504.

Castagnino, M. and Lombardi, O. (2004). "The generic nature of the global and non-entropic arrow of time and the dual role of the energy-momentum tensor." *Journal of Physics A* (*Mathematical and General*), 37: 4445–4463.

Castagnino, M. and Lombardi, O. (2009). "The global non-entropic arrow of time: From global geometrical asymmetry to local energy flow." *Synthese*, 169: 1–25.

Castagnino, M., Lombardi, O., and Lara, L. (2003). "The global arrow of time as a geometrical property of the universe." *Foundations of Physics*, 33: 877–912.

Costa de Beauregard, O. (1980). "CPT invariance and interpretation of quantum mechanics." *Foundations of Physics*, 10: 513–530.

Dasgupta, S. (2016). "Symmetry as an epistemic notion (twice over)." *The British Journal for Philosophy of Science*, 67: 837–878.

Davies, P. C. W. (1974). *The Physics of Time Asymmetry*. Berkeley: University of California Press.

Davies, P. C. W. (1994). "Stirring up trouble." In J. J. Halliwell, J. Pérez-Mercader, and W. H. Zurek (eds.), *Physical Origins of Time Asymmetry*, Cambridge: Cambridge University Press, pp. 119–130.

Dürr, D. and Teufel, S. (2009). *Bohmian Mechanics. The Physics and Mathematics of Quantum Theory*. Dordrecht: Springer.

Earman, J. (1974). "An attempt to add a little direction to 'The problem of the direction of time'." *Philosophy of Science*, 41: 15–47.

Earman, J. (1986). "The problem of irreversibility." *PSA: Proceedings of the Biennial Meeting for the Philosophy of Science Association*, 2: 226–233.

Earman, J. (1989). *World Enough and Space-Time: Absolute Versus Relational Theories of Space and Time*. Cambridge, MA: MIT Bradford.

Earman, J. (2002). "What time reversal invariance is and why it matters." *International Studies in the Philosophy of Science*, 16: 245–264.

Earman, J. (2006). "The 'Past Hypothesis': Not even false." *Studies in History and Philosophy of Modern Physics*, 37: 399–430.

Earman, J. (2008). "Pruning some branches from 'branching spacetimes'." In D. Dieks (ed.), *The Ontology of Spacetime II*. Amsterdam: Elsevier, pp. 186–205.

Earman, J. and Norton, J. (1987). "What price spacetime substantivalism? The hole story." *The British Journal for the Philosophy of Science*, 38: 515–525.

Earman, J., Wüthrich, C., and Manchak, J. B. (2021). "Time machines." In E. N. Zalta (ed.), *The Stanford Encyclopedia of Philosophy* (Winter 2021 Edition), https://plato.stanford.edu/archives/win2021/entries/time-machine/.

Ehrenfest, P. and Ehrenfest, T. (1959). *The Conceptual Foundations of the Statistical Approach in Mechanics*. Ithaca, NY: Cornell University Press (original 1912).

Emery, N., Markosian, N., and Sullivan, M. (2020). "Time." In E. N. Zalta (ed.), *The Stanford Encyclopedia of Philosophy* (Winter 2020 Edition), https://plato.stanford.edu/archives/win2020/entries/time/.

Feynman, R. P., Leighton, R. B., and Sands, M. (1964). *The Feynman Lectures on Physics*. New York: Addison-Wesley.

Gasiorowicz, S. (1966). *Elementary Particle Physics*. New York: John Wiley and Sons.

Grünbaum, A. (1963). *Philosophical Problems of Space and Time*. New York: Alfred A. Knopf.

Gryb, S. and Thébault, K. (2016). "Time remains." *The British Journal for the Philosophy of Science*, 67: 663–705.

Haag, R. (1996). *Local Quantum Physics: Fields, Particles, Algebras*. Berlin: Springer.

Halliwell, J. J. (1994). "Quantum cosmology and time asymmetry." In J. J. Halliwell, J. Perez-Mercader, and W. H. Zurek (eds.), *Physical Origins of Time Asymmetry*, Cambridge: Cambridge University Press, pp. 369–389.

Hawking, S. W. and Ellis, G. F. R. (1973). *The Large Scale Structure of Space-Time*. Cambridge: Cambridge University Press.

Horwich, P. (1987). *Asymmetries in Time*. Cambridge, MA: MIT Press.

Lineweaver, C. H. (2014). "The entropy of the universe and the Maximum Entropy Production Principle." In R. Dewar, C. Lineweaver, R. Niven, and K. Regenauer-Lieb (eds.), *Beyond the Second Law: Entropy Production and Non-equilibrium Systems*. Berlin: Springer, pp. 415–427.

Loewer, B. (2012a). "The emergence of time's arrows and special science laws from physics." *Interface Focus*, **2**: 13–19.

Loewer, B. (2012b). "Two accounts of laws and time." *Philosophical Studies*, 160: 115–137.

Lombardi, O. and López, C. (2023). "Entropy and time." In W. Freeden and M. Z. Nashed (eds.), *Frontiers in Entropy Across the Disciplines*. Singapore: World Scientific, pp. 19–53.

López, C. (2019). "Roads to the past: How to go and not to go backward in time in quantum theories." *European Journal for Philosophy of Science*, 9: 1–30.

López, C. (2021a). "The physics and metaphysics of time reversal in standard quantum mechanics." *Synthese*, 199: 14267–14292.

López, C. (2021b). "Three facets of time-reversal symmetry." *European Journal for Philosophy of Science*, 11: 1–19.

López, C. (2023). "Should physical symmetries guide metaphysics? Two reasons why they should maybe not." *European Journal for Philosophy of Science*, 13: #23.

López, C. and Lombardi, O. (2019). "Space-time symmetries in quantum mechanics." In O. Lombardi, S. Fortin, C. López, and F. Holik (eds.), *Quantum Worlds. Perspectives on the Ontology of Quantum Mechanics*. Cambridge: Cambridge University Press, pp. 269–293.

Matthews, G. (1979). "Time's arrow and the structure of spacetime." *Philosophy of Science*, 46: 82–97.

Maudlin, T. (2002). "Remarks on the passing of time." *Proceedings of the Aristotelian Society*, 102: 237–252.

Maudlin, T. (2007). "On the passing of time." Chapter 4 in *The Metaphysics within Physics*. Oxford: Oxford University Press.

Misner, C. W., Thorne, K. S., and Wheeler, J. A. (1973). *Gravitation*. San Francisco: Freeman and Co.

North, J. (2008). "Two views on time reversal." *Philosophy of Science*, 75: 201–223.

North, J. (2009). "The 'structure' of physics: A case study." *Journal of Philosophy*, 106: 57–88.

Norton, J. (2016). "The impossible process: Thermodynamics reversibility." *Studies in History and Philosophy of Modern Physics*, 55: 43–61.

Nozick, R. (2001). *Invariances: The Structure of the Objective World*. Cambridge, MA: Harvard University Press.

Penrose, R. (1989). *The Emperor's New Mind*. Oxford: Oxford University Press.

Penrose, R. (2004). *The Road to Reality: A Complete Guide to the Laws of the Universe*. New York: Vintage Books.

Popper, K. (1956a). *Nature*, 177: 538; (1956b). *Nature*, 178: 382; (1957). *Nature*, 179: 1297; (1958). *Nature*, 181: 402; (1965). *Nature*, 207: 233; (1967a). *Nature*, 213: 320; (1967b). *Nature*, 214: 322.

Price, H. (1996). *Time's Arrow and Archimedes' Point: New Directions for the Physics of Time*. Oxford: Oxford University Press.

Price, H. (2004). "On the origins of the arrow of time: Why there is still a puzzle about the low-entropy past." In C. Hitchcock (ed.), *Contemporary Debates in Philosophy of Science*. London: Blackwell, pp. 219–239.

Racah, G. (1937). "Sulla simmetria tra particelle e antiparticelle." *Il Nuovo Cimento*, 14: 322–326.

Rechel, E. (1947). "The reversible process in thermodynamics." *Journal of Chemical Education*, 24: 298–301.

Reichenbach, H. (1956). *The Direction of Time*. Berkeley: University of California Press.

Roberts, B. (2017). "Three myths about time reversal invariance." *Philosophy of Science*, 84: 315–334.

Roberts, B. (2022). *Reversing the Arrow of Time*. Cambridge: Cambridge University Press.

Sachs, R. (1987). *The Physics of Time Reversal*. London: University of Chicago Press.

Sakurai, J. J. (1994). *Modern Quantum Mechanics*. Revised Edition. Reading, MA: Addison- Wesley.
Savitt, S. F. (1996). "The direction of time." *The British Journal for the Philosophy of Science*, 47: 347–370.
Schutz, B. F. (1980). *Geometrical Methods of Mathematical Physics*. Cambridge: Cambridge University Press.
Sklar, L. (1974). *Space, Time, and Spacetime*. Berkeley: University of California Press.
Sklar, L. (1993). *Physics and Chance*. Cambridge: Cambridge University Press.
Torretti, R. (1983). *Relativity and Geometry*. Oxford: Pergamon Press.
Uffink, J. (2001). "Bluff your way in the second law of thermodynamics." *Studies in History and Philosophy of Science*, 32: 305–394.
Visser, M. (1995). *Lorentzian Wormholes. From Einstein to Hawking*. New York: Springer.
Wallace, D. (2017). "The nature of the past hypothesis." In K. Chamcham, J. Silk, J. Barrow, and S. Saunders (eds.), *The Philosophy of Cosmology*. Cambridge: Cambridge University Press, pp. 486–499.
Watanabe, S. (1955). "Symmetry of physical laws. Part 3: Prediction and retrodiction." *Reviews of Modern Physics*, 27: 179–186.
Weyl, H. (1952). *Symmetry*. Princeton, NJ: Princeton University Press.

12

The Seeds of Cosmic Structure, the Black Hole Information Puzzle, and the Entropic Arrow of Time

DANIEL SUDARSKY AND ELÍAS OKÓN

12.1 Introduction

The study of the interface between quantum theory and gravitation often takes the form of the search for a theory of quantum gravity; and although great strides have been made along various research lines adopted for that search, it is fair to say that, at this time, there is no fully satisfactory version of such a theory. Generally, the search for quantum gravity is viewed as independent of the conceptual problems afflicting quantum theory, namely, the lack of ontological clarity and the so-called measurement problem. There are, though, some exceptions to that general attitude. Several works exploring the interface between quantum theory and gravitation, in which conceptual quantum aspects are central to the discussion, have appeared in the literature over the years. Moreover, a few meetings devoted to such questions have taken place, specially over the last few years. In fact, the event that led to this book is, in a sense, one such example – even though, sometimes, questions regarding the nature of time are viewed as separated from the general relativistic context and, thus, from its intricate connection with gravitation. This is understandable because, from the philosophical point of view, the nature of time involves a large number of issues beyond those that arise in physics (issues which we do not discuss here).

In this work, we describe a research project that combines a particular strategy for the exploration of the gravity-quantum interface, with a specific point of view regarding foundational aspects of quantum theory. Moreover, we explain how the effort to adapt one to the other has led to a peculiar outlook regarding the entropic arrow of time. Our project takes general relativity (GR) as the (tentative) preferred description of gravitation and spacetime structure, while adopting an agnostic attitude toward the various existing approaches toward the construction of a true quantum theory of gravity. We, of course, acknowledge that, in all likelihood, a full theory of quantum gravity will be found and required but, at the same time, recognizing that we might be far from that point. Moreover, given that GR works rather well in the realms it has been tested, we work under the assumption that any reasonable theory of quantum gravity must have GR as a suitable approximation, under the appropriate circumstances. Regarding our treatment of matter, we take it as evident that it requires a quantum description and that the classical realm must be viewed only as an approximation – which we take, for the sake of our discussion, to be under relatively good control. In that sense, we take quantum field theory and, in particular, its curved space-time version, to provide our current best characterization of matter in general, and the standard model of particle physics as its concrete realization in our world.

These considerations drive us to adopt semiclassical gravity as the basic framework for our analysis. We cannot, then, ignore the objections against such a framework, particularly those raised by the experiment and analysis in Page and Geilker (1981). However, it has been pointed out that the issue is closely connected with the conceptual difficulties in quantum theory (Carlip 2008, Huggett and Callender 2001, Mattingly 2005, 2006), which takes us to confront such issues head on. For this, we adopt the approach based on spontaneous collapse theories. It is well-known, however, that such frameworks display a lack of conservation of energy when a collapse is involved, which, as observed in Page and Geilker (1981), seems inconsistent with semiclassical gravity. The issue is substantially mitigated by taking semiclassical gravity as approximate, as we do, and by observing that, as discussed in Maudlin et al. (2020), the problem of nonconservation seems to appear, in one guise or another, in all open avenues to deal with the measurement problem. As a result, it seems that our choice is not forcing us into a problem that could otherwise disappear.

Throughout, we try to be as conservative as possible, giving priority to well-established theories when possible, aiming to find, often in connection with concrete examples, aspects of the various theoretical frameworks that give raise to tensions to be explored. We believe that, given the sheer complexity of the issues that arise in discussing a theory of spacetime itself, and the intrinsic difficulties of trying to discuss physics without a spacetime framework – together with the tendency to revert to our intuitions when facing the peculiar nature of the conceptual conundrums characteristic of quantum physics – makes essential the adoption of relatively clear postures as a means to prevent confusion and distraction from interfering with the sought-out progress.

Our chapter is organized as follows. In Section 12.2, we describe in more detail our general approach, including a formalism that has been under development for some years devoted to the application of collapse models to semiclassical settings. Then, we present two specific scenarios to which we have applied such a formalism: the inflationary account for the emergence of the seeds of cosmic structure, in

Section 12.3, and the black hole information puzzle, in Section 12.4. These explorations have provided us with interesting lessons which, in Section 12.5, we employ to deal with the problem that concerns us most in this chapter, the nature and origin of the entropic arrow of time. We show, in particular, how this approach offers an attractive way to implement some of the conjectures on the subject introduced by Roger Penrose in Penrose (1979). We close, in Section 12.6, with a discussion.

12.2 The Program

It seems quite difficult to propose a physical theory that does not contain some basic notions intrinsically tied to spacetime. Proponents of theoretical frameworks, where spacetime is supposed to emerge, must face this difficulty. This is so even when relying on classical notions, such as points, areas, volumes, curvature, and so on; but if the theory is to be formulated in a purely quantum mechanical language from the start, the emergence analysis becomes even more complex, as one can expect a rigorous treatment to require adopting a clear position regarding the conceptual difficulties of quantum theory.

In our opinion, a failure to acknowledge these issues often leads to confusion, as avoiding (at least a temporary) commitment to a definite position regarding the difficult interpretative issues of quantum theory, removes self-consistency constrains or allows for specific words to change their meaning in the middle of the discussion. One rather common example is provided by the word "fluctuation," which is used to reflect various notions that are, in principle, quite distinct. For instance, on one hand, there are *quantum uncertainties*, which, according to the standard interpretation, represent levels of indeterminacy or lack of actual values. On the other, there are *stochastic fluctuations*, which are associated with either ensembles of similar systems (say, the grades of students in a class), or pertain to a single extended system and the spatiotemporal variation of a local quantity in it (e.g., the water level in a lake). This kind of confusion often occurs in connection with another prevalent source of misunderstandings, namely, the failure to distinguish between *proper* and *improper* mixtures.

Many of these problems can often be traced to the widespread propensity among physicists to ignore the elephant in the room: the *measurement problem* in quantum theory. In practice, the issue is reflected in the fact that the theory contains two rules determining the dynamics of the quantum state: the deterministic, time-reversible and unitary evolutionary rule, provided by the Schrödinger equation, and the stochastic, irreversible and nonunitary reduction rule associated with a measurement process. The problem, of course, is not so much that there are two rules, but that there is no unambiguous recipe specifying which one applies in each circumstance – that is, precisely determining what kind of process constitutes a measurement. The issue has been faced by the community by adopting a variety of postures and strategies. Valuable guidance regarding the available options can be extracted from the result in Maudlin (1995), showing the following three premises to be, in conjunction, inconsistent:

1. The physical description given by the quantum state is complete.
2. Quantum evolution is always unitary.
3. Measurements always yield definite results.

The negation of 1 leads to hidden-variable theories, that of 2 to spontaneous collapse models and that of 3 to Everettian interpretations. It is worth pointing out that, contrary to widespread belief, the measurement problem is not solved by decoherence (Okon and Sudarsky 2016a). Next, we must discuss the way in which we approach the exploration of the quantum gravity regime. We refer to it as a *top-down* approach, in contrast with the more traditional *bottom-up* one, which starts by assuming one has a fundamental theory of quantum gravity at hand (say, string theory, loop quantum gravity, or causal sets) and then works in attempts to apply such theory to deal with concrete problems, usually in regimes deemed of interest for representing concrete aspects of the "world out there." The regimes more often studied in these specific applications involve aspects of cosmology, black hole physics, and so on. The top-down approach involves adopting an agnostic posture regarding the nature of the fundamental theory of gravitation, and thus spacetime, and focuses instead in pushing the application of our existing, well-tested theories to situations that seem to lie just beyond their standard domain of application. The idea is that, in so doing, one might be forced to introduce some relatively mild modifications which, if successful, could provide valuable clues about the nature of the more fundamental theory.

More specifically, the idea is to use GR and quantum field theory in curved spacetimes (i.e., semiclassical gravity) to address questions commonly expected to lie beyond their applicability. To do so, we will introduce minimal modifications to the general framework, as required by the specific problem at hand, all this while taking a definite position regarding how to deal with the measurement problem. Regarding the latter, we will focus on spontaneous collapse theories, on which there is a vast amount of work in the last 40 years (see Bassi and Ghirardi 2003 and Bassi et al. 2013 for overviews). The basic idea is to construct a single dynamical equation to encompass both the unitary evolution and the collapse process.

The first viable theory of this sort is known as GRW, Ghirardi et al. (1986), and involves discrete spontaneous collapses of the wave function, separated by continuous periods in which the system is governed by the Schrödinger equation. Such a model introduces, on top of the Schrödinger equation, spontaneous reductions with rate λ, driving the state toward eigenstates of the position operator (smeared on scale r_c). The changes introduced are supposed to be very small when only a few particles are involved, but to become large when something like 10^{23} particles are entangled (and delocalized). For this, λ is chosen to be small enough, not to conflict with tests of quantum mechanics in the domain of subatomic physics, but big enough to result in rapid localization of "macroscopic objects." Ghirardi et. al. (1986) suggested $\lambda \sim 10^{-16}\text{sec}^{-1}$ and $r_c \sim 10^{-5}\text{cm}$. These theories address the measurement problem successfully and are empirically viable (at least in nonrelativistic regimes). They have been subject to experimental tests and some attractive specific versions have recently been ruled out.

Another interesting collapse model is Continuous Spontaneous Localization (CSL), Pearle (1989), which replaces the discontinuous jumps by a continuous stochastic evolution. The theory is defined by a modified Schrödinger equation, whose solution is

$$|\psi, t\rangle_w = \hat{\mathcal{T}} e^{\int_0^t dt' \left[i\hat{H} + \frac{1}{4\lambda}[w(t') - 2\lambda\hat{A}]^2\right]}|\psi, t, \tag{12.1}$$

where $\hat{\mathcal{T}}$ is the time-ordering operator, $\hat{A}$ the so-called collapse operator and $w(t)$ a real white-noise stochastic function whose probability is given by

$$PDw(t) \equiv {}_w\langle\psi, t|\psi, t\rangle_w \prod_{t_i=0}^{t} \frac{dw(t_i)}{\sqrt{2\pi\lambda/dt}}. \tag{12.2}$$

As in GRW, the collapse operator is taken to be related to a smeared position operator. Thus, in the context of many particle quantum mechanics (as a step toward quantum field theory), one would use the smeared mass density operator, namely

$$\hat{M}(x) = \sum_i m_i \int d^3y \hat{\Psi}_i(y) e^{-(\|x-y\|/r_c)}, \tag{12.3}$$

where the sum is over particle species. Moreover, as argued in Pearle and Squires (1994), there are good reasons to believe that λ, rather than a universal constant, should depend on the particle's mass.

In order to complete the theory, one must specify its ontology, that is, make explicit the connection between the formalism and what the theory says that exists "out there in the world." One of the most favored ontologies, adopted when working with nonrelativistic spontaneous collapse theories, is the so-called mass-density ontology, which takes the corresponding nonrelativistic version of the expectation value of $\hat{M}(x)$ as providing the required connection between the formalism and the world.

Next, we expand on other aspects of our treatment. Spacetime will be provisionally described in classical terms – something that can be expected to be a good approximation in regimes where curvatures are small (i.e., $\ll(1/l_{Planck})^2$) and quantum uncertainties are "not too large."[1] Matter, on the other hand, is to be treated quantum mechanically (more specifically, when possible, we will use quantum field theory in curved spacetimes). These choices quite naturally lead to the semiclassical Einstein equation, in which classical spacetime is sourced by the (renormalized) expectation value of the energy–momentum tensor for the relevant state of the matter fields. Since the Einstein tensor is identically divergence-free, the nonconservation of the expectation value of the energy–momentum tensor within collapse theories would seem to lead to inconsistencies.[2] However, the fact that we do not take these equations to be fundamental allows us to move forward.

As noted from the start, we regard the theoretical framework we are working with not as fundamental, as would be the case if one adopted a bottom-up approach, but as an approximate description with limited applicability. Thus, it is natural to take the view that, during collapses, the semiclassical Einstein equations are not valid. This would be analogous to studying, say, the Navier–Stokes equations for describing a fluid, but having in mind that such equations are not fundamental, but merely effective. In that case, one should not find it surprising that there are situations in which those equations fail to hold. Consider, for instance, a wave breaking on a shore, generating foam, and other phenomena whose explanation would have to rely on the molecular dynamics underlying the very nature of the fluid. Clearly these effects are not something that one could expect the Navier–Stokes equations to account for. In an analogous way, we must regard semiclassical gravity as an effective description and expect that there will be situations in which the description will fail and where, presumably, a full theory of quantum gravity would be required to account for the phenomena. One such situation would correspond to what, at the effective level, we describe as a spontaneous collapse of the quantum state of the matter fields.

At the formal level, we rely on the notion of *Semiclassical Self-consistent Configuration* (SSC), defined as follows. The set $\{g_{\mu\nu}(x), \hat{\varphi}(x), \hat{\pi}(x), \mathcal{H}, |\xi\rangle \in \mathcal{H}\}$ represents a SSC if and only if $\hat{\varphi}(x)$, $\hat{\pi}(x)$ and $\mathcal{H}$ corresponds to a quantum field theory over a spacetime with metric $g_{\mu\nu}(x)$ and, moreover, the state $|\xi\rangle$ in $\mathcal{H}$ is such that $G_{\mu\nu}[g(x)] = 8\pi G\langle\xi|\hat{T}_{\mu\nu}[g(x), \hat{\varphi}(x), \hat{\pi}(x)]|\xi\rangle$.

[1] The question of when such fluctuations are too large is, in itself, a rather nontrivial question.

[2] Moreover, as mentioned above, in Maudlin et al. (2020) it is shown that all reasonable approaches to the measurement problem seem to lead to violations of $\nabla^a T_{ab} = 0$, for whatever object one tries to assign the role of $T_{ab}(x)$ within such an approach. That means that, for any of the available paths to deal with the measurement problem, this issue would arise.

Regarding collapse, we take them to be not simple jumps from state to state within an SSC, but transitions from one SSC to another SSC. Thus, a spontaneous collapse on the quantum state of the matter field will be accompanied by a sudden change in the spacetime metric. Establishing exactly how this is supposed to occur implies providing matching conditions for spacetime and rules for determining – assuming a standard type of collapse theory, involving jumps within a given Hilbert space – corresponding states in the new Hilbert space. It must be noted that, technically, all this is highly nontrivial (a preliminary proposal in this direction was outlined in Diez-Tejedor and Sudarsky 2012) – and things only get more complicated when considering continuous collapses, such as in CSL, a situation which, we can only hope, might be addressed by taking an appropriate continuous limit, starting from a version designed for theories involving discrete collapses.

In any case, in what follows we will ignore some of the complications required to do this precisely and work with effective methods that reflect the approach we have discussed. In the next couple of sections, we show how this is can be done in some concrete applications, which are in themselves connected to nontrivial difficulties in our current physical understanding.

12.3 The Seeds of Cosmic Structure

Contemporary cosmology includes inflation as one of its most attractive components. One of the biggest achievements of inflation is claimed to be the account for the emergence of the seeds of cosmic structure with the correct spectrum, as a result of "quantum fluctuations." However, at the conceptual level, the standard account is not truly satisfactory. Let us explain why.

The starting point of the analysis is a cosmological spacetime (in a particular gauge)

$$ds^2 = a^2(\eta)\{-(1+2\Psi)d\eta^2 + [(1-2\Psi)\mathrm{d}_{ij} + h_{ij}]dx^i dx^j\} \qquad (12.4)$$

and matter represented by an inflaton field, written as $\varphi = \varphi_0(\eta) + \delta\phi$, with $\delta\phi, \Psi, h_{ij}$ small perturbations containing the spatial dependencies. The background (a, ϕ_0) is treated classically and assumed to be dominated by the inflaton potential slow roll regime, so ϕ_0 changes slowly and the scale factor is approximately given by $a(\eta) = -1/(\eta H_I)$. We set $a = 1$ at the present cosmological time and assume that the inflationary regime corresponds to η in $(-\mathcal{T}, \eta_0)$ with $\eta_0 < 0$. The perturbations are treated quantum mechanically and assumed to be characterized by the vacuum state (essentially the Bunch–Davies vacuum) $|0\rangle$. The idea is that inflation dilutes all preexisting features and drives all space-dependent fields toward their vacuum states.

These quantum fields are also said to be characterized by their "quantum fluctuations." However, the sort of confusion discussed in the introduction arises as, in the standard treatment, those features are wrongly identified as the primordial inhomogeneities, which eventually evolved into all the structure in our universe: galaxies, stars planets, and so on. However, note that, according to the inflationary picture, the universe was *strictly homogeneous and isotropic*, both in the sectors described classically and quantum mechanically, and the unitary dynamics cannot change that fact. How is it, then, that this fully homogeneous and isotropic scenario leads, at later times, to the formation of galaxies, stars, planets, and life?

12.3.1 Dealing with the Problem

In our picture, spacetime is treated classically, while the inflaton field is treated using quantum field theory in curved spacetime. Thus, quantum mechanically, the zero mode of the field, $\hat{\phi}_0$, is taken to be in a highly excited (and sharply peaked) state (see Diez-Tejedor and Sudarsky (2012)), while the space-dependent modes are in the vacuum state. The quantum state of the inflaton and the spacetime metric are then assumed to satisfy the semiclassical Einstein equations:

$$G_{\mu\upsilon} = 8\pi G\langle\xi|\hat{T}_{\mu\upsilon}|\xi\rangle. \tag{12.5}$$

Under these conditions, one essentially obtains the standard behavior for the background, $a(\eta) = \frac{-1}{\eta H_I}$, and slow roll for $\langle\hat{\phi}_0\rangle$ in $(-\mathcal{T}, \eta_0), \eta_0 < 0$.

We concentrate, next, on the $\vec{k} \neq 0$ modes. For their study, we rely on an effective procedure we have checked to give equivalent results as the SSC formalism. In conformity with standard assumptions, we take, for the early stages of inflation $\eta = -\mathcal{T}$, the state to correspond to $|0\rangle$ and for the operators $\delta\hat{\phi}_k, \hat{\pi}_k$ to be characterized by Gaussian wave functions centered on 0, with uncertainties $\Delta\delta\phi_k$ and $\Delta\pi_k$. We also assume $\Psi = h_{ij} = 0$.

Next enters the collapse, which modifies the quantum state and the expectation values. We first assume that the collapse is described by a collapse theory adapted to the situation, in which collapses occur discretely and mode by mode. Our universe then corresponds to one specific realization of the stochastic evolution. We focus on the scalar metric perturbations, $\Psi(\eta, x)$, which characterize the cosmic microwave background (CMB) temperature fluctuations (and seeds of structure). The Fourier decomposition of the semiclassical Einstein equations yields

$$-k^2\Psi(\eta, \vec{k}) = \frac{4\pi G\phi_0'(\eta)}{a}\langle\hat{\pi}(\vec{k}, \eta)\rangle \tag{12.6}$$

With reasonable choices for the details of the collapse theory, agreement with observations can be achieved (Cañate et al. 2013).

In a CSL version, one must select the collapse operator. Obvious choices are the field operator or the momentum conjugate operator, with $\lambda = \tilde{\lambda} k^{\pm 1}$ fixed by dimensional considerations, but $\left(-\nabla^2\right)^{-1/4}\hat{\pi}(\vec{x})$ or $\left(\nabla^2\right)^{1/4}\hat{\phi}(\vec{x})$ can also be chosen. How to decide? We do not really know, but the fact that, in the nonrelativistic limit, the mass density seems to offer a good possibility for the collapse operator suggests exploring how these operators are related to operators constructed out of $\hat{T}_{ab}(x)$. Moreover, as we will discuss later, there are good reasons to introduce curvature-dependent coefficients.

In any event, once the choice of a collapse operator is made, the resulting prediction for the power spectrum for density inhomogeneities is

$$P_S(k) \sim \left(1/k^3\right)(1/\epsilon)\left(V/M_{Pl}^4\right)\tilde{\lambda}T. \tag{12.7}$$

Taking the GUT scale for the inflation potential and standard values for the slow-roll leads to agreement with observation for $\tilde{\lambda} \sim 10^{-5} MpC^{-1} \approx 10^{-19} sec^{-1}$. We note that the order of magnitude obtained is not very different from the one proposed in the context of GRW.

Treatments with a similar spirit can be found in Martin et al. (2012) and Das et al. (2013). We must mention, though, that the recent analysis in Martin and Vennin (2020) leads to the conclusion that the details of the CMB can be used to rule out some simple extrapolations of CSL theory, in which the collapse operator is taken as the matter density perturbation. In that case, a prediction density perturbation spectrum incompatible with observations is encountered. However, as discussed in Bengochea et al. (2020), it is not at all clear how to extrapolate the version of CSL that works in the low-energy, many-particle regime (i.e., one based on a smeared mass density collapse operator) to the semiclassical realm under consideration.

12.4 The Black Hole Information Puzzle

The discovery by Stephen Hawking that, as a result of quantum field theory effects, black holes emit thermal radiation, has had enormous repercussions in our understanding, not only of black hole physics, but also of broader aspects of the nature of our world. In particular, it is expected that, after settling into quasi-stationary configurations, black holes should start losing mass and, eventually, essentially disappear, leaving behind only thermal radiation (although there is the possibility that they could leave a small remnant, which we will ignore in this discussion). All this, however, seems to lead to a problem since quantum theory requires unitary relations between initial and final states.

One must note, though, that this demand only applies to states on Cauchy hypersurfaces. Still, people often modify the demand, requiring unitary relation between states in $\mathscr{I}^-$ and $\mathscr{I}^+$, leading to a puzzle. Such demand seems to us very hard to account for. The point is that a true paradox only arises when one assumes (among other things, as discussed in detail in Okon and Sudarsky 2018), that quantum gravity cures the singularity (see Unruh and Wald 2017, or Maudlin 2017, for alternative positions on this issue). Only in that case, that lack of unitarity displayed by the Hawking evaporation of the black hole (assuming no remnants) would seem to lead to a conflict with quantum mechanics. However, even that argument ignores something important. Even standard textbook quantum mechanics involves departures from unitarity in connection to measurements, and it is only under the assumption of purely unitary evolution that a conflict could arise.

In what follows we explore the issue from the perspective of spontaneous collapse theories, in which departures from unitary evolution are always present. We present a picture where the two kinds of departures from unitarity are unified, as first proposed in Okon and Sudarsky (2014). The proposal has been studied explicitly in the 2-D CGHS Model and schematically in 4-D, Modak et al. (2015b,a) (for further discussion, see Perez and Sudarsky 2022).

12.4.1 The Objective Collapse Point of View

As noted, we will consider a quantum-field-theory-in-curved-spacetime treatment for the matter fields. We take the field ξ as corresponding to the matter that will undergo gravitational collapse and help form the black hole, while ψ represents the quantum field initially taken to be in the vacuum state, and on which we will focus for consideration of the issue of information loss. For the unitary evolution, we will be using the Heisenberg picture, in which the state remains fixed, but the field operators depend on time (and space), $\hat{\psi}(x), \hat{\xi}(x)$. The effects of the spontaneous collapse theory will be treated as those of an interaction and for those we will use the interaction picture.

Consider first the characterization of the *in* region, before the black hole forms. There, the initial state schematically can be written as

$$|\Psi_{in}\rangle = |0_{in}\rangle_{\psi} \otimes |matt\rangle_{\xi}, \tag{12.8}$$

where $|matt\rangle_{\xi}$ represents the matter undergoing gravitational collapse. One then describes the state of a quantum field ψ at late times, in terms of degrees of freedom *inside* and *outside* of the black hole. The initial vacuum state can then be written as

$$|0_{in}\rangle_{\psi} = \sum_{F\propto} C_{F\propto} |F_{\propto}\rangle^{ext} \otimes |F_{\propto}\rangle^{int}, \tag{12.9}$$

where a particle state $F_{\propto}$ consists of an arbitrary, but *finite*, number of particles (or individual mode excitations). Tracing over the interior degrees of freedom would lead to an improper thermal state, corresponding to the Hawking flux. The complete initial state can then be written schematically as

$$|\Psi_{in}\rangle = \sum_{F\propto} C_{F\propto} |F_{\propto}\rangle^{ext} \otimes |F_{\propto}\rangle^{int} \otimes |matt\rangle_{\xi}, \tag{12.10}$$

The idea now is to consider the evolution of the initial state, employing a modified theory involving spontaneous collapse. For concreteness, we will consider a CSL-type theory. To do so, we introduce a foliation parameterized by τ, corresponding to $W^2 = const$ in the inside (with W *as* the Weyl tensor) and continue it "almost arbitrary" outside. This makes the collapse parameter an effective function of τ which, in fact, diverges as the region where the classical singularity would have been is approached.

The CSL equations can be generalized to drive collapse into a state of a joint eigenbasis of a set of commuting operators $\hat{A}^I$. For each $\hat{A}^I$ there will be one $w^I(t)$. In that case, we have

$$|\psi, t\rangle_w = \hat{\mathcal{T}} e^{-\int_0^t dt' \left[i\hat{H} + \frac{1}{4\lambda}\sum_I [w(t') - 2\lambda \hat{A}^I]^2\right]} |\psi, 0\rangle. \tag{12.11}$$

We call $\hat{A}^I$ the *set of collapse operators.* Then we make a simplifying choice: collapses lead to a state of definite number of particles in the inside region. Moreover, since we are working in the interaction picture, $\hat{H} \rightarrow 0$ in the preceding equation.

Next, we assume that the CSL collapse mechanism is amplified by the curvature of space-time. That is, that the rate of collapse λ depends on the Weyl tensor as follows:

$$\lambda(W) = \lambda_0 \left[1 + \left(\frac{W^2}{\mu}\right)^{\gamma}\right], \tag{12.12}$$

where $W^2 = W^{abcd} W_{abcd}$, $\gamma > 1/2$ is a constant and μ provides an appropriate scale (R^2 in 2-D). Therefore, in the region of interest, we have $\lambda = \lambda(\tau)$. As a result, this evolution achieves, in the finite time to the singularity, what ordinary CSL achieves in infinite time, that is, to drive the state to one of the eigenstates of the collapse operators.

Then, the effect of CSL on the initial state is to drive it to one of the eigenstates of the joint number operators. Thus, at a hypersurface Σ very close to the singularity, the state will be

$$|\Psi_{in,\tau}\rangle = NC_{F\propto}|F_{\propto}\rangle^{ext} \otimes |F_{\propto}\rangle^{int} \otimes |matt\rangle_{\xi}. \tag{12.13}$$

Note that there is no summation and that the state is pure. However, due to the stochastic nature of the evolution, we do not know which one.

Let us now consider the role of quantum gravity. As discussed earlier, we will assume that it resolves the singularity and leads, on the other side, to a reasonable spacetime. Moreover, we will assume that it does not lead to large violations of the basic spacetime conservation laws. As a result of these assumptions, the effects of quantum gravity can be represented by the transformation

$$NC_{F\propto}|F_{\propto}\rangle^{ext} \otimes |F_{\propto}\rangle^{int} \otimes |matt\rangle_{\xi} \rightarrow NC_{F\propto}|F_{\propto}\rangle^{ext} \otimes |0^{post-sing}\rangle, \tag{12.14}$$

where $|0^{post-sing}\rangle$ represents a zero-energy momentum state, corresponding to a trivial region of spacetime (we are ignoring possible small remnants). We end up, then, with a pure quantum state but, as we explained earlier, we do not know which one.

Let us consider, then, an ensemble of systems prepared in the same initial state, described by the pure density matrix

$$\rho(\tau_0) = |\Psi_{in}\rangle\langle\Psi_{in}|. \tag{12.15}$$

Now, let us consider the CSL evolution of this density matrix, up to the hypersurface just before the singularity. Finally, let us make use of our assumptions about the effects of quantum gravity. The density matrix characterizing the ensemble after the would-be singularity is

$$\rho^{Final} = N^2 \sum_F e^{-\frac{E_F}{T}} |F\rangle^{out} \otimes |0^{post-sing}\rangle \langle F|^{out} \otimes \langle 0^{post-sing}| \tag{12.16}$$

$$= |0^{post-sing}\rangle\langle 0^{post-sing}| \otimes \rho^{out}_{Thermal}. \tag{12.17}$$

We see, then, that, in the end, the ensemble is described by a proper thermal state on future null infinity, followed by an empty region. Therefore, information was in fact lost as a result of the general quantum evolution, but there is nothing paradoxical at all about that fact.

12.5 The Entropic Arrow of Time

It has been argued extensively that, in order to account for the thermodynamic arrow of time in the context of time-reversal-invariant laws of physics, one needs to assume a very special initial state of the universe. In Penrose (1979), Penrose has proposed that such demand can take the form of an initial condition law restricting the initial state to be one with vanishing Weyl curvature (while allowing arbitrarily large R). His arguments are motivated by the observation that, at late times, the entropy of the universe seems to be connected with very large black holes, while there is no indication of large white holes in the early universe – despite the very large curvatures associated with that regime. That kind of "initial condition" law seems quite different from all other laws we have encountered before (except, perhaps, for the constraint equations that appear in theories involving gauge invariance). The considerations described earlier, involving the measurement problem, the origin of structure and black hole information puzzle suggest a different possibility.

Let us look again at the CSL evolution law:

$$|\psi, t\rangle_w = \hat{\mathcal{T}} e^{-\int_0^t dt' \left[i\hat{H} + \frac{1}{4\lambda}[w(t') - 2\lambda \hat{A}]^2\right]} |\psi, 0\rangle. \tag{12.18}$$

Ordinarily, one considers that, in generic circumstances, the evolution of large systems would be dominated by the Hamiltonian component (with the collapse sector becoming most relevant in situations where only a few degrees of freedom play a crucial role). However, if λ grows with $W^2 = W^{abcd}W_{abcd}$, as we suggested earlier, then, in regimes where W^2 is large, the large-scale evolution would be practically random. As a result, we have a picture in which we can have a universe evolving randomly for an indefinite "lapse of time," until some point where, just by chance, there is a jump into a state of an almost vanishing W^2. Thereafter, the evolution would be orderly and dominated by the Hamiltonian dynamics, with relatively small stochastic modifications (this idea was first put forward in Okon and Sudarsky 2016b).

In such a universe, structure such as galaxies, stars, planets and life would only appear in the orderly regime, and creatures studying it would look back and find their past characterized by a regime with extremely small Weyl curvature. This would be a kind of dynamical realization of Penrose's proposal, framed within a unified scheme, capable of accounting for various other open issues in physics. We find this quite attractive. Surprisingly, on the other hand, the emerging picture cannot but remind us of some more traditional accounts of creation, like those in Genesis or Hesiod's *Theogony*.

12.6 Discussion

As we have seen, the approach we have been following seems to be quite promising in addressing various issues that arise in the gravity–quantum interface. However, it is worth pointing out that, at the same time, it has given rise to certain concerns. One of the most important relates to the issue of violation of the conservation of energy (and more specifically the local conservation of energy-momentum). The issue was already noted early in Banks et al. (1984), but further analysis in Unruh and Wald (1995) indicated the initial worries to be exaggerated. In fact, various versions of dynamical collapse theories have been constructed to ensure compatibility with experimental bounds. Still, as we noted, there is a potential problem of inconsistency, which we have been working on with the use of the SSC and gluing formalism. Recent progress along these lines has been obtained in Juárez-Aubry et al. (2018), Juárez-Aubry and Sudarsky (2020), and Kay et al. (2023). On the other hand, the resulting formalism is rather complicated and dealing with more than a few collapse events often becomes completely impractical. It is important to point out, though, that recent work seems to suggest that "cumulative effects" of $\nabla^a \langle \hat{T}_{ab} \rangle \neq 0$ might account for the existence of dark energy (see Josset et al. 2017, Perez and Sudarsky 2019, Perez et al. 2021, and Perez and Sudarsky 2021).

Another key issue is the construction of a fully general-relativistic collapse theory. Some proposals in this direction have been put forward in Bedingham et al. (2016) and a similar analysis of black hole evaporation and information loss has been carried out using a relativistic version of dynamical collapse theory developed by Bedingham (2011).

There are, of course, other open issues that require resolution. For instance, one should ensure that the unwanted foliation dependence is removed from the framework. In particular, without an appropriately defined prescription, the essential quantity $\langle \hat{T}_{ab}(x) \rangle$ could depend on the choice of hypersurface passing through x simply because, in the context of a spontaneous collapse theory, different quantum states would be associated with different hypersurfaces and infinitely many such hypersurfaces go through the event x. It is natural to expect, though, that such problems would be eliminated by the construction of a fully relativistic collapse dynamics.

One more issue that must be dealt with in implementing the program we have outlined is the fact that the expectation value of the energy–momentum tensor is an object that requires renormalization. This is something well-understood in situations in which the spacetime geometry is given (often as a background) and requires, among other things, that the state in question be a Hadamard state. In our case, this needs to be viewed as a requirement on the collapse theory, Juárez-Aubry et al. (2018), but becomes much more complex in situations in which one needs to solve at the same time for the spacetime, as would be the case in, say, constructing the SSC to which the system collapses. Relevant progress in dealing with this question has recently been achieved and reported in Kay et al. (2023). Another open question relates to the nature of the universal collapse operator. As noted, in the nonrelativistic context, a natural choice seems to be the (smeared) mass density operator and it is not clear what would make a suitable general-relativistic substitute for that. One might guess that the role should be played by the energy–momentum tensor, but complications are likely to arise (besides those already mentioned) from the fact that the various components of that tensor do not commute.

On the other hand, it should be noted that the program has much more potential. There is the natural resolution of the lack of primordial gravity waves generated during inflation (which have been searched for in the form of B-modes, without any evidence for them so far, León et al. 2017, 2018). The program also seems to offer a simple path to address the problem of eternal inflation, Lechuga Soliz and Sudarsky (2023), and to account for the anomalous low power at low *l*, León and Sudarsky (2015, 2012). Moreover, the introduction of collapses could diffuse the problem of time in canonical quantum gravity, as discussed in Okon and Sudarsky (2014). There is a large amount of work ahead to continue the exploration of this line of research. Of course, our approach could, in the future, be shown to be unviable. However, as noted by Sir Francis Bacon, when considering the scientific enterprise in general, "truth emerges more readily from error than from confusion." We believe that ignoring the conceptual problems of quantum mechanics in the application of the theory to certain domains can be a serious source of confusion, particularly when referring to situations beyond the laboratory, as the ones considered here.

References

Banks, T., Susskind, L., and Peskin, M. E. (1984). "Difficulties for the evolution of pure states into mixed states." *Nuclear Physics*, 44 B (234): 125.

Bassi, A., and Ghirardi, G. (2003). "Dynamical reduction models." *Physics Reports*, 379:257–426.

Bassi, A., Lochan, K., Satin, S., Singh, T., and Ulbricht, H. (2013). "Models of wave-function collapse, underlying theories, and experimental tests." *Review of Modern Physics*, 85: 471.

Bedingham, D., Modak, S. K., and Sudarsky, D. (2016). "Relativistic collapse dynamics and black hole information loss." *Physical Review D*, 94(4):045009.

Bedingham, D. J. (2011). "Relativistic state reduction dynamics." *Foundations of Physics*, 41: 41.

Bengochea, G., León, G., Pearle, P., and Sudarsky, D. (2020). "Discussions about the landscape of possibilities for treatments of cosmic inflation involving continuous spontaneous localization models." *European Physical Journal C*, 80: 1021.

Carlip, S. (2008). "Is quantum gravity necessary?" *Classical Quantum Gravity*, 25: 154010.

Cañate, P., Pearle, P., and Sudarsky, D. (2013). "Continuous spontaneous localization wave function collapse model as a mechanism for the emergence of cosmological asymmetries in inflation." *Physical Review D*, 87: 104024.

Das, S., Lochan, K., Sahu, S., and Singh, T. P. (2013). "Quantum to classical transition of inflationary perturbations: Continuous spontaneous localization as a possible mechanism." *Physical Review D*, 88:085020.

Diez-Tejedor, A., and Sudarsky, D. (2012). "Towards a formal description of the collapse approach to the inflationary origin of the seeds of cosmic structure." *JCAP*, 045:1207.

Ghirardi, G., Rimini, A., and Weber, T. (1986). "Unified dynamics for microscopic and macroscopic systems." *Physical Review D*, 34:470–491.

Huggett, N., and Callender, C. (2001). "Why quantize gravity (or any other field for that matter)?" *Philosophy of Science*, 68(3): S382–S394.

Josset, T., Perez, A., and Sudarsky, D. (2017). "Dark energy as the weight of violating energy conservation." *Physical Review Letters*, 118: 021102.

Juárez-Aubry, B. A., Kay, B. S., and Sudarsky, D. (2018). "Generally covariant dynamical reduction models and the hadamard condition." *Physical Review D*, 97: 025010.

Juárez-Aubry, Miramontes, T., and Sudarsky, D. (2020). "Semiclassical theories as initial value problems." *Journal of Mathematical Physics*, 61: 032301.

Kay, B. S., Juárez-Aubry, B. A., Miramontes, T., and Sudarsky, D. (2023). "Semiclassical theories as initial value problems spontaneous collapse theories, and the initial value formulation of semiclassical gravity." *Journal of Cosmology and Astroparticle Physics*, 01:40.

Lechuga Soliz, R. L., and Sudarsky, D. (2023). "On the issue of eternal inflation." *arXiv:2308.01383*.

León, G., Majhi, A., Okon, E., and Sudarsky, D. (2017). "Reassessing the link between b-modes and inflation." *Physical Review D*, 96:101301(R).

León, G., Majhi, A., Okon, E., and Sudarsky, D. (2018). "Expectation of primordial gravity waves generated during inflation." *Physical Review D*, 98(2):023512.

León, G., and Sudarsky, D. (2012). "Novel possibility of nonstandard statistics in the inflationary spectrum of primordial inhomogeneities." *Sigma*, 8:024.

León, G., and Sudarsky, D. (2015). "Origin of structure: Statistical characterization of the primordial density fluctuations and the collapse of the wave function." *Journal of Cosmology and Astroparticle Physics*, 06:020.

Martin, J., and Vennin, V. (2020). "Cosmic microwave background constraints cast a shadow on continuous spontaneous localization models." *Physical Review Letters*, 124:080402.

Martin, J., Vennin, V., and Peter, P. (2012). "Cosmological inflation and the quantum measurement problem." *Physical Review D*, 86: 103524.

Mattingly, J. (2005). "Is quantum gravity necessary?" In Kox, A. J., and Eisenstaedt, J., editors, *The Universe of General Relativity*, pages 325–338. Cham: Springer Birkhäuser.

Mattingly, J. (2006). "Why Epply and Hannah's thought experiment fails." *Physical Review D*, 73: 064025.

Maudlin, T. (1995). "Three measurement problems." *Topoi*, 14.

Maudlin, T. (2017). "(Information) paradox lost." DOI: https://doi.org/10.48550/arXiv.1705.03541.

Maudlin, T., Okon, E., and Sudarsky, D. (2020). "On the status of conservation laws in physics: Implications for semiclassical gravity." *Studies in History and Philosophy of Modern Physics*, 69: 67–81.

Modak, S. K., Peña, I., and Sudarsky, D. (2015a). "Loss of information in black hole evaporation with no paradox." *General Relativity and Gravity*, 47: 120.

Modak, S. K., Peña, I., and Sudarsky, D. (2015b). "Non-paradoxical loss of information in black hole evaporation in collapse theories." *Physical Review D*, 91(12): 124009.

Okon, E., and Sudarsky, D. (2014). "Benefits of objective collapse models for cosmology and quantum gravity." *Foundations of Physics*, 44: 114–143.

Okon, E., and Sudarsky, D. (2016a). "Less decoherence and more coherence in quantum gravity, inflationary cosmology and elsewhere." *Foundations of Physics*, 46: 852–879.

Okon, E., and Sudarsky, D. (2016b). "A (not so?) novel explanation for the very special initial state of the universe." *Classical Quantum Gravity*, 33: 225015.

Okon, E., and Sudarsky, D. (2018). "Losing stuff down a black hole." *Foundations of Physics*, 48: 411–428.

Page, D. N., and Geilker, C. D. (1981). "Indirect evidence for quantum gravity." *Physical Review Letters*, 47:979.

Pearle, P. (1989). "Combining stochastic dynamical state-vector reduction with spontaneous localization." *Physical Review A*, 39:2277–2289.

Pearle, P., and Squires, E. (1994). "Bound state excitation, nucleon decay experiments, and models of wave function collapse." *Physical Review Letter*, 73:1.

Penrose, R. (1979). "Singularities and time-asymmetry." In Hawking, S. W., and Israel, W. (eds.), *General Relativity: An Einstein Centenary Survey*. Cambridge: Cambridge University Press, pp. 581–638.

Perez, A., and Sudarsky, D. (2019). "Dark energy from quantum gravity discreteness." *Physical Review Letters*, 122: 221302.

Perez, A., and Sudarsky, D. (2021). "Cosmological constraints on unimodular gravity models with diffusion." *General Relativity and Gravity*, 53: 40.

Perez, A., and Sudarsky, D. (2022). "A dialog on the fate of information in black hole evaporation." In *Special Topic Collection Celebrating Sir Roger Penrose's Nobel Prize*. AVS Quantum Science (AIP Press).

Perez, A., Sudarsky, D., and Wilson-Ewing, E. (2021). "Resolving the h_0 tension with diffusion." *General Relativity and Gravity*, 57: 7.

Unruh, W. G., and Wald, R. M. (1995). "On evolution laws taking pure states to mixed states in quantum field theory." *Physical Review D*, 52: 2176.

Unruh, W. G., and Wald, R. M. (2017). "Information loss." *Reports on Progress in Physics*, 80: 092002.

13

Finding Time for Wheeler–DeWitt Cosmology

NICK HUGGETT AND KARIM THÉBAULT

13.1 Introduction

Time may change me, but I can't trace time. But could the *trace of time* itself change? Might the *structure of time* be *emergent*? And, if so, could such temporal structure not *de*-emerge? And then even *re*-emerge again? In this chapter we will consider, in detail, the formal and physical features of a quantum cosmological model due to Kiefer (1988) and Kiefer and Zeh (1995) which displays precisely such a remarkable emergent temporal structure.

By temporal structure we mean the basic features of time that provide us with a temporal betweenness relation, or chron*ordinal* structure; a quantitative duration relation between events, or chrono*metric* structure; and temporal directionality, or chrono*directed* structure. By emergent we mean derivation under some approximation.

Our central goal will be to explore, in the context of a concrete model, a cluster of challenges to the temporal emergentist story that each seek to undermine the relevant approximate derivations on the grounds that they assume what they are aiming to show. We will consider a specific challenge in the context of the appeal to the Born–Oppenheimer (BO) and Wentzel–Kramers–Brillouin (WKB) approximations in the approximate derivation of chronordinal and chronometric structures due to Chua and Callender (2021), and the idea of "temporal double standards" due to Price (1996) in the context of the derivation of chronodirected structure.

The chapter is organized as follows: Section 13.2 provides more details regarding what we mean by the structure and emergence of time. Section 13.3 contains an extended discussion of explicit quantum cosmological models in context of the emergence of time. In Section 13.3.1, we consider the application of the Born–Oppenheimer approximation to a simple quantum cosmological model, in comparison with its standard application in molecular chemistry. In Section 13.3.2, we then consider the approximate derivation of chronordinal and chronometric structure in the semiclassical approximation. Finally, in Section 13.3.3, we examine the potential for failure of the semiclassical approximation to be remedied via a decoherence-based argument that contains a mechanism for the potential emergence of chronodirected structure. Section 13.4 considers skeptical arguments that challenge the cogency of the temporal emergentist story based upon the idea of temporal double standards, or assuming the temporal structure that you are trying to derive. In particular, we provide reconstructions of double standards arguments regarding the emergence of chronordinal and chronometric structure due to Chua and Callender (2021) and regarding the emergence of chronodirected structure due to Price (1996). In each case, the relevant charge is found to fail. The model thus provides a self-consistent account of the emergence of chronordinal, chronometric and chronodirected structure. Residual concerns relate to explanatory rather than consistency considerations.

13.2 The Structure and Emergence of Time

Even restricting attention to mechanics, physical time is not a unitary concept; rather, we can distinguish various properties and structures. So doing will help us analyze the emergence of time in quantum cosmology later in the chapter. They are partially ordered, so we will label them "grades" of temporal structure.[1] They should be thought of as structuring spacetime in the first place, though here we will define them in terms of relations between "events" in a *linear order*. In the case of Friedmann–Lemaitre–Robinson–Walker (FLRW) universes that concerns us here, these "events" are the well-defined instants of "cosmological time." While in more general relativistic spacetimes, the corresponding relations can be defined as *invariant structures* for points on *timelike curves*, in terms of the causal and metrical structures of spacetime, or as *surplus structures* based upon a space-like foliation (Gryb and Thébault 2024).

The first grade of temporal structure that we consider orders sets of events with respect to a three-place relation of temporal "betweenness": such a "chronordinal" structure is that of an *undirected temporal line*, which McTaggart named "C-series time."[2] That is, given three chronordinally structured events, a, b, c, it is determinate, for example, that b is *temporally between* a and c but not (necessarily) whether it is *before* or *after* a. We assume that a is not also between a and

[1] This account of the structure of time builds on Gryb and Thébault (2024) which focuses on analysis of "chronordinal" and "chronometric" structure in the context of the so-called problem of time. For a fascinating and wide-ranging philosophical investigation of temporal structure we highly recommend Newton-Smith (1982).

[2] For more general discussion on the status of time ordering structure we strongly recommend the work of Farr (2012, 2016, 2020). Building upon the ideas of McTaggart (1908), Reichenbach (1956) and Black (1959), Farr advocates a "C theory of time" in which there exists an undirected causal relation sufficient to define a partial ordering on the space of events.

c (ruling out cyclic time) though, of course, one could consider relaxing that assumption.[3]

The second grade of temporal structure that we shall consider is *chrono-metric*: the temporal distance relations between events. For every pair $\{a, b\}$ of chronordinally structured events there is also a unique (since we consider noncyclic time) nonnegative real number, the *duration* of the interval between them, τ_{ab} (with $\tau_{ab} = 0$ for $a = b$), satisfying the "triangle equality": for any triple of events a, b, c, the sum of two of the durations equals the third, $\tau_{ac} = \tau_{ab} + \tau_{bc}$, say.

However, we can strengthen chronordinal structure in another way, by imposing *chronodirected* structure on chronordinally structured events: the relation of "not-later-than" between pairs, $a \le b$ (b is between a and c, if $c \le a \wedge c \le b \wedge b \le a$, assuming that they are distinct). This relation gives McTaggart's B-series, our third grade of temporal structure.

Chronometric and chronodirected structures are logically independent. On the one hand, chronodirected structures are often not posited within physical theories with a temporal metric, leading to a long tradition of attempts to derive an arrow of time from special initial conditions (e.g., Callender (2021)). On the other hand, one could envision a chronodirected world in which no durations are given; times organized as a directed line with no fact of the matter about the duration between pairs of points. However, if the set of events possesses a "clock," a real-valued quantity that is strictly increasing with respect to some total ordering of time, then it will provide both chronometric and chronodirected – hence chronordinal – structure: the duration between two events is the difference between their times, and a is earlier than b just in case a's time is less than b's.

Now, when a set of objects possesses structures with the formal properties that we have described, it is still legitimate to ask whether they are in fact *temporal*: perhaps instead they refer to some other physical relations (spatial relations, for instance). We take the general attitude that a structure can be identified as temporal in virtue of the roles it plays in dynamical physical theories. We do not offer a general account of what these roles are; rather we appeal to the fact that time is already identified in existing physical theories, and in this chapter aim to pinpoint the (emergent) structures playing those roles.[4]

This is not the only attitude one could take to the temporal nature of the structures. Suppose one posits that temporal structures are metaphysically "absolute." So doing is compatible with our position, since we do not claim that temporal structures are identified *with* their roles in physical theories, but *by* them. However,

[3] Moreover, if b is between a and c, and d is between a and b, then d is between a and c. In relativistic spacetime these temporal lines can be specified invariantly as timelike curves, defined in terms of the lightcone structure; on such a curve, b is temporally between a and c, if and only if it lies in their "causal diamond."

[4] See Huggett and Wuthrich (2025) for more on such spacetime functionalism.

one might deny even our weaker claim, so that temporal structures might come apart from physical theories. For instance, suppose a temporally oriented world, in which entropy is a monotonic function; even if one identifies the orientability of time by the arrow of increasing entropy, there would still remain the question of whether that arrow points to the absolute future or past. On our view, on the contrary, it is a matter of definition that the entropic arrow points to the "future," since that is how we use the term (assuming the alignment of that arrow with those of other physical processes). There is no further issue once we assume that temporal structures are identified by their roles in physical theory.[5] In short, once we have identified an emergent arrow that aligns with familiar temporal arrows, we are done; we do not need to – and so won't! – argue that it is future directed.

Having clarified what we will mean by temporal structure, let us next explain what we will mean by "emergence" of such a structure in a quantum cosmological model. We are interested in situations in which the following three features obtain:

1. A given temporal structure is not explicitly specified in the basic equations of a quantum cosmological model.
2. There exists a formal derivation of the relevant structure, or an approximation to such a structure, in some limit of the model.
3. The formal derivation can be physically justified in low-energy and/or semiclassical regimes.

This approach is along the lines of the view of emergence, common in physics and becoming more popular in philosophy, in which emergence is compatible with reduction. We will assume, and not argue for, the basic structure of such an approach in what follows.[6]

Our focus in the chapter will be a package of approaches to quantum cosmology, developed both separately and in collaboration by the physicists Claus Kiefer and Dieter Zeh, in which all three grades of temporal structure (putatively) emerge in our sense.[7]

13.3 Emergence of Temporal Structure in Quantum Cosmology

> [T]he "time ordering of events" is a notion devoid of meaning [. . .] the concept of spacetime and time itself are not primary but secondary ideas in the structure of physical theory. These concepts are valid in the classical approximation. However, they have neither meaning nor

[5] This attitude is compatible with asking how posited temporal structures that do not play a role in physical theory relate to those that do. For instance, does entropy increase or decrease with respect to growth of the "growing block"?

[6] For more details see Butterfield (2011), Huggett and Wuthrich (2025), Huggett (2021), and Palacios (2022).

[7] See in particular Zeh (1986), Kiefer (1988), Zeh (1989), Kiefer and Zeh (1995), and Kiefer (2012).

> application under circumstances when quantum-geometrodynamic effects become important. Then one has to forgo that view of nature in which every event, past, present, or future, occupies its preordained position in a grand catalog called "spacetime." There is no spacetime, there is no time, there is no before, there is no after. The question what happens "next" is without meaning.
>
> (Wheeler 1968, p. 1124)

The full gravitational Wheeler–DeWitt equation (DeWitt 1967) is a heuristic semi-mathematical expression that results from informal application of the Dirac constraint quantization algorithm (Dirac 1964) to the Hamiltonian formulation of general relativity (Dirac 1958; Arnowitt et al. 1960). Famously, as indicated by Wheeler himself in the quote above, the equation does not contain any extrinsic temporal structure. Its formal and physical justification, both in general and in specific simple cosmological applications is open to dispute. We will not enter into these debates here.[8] Rather, we will assume the general pattern of argument leading to the Wheeler–DeWitt equation to be well enough justified and consider the explicit details of a particular model resulting from the quantization of symmetry-reduced "mini-superspace" (MSS) cosmology.

Explicit model building in the context of quantum cosmology is usually based upon the quantization of classical, finite-dimensional FLRW cosmologies, in which spacetime can be foliated into homogeneous and isotropic spatial slices (or their generalization into Bianchi spacetimes, with homogeneous but anisotropic spatial slices or "midi-superspace" with infinite degrees of freedom and only local symmetry). The only variables are the scale factor a for the spatial geometry, capturing the gravitational degrees of freedom, and the matter degree of freedom, ϕ, describing a spatially homogeneous scalar field. (The story about the emergence of time in quantum cosmology we describe is *in principle* extendable to a full theory of quantum gravity able to recover time in standard model cosmology in full, even if *in practice* this would present a fearsome challenge). The space parameterized by these quantities is MSS, and the standard form of the Wheeler–DeWitt equation in quantum cosmology (QC) is in the MSS representation: an equation for a wavefunction $\Psi(a, \phi)$.[9] This formulation brings the disappearance of time in quantum gravity into sharp relief: what kind of universe is described by a solution in

[8] For discussion see Gryb and Thébault (2011, 2014); Gryb and Thébault (2016a,b, 2017); Gryb and Thébault (2018); Gryb and Thébault (2024). Note that a full understanding of Wheeler's assertion (which will not be necessary here) requires consideration of the "problem of time."

[9] Analysis of these models goes back to DeWitt (1967) and includes notable work by Blyth and Isham (1975); Gotay and Isenberg (1980); Hartle and Hawking (1983); Vilenkin (1984); Kuchař and Ryan (1989); Bojowald (2001); Ashtekar et al. (2006). See Ashtekar and Bianchi (2021); Gielen and Menéndez-Pidal (2022) and citations therein for recent work. In what follows we will focus on the work on Claus Kiefer and provide full citations. Further work includes study of symmetry reduced but infinite dimensional midi-superspace models and so-called Gowdy models that feature gravitational waves. See for example, Barbero G and Villaseñor (2010), (Ashtekar and Singh 2011, §6), Tarrío et al. (2013); de Blas et al. (2017).

which $\Psi(a,\phi)$ has support across a large swathe of MSS? It seems to describe amplitudes for all kinds of spatial slices, and no evolution through them (see Warrier [2022] for a recent discussion of such issues) – returning us to the quotation from Wheeler.

13.3.1 Wheeler–DeWitt meets Born–Oppenheimer

Even in such a restricted context as MSS, finding solutions to the relevant equations requires semiclassical approximations; typically one or both of the WKB-approximation and the BO-approximation.[10] Here we consider some of the explicit details of such an approach to the quantization of a MSS model with no cosmological constant, found in Kiefer (1988). The Wheeler–DeWitt equation of the model is

$$\hat{H}\Psi(\alpha,\phi) = \left[\frac{2}{3\pi m_p^2}\frac{\partial^2}{\partial\alpha^2} - \frac{\partial^2}{\partial\phi^2} - \frac{3\pi m_p^2}{2}ke^{4\alpha} + m^2e^{6\alpha}\phi^2\right]\Psi(\alpha,\phi) = 0, \tag{13.1}$$

where $k \in \{-1,0,+1\}$ is the sign of the spatial curvature, the scale factor a (relative to some reference scale a_0 which we set to 1) is written in terms of a logarithmic variable $\alpha = \ln a$, the matter content is given entirely by a homogeneous scalar field φ with mass m, and $m_p \gg m$ is the Planck mass.[11]

Our Wheeler–DeWitt equation has the form

$$(\hat{T}_1 + \hat{T}_2 + \hat{W})\Psi(x_1,x_2) = E, \tag{13.2}$$

with $E = 0$. The Hamiltonian is the sum of the "kinetic energies" ($\sim \partial^2/\partial x_i^2$) of two subsystems, with a potential term that describes their interaction. Moreover, for a realistic field, the masses of the two subsystems are very different: the Planck mass of the gravitation field is much greater than that of the matter field, $m_p \gg m$. Thus the equation is formally analogous – with $E = 0$ – to the time-independent Schrödinger equation for the kind of molecular system, containing heavy nuclei and light electrons, for which the Born–Oppenheimer (BO) approximation was developed (Born and Oppenheimer 1927; Born et al. 1955). However, we will see that the rationale for its use in quantum cosmology is importantly different from that in atomic physics, a point that is relevant to addressing recent criticisms of derivations like Kiefer's.

10 Early work is Banks (1985), Halliwell and Hawking (1985), Kiefer (1988), and Vilenkin (1989). See Kiefer (2005, 2013), Kuchař (2011), Kiefer and Peter (2022), and Maniccia et al. (2022) for reviews.

11 Note that our treatment here, mainly following Kiefer (1988), does not include full details of the quantum formalism. A fully rigorous presentation of this model would require explicit definition of the Hilbert space and operators and would encounter issues with self-adjointness related to the $a \to 0$ singularity. Neglect of these features is justified here since the behavior we are interested in is far away from the singular "region." See Thébault (2023).

In the molecular case, the BO approximation can be understood as a three-step method (see Huggett et al [2024] for a more complete account):

Step 1: solve the (normalized) eigenvector equation for the "reduced Hamiltonian":

$$\left(\hat{T}_2 + \hat{W}(x_1)\right)\psi_n(x_1;x_2) = \lambda_n(x_1)\psi_n(x_1;x_2). \tag{13.3}$$

This equation treats the "heavy" subsystem as if it were fixed at a definite location x_1, and the "light" subsystem as moving in the resulting potential $\hat{W}(x_1)$; for instance, the equation could describe how electrons would move for a certain classical configuration of nuclei.[12] As a result (13.3) is parameterized by x_1, including the eigenvalues (see Figure 13.1) and solutions: hence here $\psi_n(x_1;x_2)$ is a function of x_2 (only), of a solution for the potential $\hat{W}(x_1)$ (hence the semicolon).

Step 2: since the $(\hat{T}_2 + \hat{W}(x_1))$ are commuting Hermitian operators for the full system, the $\psi_n(x_1;x_2)$ form an orthonormal basis for states $\Psi(x_1,x_2)$, provided that the observable is nondegenerate. Thus, any solution to (13.3) can be written

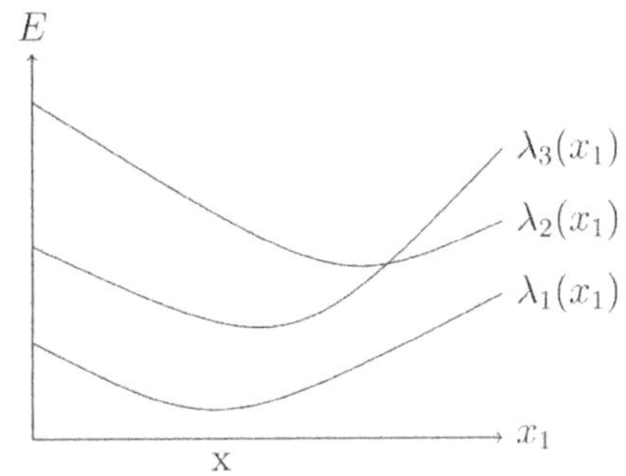

Figure 13.1 The eigenvalues λ_n of the reduced Hamiltonian, as (hypothetical) functions of the heavy degrees of freedom x_1. In the region around $x_1 = x$ the first three electronic energy levels can be seen to be widely separated: specifically, by far more than the kinetic energy of the nuclei. This is the condition for stable molecules, and for the BO approximation.

12 This step is typically given a heuristic gloss in terms of treating the "heavy" subsystem as if it were "fixed" or "clamped" at a definite location x_1, and the "light" subsystem as moving in the resulting potential $\hat{W}(x_1)$. Physically, however, there is no sense in which the nuclei are literally fixed at points, since they are quantum objects. Rather the step should be understood as a formal move in an approximation scheme, and the "fixed" description a heuristic gloss, which is ultimately both unphysical and unnecessary.

$$\Psi(x_1, x_2) = \sum_n c_n \theta_n(x_1) \psi_n(x_1; x_2) \tag{13.4}$$

for some (normalized) coefficients $\theta_n(x_1)$. This equation is exact, but now one makes the first approximation and makes a *separation ansatz*:

$$\Psi(x_1, x_2) \approx \theta_n(x_1) \psi_n(x_1; x_2). \tag{13.5}$$

The rationale for this separation is that because of the separation of masses, there will be far more kinetic energy in the light subsystem than in the heavy one, and $E \approx \lambda_n$. In particular, in the regime of stable molecules we have energies and atomic configurations for which the separations between λ_ns for different values of n are much greater than the kinetic energy of the heavy subsystem: $|\lambda_m - \lambda_n| \gg T_1$. But then (see Appendix [1]) a superposition of ψ_n cannot be an eigenvector of total energy, and (13.5) holds.

Step 3: it follows from (13.5) (see Appendix [2]) that solutions to (13.3) have small variation with respect to x_1: for molecules, that the electron wavefunction is unchanged by small changes of the nuclei position parameters:

$$\frac{\partial \psi_n(x_1; x_2)}{\partial x_1} \approx 0. \tag{13.6}$$

Using this *adiabatic approximation*, an equation for $\theta_n(x_1)$ can be found from (13.2), (13.3), and (13.4):

$$(\hat{T}_1 - (E - \lambda_n))\theta_n(x_1) = 0 \tag{13.7}$$

In sum, for molecules, the overall BO method is to solve (13.3) and (13.7), and then insert into (13.5) to yield an approximate solution to (13.2). This is a formal technique to construct approximate solutions to a partial differential equation within a well-controlled regime of validity. The approximation requires neither time-dependent dynamics nor "fixed" nuclear positions for its application.[13]

For the MSS Wheeler–DeWitt equation, the rationale – hence method – is somewhat different. In particular, we are not interested in finding solutions to

[13] The second point is significant for the interpretation of Born-Oppenheimer in the context of the reduction-emergence debate and will be considered in detail in Huggett, Ladyman and Thébault (2024). See Woolley and Sutcliffe (1977); Woolley (1978); Claverie and Diner (1980); Hendry (1998, 2006, 2010a,b, 2017) Fortin and Lombardi (2021); Accorinti and González (2022); Chang (2015); González et al. (2019); Cartwright (2022); Scerri (2012); Hettema (2017); Franklin and Seifert (2020); Seifert (2020, 2022).

(13.2) for *different* values of E, but only for $E = 0$. As a result, we are not interested in solving (13.3) for values of λ_n – the energy of the light subsystem – *widely separated* with respect to the kinetic energy of the heavy subsystem, T_1, since these would correspond to different energy levels for the total system. We are concerned with an energy spectrum of a different character than that depicted in Figure 13.1. Hence, (a) we *cannot* justify the separation ansatz (13.5) of Step 2, since it depends on such energy gaps, and we instead look for superpositions of the form (13.4). This turns out to be important, since finding solutions that plausibly describe classical spacetimes requires forming states from such superpositions, as we shall soon see. Then, (b) we cannot justify the adiabatic approximation of Step (3) (13.6) in terms of the separation ansatz, and a different justification must be given to obtain it and hence (13.7).

That is to say that, for quantum cosmology, the BO method is *first* to follow Step 1, and find (normalized) solutions to (13.3) for the light – field – subsystem

$$\left(-\frac{\partial^2}{\partial\phi^2} - ke^{4\alpha} + m^2e^{6\alpha}\phi^2\right)\psi_n(\alpha;\phi) = \lambda_n(\alpha)\psi_n(\alpha;\phi), \tag{13.8}$$

where x_1 and x_1 are replaced by α and ϕ, respectively. As before, both the "potential" $\hat{W}(\alpha) = -ke^{4\alpha} + m^2e^{6\alpha}\phi^2$ and the reduced energy eigenfunctions $\lambda(\alpha)$ are parameterized by the heavy coordinate α, which is considered to be a "fixed" c-number parameter rather than a wavefunction variable: *as if* the field were propagating in a fixed spatial geometry α. The (exact) solutions are given in Kiefer (1988, (5.15)). As before, they form a basis (13.4) for the joint system, and we can write any state as a superposition:

$$\Psi(\alpha,\phi) = \sum_n c_n\theta_n(\alpha)\psi_n(\alpha;\phi). \tag{13.9}$$

Then *second*, without making a separation ansatz, one follows Step 3, and makes the adiabatic approximation

$$\frac{\partial\psi_n(\alpha;\phi)}{\partial\alpha} \approx 0 \tag{13.10}$$

and uses the Wheeler–DeWitt equation (13.1), (13.8), and (13.9) to derive an equation for the wavefunction of the "heavy" degree of freedom:

$$\left[\frac{\partial^2}{\partial^2\alpha} + \lambda_n(\alpha)\right]\theta_n(\alpha) = 0. \tag{13.11}$$

Of course, compared to (13.7) we have $E = 0$. Moreover, one can further show that products $\theta_n \psi_n$ are solutions of the Wheeler–DeWitt equation (13.1), as of course are their superpositions, by linearity.

This equation one solves under a semiclassical WKB approximation with wave-packets peaked about classical trajectories for α; we are ultimately interested in understanding the emergence of classical spacetime. We will not investigate here the application of the WKB approximation in the same detail as the Born–Oppenheimer method, but rather refer the reader to Kiefer (1988, eq. 5.23) for the explicit expression for $\theta_n(\alpha)$. It is important for our later analysis that in general terms, the validity of WKB approximation can be understood in terms of a restriction on the *time-independent* characteristic functional of the Hamilton–Jacobi formalism. This, in turn, can be shown to imply that the approximation is valid for regimes in which the de Broglie wavelength is small compared to the characteristic distance over which the spatial potential varies (Sakurai and Commins 1995: 112–116). (Or, for the Wheeler–DeWitt equation, for regimes in which the wavelength of $\Psi(\alpha, \phi)$ with respect to α is small compared to the distance over which $W(\alpha)$ varies.)

What of the required justification of the BO approximation (13.10)? (Recall, in the molecular case it was justified by the separation ansatz, itself justified by the large electronic energy gaps, which does not now hold.) In the *third* step, once one has solved for $\psi_n(\alpha; \phi)$ and $\theta_n(\alpha)$, then one can simply check by explicit calculation whether (13.10) holds. Kiefer (1987, (5.24)) identifies the regime of validity as

$$n \ll m^{3\alpha}, \tag{13.12}$$

where n is the excitation level of the reduced eigenvalue problem (13.8). When this condition is satisfied, the adiabatic approximation holds, and the Born–Oppenheimer method is formally valid. The functions $\theta_n(\alpha)$ and $\psi_n(\alpha; \phi)$ obtained satisfy the equations for the light and heavy parts – (13.7) and (13.8), respectively – derived from (13.1) and the adiabatic approximation: so their products are approximate solutions of the Wheeler–DeWitt equation. That is, (13.10) is an ansatz whose validity can be – *and is* – checked after putative solutions are found. We will return to this point in what follows.

13.3.2 Approximate Derivation of Chronordinal and Chronometric Structure

With the general solution found (in the specified regime), the next step is to fix boundary conditions and construct explicit solutions for an emergent FLRW model in which the appropriate semiclassical behavior is obtained. In particular, we require that we: (i) have a well-posed boundary problem for the wavefunction;

and (ii) it is possible to construct quantum states that form narrow wave-packets that follow the classical solution according to the Ehrenfest relation (i.e., with the centroid of the wave-packet approximating the classical behavior). Significantly, the emergence of such semiclassical behavior is not guaranteed merely by the validity of the WKB-approximation.

In Kiefer's model, one finds that the spatially flat, $k = 0$, case allows the formulation of a well-posed Cauchy problem for "initial" data surfaces $\alpha_0 = constant$. The full wavefunction (13.4) is uniquely determined by a choice of $\Psi(\alpha_0, \phi)$ and $\frac{\partial}{\partial\alpha}\Psi(\alpha, \phi)|_{\alpha=\alpha_0}$. For this case, Kiefer shows that (13.11) can be solved under the WKB approximation and that, within this approximation, it is possible to construct a wave-packet $\Psi(\alpha, \phi)$ that follows the classical trajectory with minimal dispersion. Specifically, we have a wave-packet whose support lies along a classical FLRW trajectory in MSS; which is uniquely determined by an "initial" wavefunction $\Psi(\alpha_0, \phi)$; and which is a Gaussian state with respect to ϕ. Such states are the quantum MSS versions of the quantum coherent states considered to describe approximately classical systems. Then, the solution $\Psi(\alpha_0, \phi)$ to the Wheeler–DeWitt equation (13.1) remains (approximately) Gaussian – hence classical – in ϕ as α varies, and the peak of the Gaussian follows an FLRW trajectory $\phi(\alpha)$ as per the Ehrenfest relation (Kiefer 1988: 1768–1769); compare Kiefer (1990).

The spatially flat case allows us to identify the features required for a quantum variable within a Wheeler–DeWitt type model to be identified as an approximate internal time – that is, to play a "classical-time-parameter-like" functional role in the emergent dynamics, as discussed in Section 13.2. The first is for it to be possible to interpret the specification of boundary conditions on the wavefunction as initial (or initial and final) "time" boundary problems. That is, we should be able to fully specify the wavefunction via conditions of the relevant variable and then construct determinate "evolution" of the wavefunction. As we noted, α has this feature. The second feature is that the variable should define a metric for points along this trajectory as, again, α does: $\tau_{ab} \equiv |\alpha_a - \alpha_b|$. Hence, in light of the discussion of Section 13.2, α provides chronometric (hence chronordinal) structure, provided it plays an appropriate temporal role in the emergent dynamics. If Kiefer's model is interpreted as intended, so that α and ϕ are the spatial scale factor and a scalar field, respectively, then we say yes: classical FLRW physics is recovered, in which α does play the role of time.

Note that in this interpretation ϕ and α are treated in very different ways: ϕ is treated as classical in virtue of being in quantum state corresponding to a classical field. But the wavefunction is not localized with respect to α (even for fixed ϕ) and so is not treated as classical in an approximation to a quantum state. Instead, it is postulated to correspond to the observed classical time and scale factor; such a correspondence cannot be derived from existing physical principles. That such a new physical postulate will typically be required for classical spacetime to emerge from quantum gravity is argued in Huggett and Wuthrich (2025, chapters 1–2) and Huggett (2021); without it the derivation is purely formal, lacking in "physical salience." Since chronometrical structure does not entail chronodirected structure, this model does not contain an arrow: no basis to say whether the universe is expanding or contracting; for that we need to consider a different model, as we will in Section 13.3.3.

Whether or not these features obtain in the model depends on the satisfaction of the conditions that allow for the BO and WKB approximations to be applied. In particular, by having a wave packet approximately peaked along the classical trajectory while the condition (13.10) holds, we know we can label the waveforms in $\{\alpha, \phi\}$ by a parameter that tracks a change that will be, at least locally in configuration space, "continuous" and single valued, namely α. By contrast, ϕ does not play the functional role of time in the model since the failure of the analogue of condition (13.10) means that the wave-packet can very rapidly vary along the ϕ-axis in $\{\alpha, \phi\}$ space.

It is worth noting that our characterization of the requirements for a quantum variable to play a functional role as semiclassical time does not include the ability to recover an explicit Schrödinger-like equation with time dependence with respect to the internal time. One could add such a further role, of course, and this additional role does feature in various semiclassical time approaches to the Wheeler–DeWitt equation, see Kiefer (2012: 179) and references therein. As noted by Chua and Callender (2021) this is a plausible sufficient condition for a variable to play the functional role of time. However, it is clearly not a necessary condition as shown by the case in hand where such an equation is not derived.

In sum, the overall result is that from the timeless Equation (13.1), we have extracted an (approximate) diachronic structure whereby the internal α degree of freedom plays the functional role of time in the sense of chronometric and chronordinal structures.

13.3.3 Approximate Derivation of Chronodirected Structure

Of particular relevance to our present discussion is that models of (13.1) display a *breakdown* of the semiclassical approximation in the case of a closed, $k = +1$, universe. Classical FLRW closed universes with vanishing cosmological constant expand from a big bang then recollapse at a big crunch: the trajectory that we want our Wheeler–DeWitt wavefunction to approximate is a curve in MSS that starts and ends at $\alpha \to -\infty$, and reaches a maximum value of α. Therefore, to find a solution to the $(k = +1)$ Wheeler–DeWitt equation one therefore (i) specifies an "initial,"

$\alpha_0 = constant$, state that is a superposition of expanding *and recollapsing* components; (ii) imposes the boundary condition $\Psi \rightarrow 0$ as $\alpha \rightarrow \infty$; and (iii) seeks dual trajectories approaching a common point as α grows. Such a solution, found by Kiefer, is shown in Figure 13.2: particularly the upper plot, of the wave-packet for smaller values of α. As before, in such a solution α would provide chronometrical, hence chronordinal structure; though, obviously, that α grows from both the "beginning" and the "end" of the trajectory will affect its identification as time, as we will discuss in Section 13.4.2.

The crucial point for now, however, is that (ii) *prevents* the construction of narrow wave-packets with minimal dispersion that would allow us to represent classical trajectories for sufficiently large values of α (Kiefer 1988: 1770; 2012: 268–269). In such circumstances, semiclassical behavior of the quantum state is not obtained, and the putative emergence of temporal structure fails. The failure of putative emergence of temporal structure in the $k = +1$ model is vividly illustrated in the lower plot of Figure 13.2 which extends the same wave-packet for larger values of α.

Fascinatingly, according to Kiefer (2012; §8.1.2), the breakdown of the semiclassical limit in this model is the reason for which we must invoke the role of the "environment" in inducing semiclassical behavior for *any* part of the cosmological evolution. At this point there is an obvious conceptual challenge. On standard approaches to cosmology the universe must be considered a closed system.[14] Thus, from this perspective, the role of the environment in a quantum cosmological context must be played by an endogenous system.

The approach followed in pioneering work by Kiefer and Zeh is to consider small inhomogeneous degrees of freedom as an endogenous environment and then apply an *internal* version of standard decoherence arguments (Zeh 1986, 1989; Kiefer 1987, 1988, 2012; Kiefer and Zeh 1995). Recall that the FLRW cosmology is formulated in terms of variables defined on spatial scales in which the universe can be assumed to be homogeneous and isotropic. These "hydrodynamic" variables accurately capture the bulk behavior of classical spacetimes and are the starting point for building Wheeler–DeWitt-type quantum cosmological models. In reality, of course, the universe is not, thankfully, homogeneous and isotropic on all scales.

A first step toward de-idealization of a FLRW-type model is to include small inhomogeneities in the classical equations of motion, specifically higher moments of the fields. This is analogous to considering hydrodynamics of bulk fluid variables

[14] The closed system view is not a physical or logical necessity in cosmology. In particular, open classical statistical systems are standardly defined in terms of conservation of phase space measure and thus the existence of empirically adequate cosmological models based upon contact dynamics which display measure compression indicates that one may model the universe as an open system in the classical domain at least. See Sloan (2021); Cuffaro and Hartmann (2021); Bravetti et al. (2022); Sloan (2023); Ladyman and Thébault (2023).

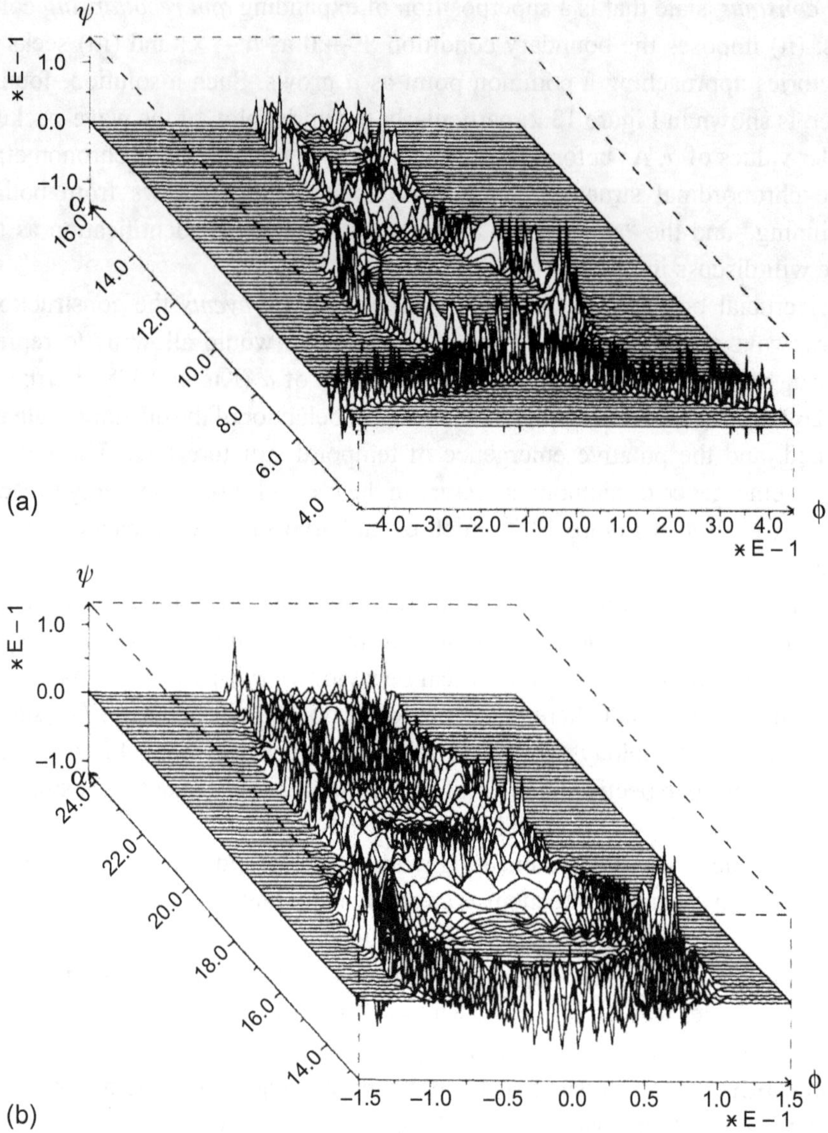

Figure 13.2 The wave-packet as a function of the geometric degrees of freedom ($a = e^{\alpha}$) and matter degree of freedom (ϕ) in the case of a closed, $k = +1$, universe. The upper figure (a) shows the region in which a classical trajectory can be recognized with a playing the role of an approximate time parameter. The lower figure (b) shows that at increasing values of a, the wave packet begins to spread around the classical trajectory. It is no longer possible to use a as an approximate time parameter and thus the semiclassical emergence of temporal structure fails. (Note the different scales of the plots, which overlap for $14 \leq \alpha \leq 16$.) Reprinted figures with permission from Kiefer, C., *Physical Review D, 38,* 1770, Copyright 1988 by the American Physical Society.

together with first-order fluctuations. This approach can be transferred to the quantum context also. In particular, following Kiefer (2012: 347), if we write the small inhomogeneous degrees of freedom as $\{x_i\}$ we can arrive at a Wheeler–DeWitt equation of the following form:

$$\left[\frac{\partial^2}{\partial^2\alpha} + \sum_i \frac{\partial^2}{\partial^2 x_i} + V_i(\alpha, x_i)\right]\Psi = 0, \tag{13.13}$$

where the potentials $V_i(\alpha, x_i) \to 0$ for $\alpha \to -\infty$. The form of the potentials means that the Wheeler–DeWitt equation possesses an asymmetry with respect to intrinsic time α. Furthermore, since the potential is such that the coupling between the bulk variables and inhomogeneities tends toward zero for $\alpha \to -\infty(a \to 0)$ one can impose a *Simple Boundary Condition* on the packets of the following form:

$$\Psi \underset{\alpha\to-\infty}{\to} \psi_0 \prod_i \psi_i(x_i). \tag{13.14}$$

In other words, we assume an "initial state" which (asymptotically) takes the form of a product state in which the bulk variables and inhomogeneities have vanishingly small entanglement.

We then have that if Equation (13.14) is taken as an "initial" condition, the Wheeler–DeWitt equation will, through the form of the potential, lead to a wave-function with an *intrinsic arrow of time*. This would be to derive chronodirected structure in precisely the sense that we described. The crucial formal feature upon which such claims can be justified is that if the intrinsic dynamics of Equation (13.13) are combined with the simple boundary condition, Equation (13.14), for increasing α, we get increasing entanglement between α and the other modes.

Kiefer (2012) (see §10.3 in particular) interprets this process in terms of an increase in *local entanglement entropy* as defined with reference to a subset of "relevant" degrees of freedom, $\{y_i\}$, where one defines

$$S(\alpha, \{y_i\}) = -k_B tr\cdot(\rho ln\rho), \tag{13.15}$$

where ρ is the reduced density matrix corresponding to α and $\{y_i\}$.

The final step is then to *interpret* the von Neumann entropy of the reduced density matrix as providing a chronodirected structure. That is, we understand the increase in $S(\alpha, \{y_i\})$ for increasing α as demarcating earlier from later times. As such, it appears that we have a means to derive not just chronordinal and chronometric structure but chronodirected structure also, all starting from a basic equation without an extrinsic time parameter. It is, of course, open to question at this point *why* we interpret the von Neumann entropy as providing chronodirected structure. In essence, what we have done is simply stipulate that the arrow of time is equivalent to the von Neumann entropy. Much more is required of the derivation of an arrow of time, including connecting the direction of the von Neumann entropy of the reduced density matrix to the entropy arrows of other physical systems. We will not enter into detailed discussion of this point here, though we will return to related issues in Section 13.4.2.

Let us conclude our discussion by isolating the three crucial assumptions that have gone into the derivation of chronodirected structure in quantum cosmology. The assumptions are:

- *Simple Boundary Condition*: The quantum state of the universe, including bulk degrees of freedom and inhomogeneities, at some boundary (initial and final) value of the internal time parameter, can be written *approximately* as a product state, that is, such that the bulk variables and inhomogeneities have vanishingly small entanglement.
- *Relevant Coarse-Graining*: The degrees of freedom representing inhomogeneities can be divided into relevant and irrelevant subsets and we can represent the state of the universe by a reduced density matrix where the irrelevant subsets have been traced out.
- *Entropic-Arrow*: Chronodirected structure can be defined via the behavior of the von Neumann entropy of the reduced density matrix such that an increase in the von Neumann entropy of the reduced density matrix is sufficient to establish a future-orientated chronodirected structure.

Forms of these assumptions will be familiar from analogous derivations found in statistical mechanics and quantum mechanics. We will return to these connections and the putative justification of the assumptions in Section 13.4.2.

13.4 Temporal Double Standards

In this section we will consider challenges to various aspects of the approximate derivation of temporal structure detailed in the previous section. The particular focus is a generalization of the idea of "temporal double standards" – due to Price (1996) – in which there is a prima facie worry regarding the cogency of some derivation of temporal structure in physical theory due to implicit appeal to or assumption of such structure within the derivation. We will turn to the specific details of Price's worry relating to the derivation of chronodirected structure in the context of the Kiefer–Zeh approach in Section 13.4.2. In Section 13.4.1, we consider a specific challenge to the cogency of the work of Kiefer due to Chua and Callender (2021).

13.4.1 Time from No Time?

The reader may have noticed that our presentation of the BO method, for both molecules and quantum cosmology – unlike typical pedagogical presentations –

avoided appeal to time-dependent physics: for instance, nothing depended on the fact that electrons, being lighter, respond more quickly to perturbations than nuclei. This approach was taken with an eye to responding to a pair of critical challenges due to Chua and Callender (2021).[15]

The general form of this challenge is concisely stated by the Chua and Callender as follows:

> Programs in quantum gravity often claim that time emerges from fundamentally timeless physics. In the semiclassical time program, time arises only after approximations are taken. Here we ask what justifies taking these approximations and show that time seems to sneak in when answering this question. This raises the worry that the approach is either unjustified or circular in deriving time from no-time. (Chua and Callender 2021:1172)

There are several aspects to this challenge, which we should consider systematically. First, we will have to consider their critique of both the BO and WKB approximations. Second, it is unclear whether the time allegedly sneaking in is external or internal time, since the authors do not explicitly make the distinction in their critique (although they do talk about a "background time metric" as we will see later). If the former then, since there is no external time in canonical quantum gravity, the charge is the devastating one that derivation is incoherent. If the latter, the concern seems weaker: for if one assumes emergent time in one's solutions, and then finds solutions consistent with that assumption, one has shown the existence of emergent time solutions. One has shown that time *can* emerge from no time, but not that it should be expected; something important has been achieved, though one should indeed be clear on its limits. Let us work through these issues one by one.

First then, let us consider the possibility that either the BO or WKB approximations presuppose external time, BO first. Chua and Callender claim that the approximation is "thoroughly laden with temporal notions" (2021: 1178) since the light versus heavy system distinction is a proxy for a slow versus quick distinction. "The change in the lighter subsystem happens on such a short timescale that there is not enough time for the heavier sub-system to react in that relevant timescale, and so it is effectively independent of lighter subsystems in that period of time" (1178). They assume this to be the case in the context of the molecular version of the approximation and then consider the possibility that it is also the case in the quantum cosmology version. Let us suppose the time scale in question to be extrinsic, so that the putative issue is one of incoherence.

But our explication shows that in the molecular case the central claim is incorrect (or at least unnecessary): the mass separation is rather proxy for the electronic energy level separation, so a property of solutions of the time-*in*dependent

[15] It should be noted that Chua and Callender also provide a very short sketch of a third critical challenge relating to appeal to decoherence. We here neglect such important considerations, which would inevitably lead into a discussion of the measurement problem, due to restrictions of space.

Schrödinger equation. Now it is true that textbook presentations do justify the BO approximation along the lines given by Callender and Chua; for instance, the highly influential Messiah (1962, XVIII.iii) essentially uses it to justify the explicit use of the adiabatic approximation (less explicitly, the move is also made in Born et al. (1955, §14)). But this is for heuristic convenience, not necessity, as our presentation – and indeed that of Born and Oppenheimer – shows.[16]

For molecules, the Born and Oppenheimer approximation rests upon the relative size of terms in a time-independent Hamiltonian and the consequent splitting of the energy spectrum, which implies separability and adiabaticity. These features justify the approximation, but do not require temporal considerations, though of course they imply dynamical properties too, through the time-*de*pendent Schrödinger equation. Things are somewhat different for quantum cosmology, since the splitting of eigenstates does not occur in the desired solutions; so, in this case, putative solutions obtained by the BO method must be checked formally to verify that they do indeed satisfy the conditions of the approximation – which they do, in a specified regime, as we saw. One may consider the relation of the time-independent properties to time-scales and the intuitive dynamical picture *if* the Hamiltonian can be taken as the generator of time translations; such an assumption is possible, but *unnecessary* in the molecular context and *inconsistent* in the Wheeler–DeWitt context if the Hamiltonian constraint is interpreted as a "gauge generator" as in the standard interpretation that follows Henneaux and Teitelboim (1992).[17]

These reflections point to the futility of arguing that *any* computation can only be understood by invoking a time-dependent property of a quantum system. For such a property supervenes on the Hamiltonian, and hence on the energy eigenstates and eigenvalues, and hence on the time-independent Schrödinger equations – as these examples illustrate.[18]

Therefore, because the BO approximation can be justified without appeal to temporal properties, it most certainly does not assume an appeal to an external time, and so its use in quantum cosmology is perfectly coherent. What about the WKB approximation? In this case, Chua and Callender recognize the prima facie atemporality of the approximation. In particular, the fact that the crucial feature is spatial smoothness of the potential function which can be expressed more precisely in

[16] To be clear, we do not claim originality for our demonstration, which draws on Jecko (2014) to reframe Born and Oppenheimer (1927). Note that Messiah discusses the relation between his approach and that of Born and Oppenheimer.

[17] For detailed arguments toward the view (not defended here) that the standard treatment is problematic, see Gryb and Thébault (2024).

[18] At least for time-independent Hamiltonians. A similar argument can be run for classical systems via the Hamilton–Jacobi formalism. In particular, for time-independent classical Hamiltonians, the separation ansatz for the principal functional in terms of the time-independent characteristic functional together with an *Et* term establishes an analogous supervenience claim.

terms of de Broglie wavelength being small compared to the characteristic distance over which the spatial potential varies. However, they contend:

> Still, time is present. There are many ways to see this. [. . .] In quantum mechanics, we know that the momentum operator depends only on spatial variables and not time [. . .] However, the classical momentum does depend on time since $p = m\frac{dx}{dt}$ [. . .]. The time dependence, evident when talking about velocities/momenta, becomes masked when we replace velocities with notions of wavelengths and spatial variations. Yet the time dependence is plainly there. From [the de Broglie relation and dimensional analysis] if [the scale of variation of the potential] is large, the WKB approximation will be good for long [time scales] and if small then only for short [time scales]. In standard cases WKB is thus justified via a background time metric. In the case of semiclassical time, however, there is no such background time metric, so we again face our challenge to justify the assumption without invoking time. (Chua and Callender 2021: 1179–1181)

What should we make of these claims? First, note that $m\frac{dx}{dt}$ is the classical canonical momentum only in a particular family of Newtonian mechanical theories, in a particular choice of canonical coordinates. In the general context, p is a generalized canonical momentum variable given by an element of the phase space. In particular, we have that $p \in \Gamma$ where (Γ, ω) is the symplectic manifold given by the cotangent bundle to configuration space $\Gamma = T^\star C$. Canonical momenta are only defined up to symplectomorphism, and privileging a representation in a preferred chart is inconsistent with mathematical practice regarding their representational roles; see Weatherall (2018) and Gryb and Thébault (2016). Moreover, even in an explicit coordinate representation, the momentum is given by $p = \frac{\partial L}{\partial \dot{x}}$, so we only recover the Newtonian momentum given a particular form of the Lagrangian. The required temporal structure is that of infinitesimal tangent and cotangent vectors, not a background temporal metric.

Furthermore, as with the case of Born–Oppenheimer, the intuitive justificatory story told for the quantum mechanics example that does feature temporal notions, is *unnecessary*. The WKB approximation in quantum theory can be fully justified in a time-independent context since its validity is a property of a spatial function. This, then, also holds for the case WKB in the context of Wheeler–DeWitt quantum cosmology. Here we can think of the formal requirements in terms of needing some notion of functional derivation in the space of Riemannian 3-metrics which is not itself a temporal structure.[19]

Thus, we conclude, the potentially damning complaint that the derivation of time from no time assumes the existence of an external time fails to have teeth. However, there remains the charge that the derivation is circular in the sense that it assumes an internal time. In response, we observe that the formal structure of Kiefer's derivation is that of an *ansatz*: a trial form of solution to a differential equation is

[19] Thanks to Henrique Gomes for pointing this out to us.

assumed and then tested for consistency. Separability and adiabaticity are assumed, and the solutions checked to see that they obey the condition. Indeed, Born and Oppenheimer's original derivation used the very same logic. Specifically, they had to assume that in the lowest-order solutions the nuclei were at the minima of their potentials: "*eine Annahme, die erst durch den Erfolg gerechtfertigt werden kann*" ("an assumption only justified by its success") (Born and Oppenheimer 1927: 465) (trans. Blinder). Similarly, in Kiefer's derivation we in fact find an explicit estimation of the regime of validity of the BO approximation in (13.12). Moreover, we have seen that the WKB approximation holds in some regimes, and not in others. There is no uncontrolled circularity here.

The issue at this point, rather, is that no formal or physical restriction prohibits the existence of solutions to our Wheeler–DeWitt equation, (13.1), which *do not* satisfy the conditions, and in which time is thus not emergent: Kiefer constructed special solutions with quasi-classical behavior. We would like a formal or physical justification for the necessary approximation conditions – and also the simple boundary condition, for chronodirected structure – otherwise what has been achieved is a merely a demonstration of possibility, not necessity: that some, but not all, solutions admit emergent time.[20]

A more ambitious temporal emergentist target would be the emergence of the full classical temporal structure of general relativity from the Wheeler–DeWitt equation. That is, the application of a WKB-type derivation as a bridge for the approximate derivation of the Einstein Field Equations from the Wheeler–DeWitt equation, via the relevant generally relativistic Hamilton–Jacobi equation due to Peres (1962). We leave analysis of this full story of the putative emergence of time to future work. Discussion of the relevant derivations in the physics literature can be found in Gerlach (1969), Komar (1971), Misner et al. (1974), Hartle (1993), Kiefer (2009), and Salisbury (2020).

13.4.2 Asymmetry from Symmetry?

In the previous subsection we considered potential issues for the emergence of chronordinal and chronometric structure in the context of their approximate derivation from a Wheeler–DeWitt cosmology, so without such structures at a basic level. The challenges accuse the derivation of circularity in some form –somehow assuming the very temporal structure to be derived. In the context of the derivation of chronodirected structure (with chronordinal structure already given), an

[20] A more general and more worrying possibility comes from the fact that in totally constrained systems, like general relativity, nonintegrable dynamics may lead to a quantization that does not admit a semiclassical limit (Dittrich et al. 2017). The story provided here regarding the semiclassical emergence of time would be inadmissible in such circumstances.

analogous challenge was made in the now-classic discussion of Price (1996, chapters 2–4), in which a "temporal double standards" argument pattern is repeated across multiple physical contexts from thermal physics to quantum cosmology. We can rationally reconstruct it as follows:

(1) Suppose theory T possesses chronordinal structure, with respect to which the dynamics is time symmetric: the time-reverse of any solution is another solution.
(2) So, the time-asymmetry of the world – its chronodirected structure – could only be explained from T via appeal to a special boundary condition.
(3) But to apply such a boundary condition to *only* one end of time is to indulge in a temporal double standard (which may not be obvious until we take the appropriate "view from nowhen").[21]
(4) Then, the chronodirectedness of the world has either not been explained at all or has been put in by hand based upon a temporal double standard.

In this section we will consider how the derivation of chronodirected structure in quantum cosmology deals with this powerful and important challenge.

To start with, observe a near intersection between our case study of Kiefer and Zeh in Section 13.3.3 with Price's own discussion of the putative emergence of time in the famous "no-boundary proposal" of Hartle and Hawking (1983). Following the argument pattern reconstructed earlier, Price takes particular issue with Hawking and coauthors' proposal that the no-boundary condition be applied to only one "end" of the universe, so that chronodirected structure emerges only in one temporal orientation (Hawking et al. 1993, Hawking 1994). (This is in contrast to Hawking's original position expressed in Hawking (1985), in which the arrow of time would "reverse" in a recollapsing universe and the no-boundary proposal would apply at the limit of both temporal extremities.)

> [Hawking] hasn't shown that asymmetric universes are the natural product of a symmetric theory [. . .] on the contrary, he seems to have simply assumed the required asymmetry, by taking the no boundary condition to apply to only one end of an arbitrary universe [. . .] this amounts to putting the asymmetry in "by hand." (Price 1996: 93)

We concur.

To what extent, then, does the emergence of chronodirected structure based on the simple boundary condition of Kiefer and Zeh also amount to "putting the asymmetry in by hand"? Answering this question carefully will help us obtain a clearer picture of the arrow of time in the semiclassical closed universe model.

First, with respect to (1) of Price's double standards argument, the laws are not time symmetric; indeed, they are neither (fundamentally) temporal nor symmetric!

[21] Of course, supposing that the boundary condition can consistently hold at both ends of time, as indeed it can in all the relevant cases; for instance, entropy could be low at either end of a chronordinal structure.

The Wheeler–DeWitt equation is atemporal, but with an asymmetry with respect to α in the form of the potential. Second, therefore the simple boundary condition (13.14) is not postulated as an *initial* condition, but a *boundary* condition that applies for $\alpha \to -\infty$: that the initial state is separable, in the regime in which the matter and gravitational subsystems are noninteracting. Third, in the semiclassical closed universe model $\alpha \to -\infty$ at both "ends," so in the sense of emergent time, the condition (13.14) is applied symmetrically, as Price insists against Hawking et al. Of course, given the identification of a future–past arrow with the direction of increasing (von Neumann) entropy, this double imposition of the boundary condition entails that the arrow of time points toward increasing α in *both* branches coming from $\alpha \to -\infty$.

Although he does not discuss Kiefer and Zeh's work,[22] Price (1996: 99–111) defends the possibility of a "Gold universe" – namely, one with low entropy at the start and end of time, and hence two, opposite arrows of time, one before and one after a "turn-around," just as in our case. He further argues that in such a universe one could now observe traces of events in the post-turn-around future: after all, with respect to the oppositely oriented temporal arrow at such an event, the event is in our past! However, Price's reasoning assumes a classical spacetime background, and so does not apply to the Kiefer–Zeh model, since at the turn-around, where the two branches meet, the system becomes fully quantum (Kiefer and Zeh 1995: 4150) and, as Wheeler said earlier, "there is no spacetime."

That is, the model provides a mechanism for the chronodirected structure to *disappear* as well as emerge. Since there is no classical connection between different "legs" across the turning point between expansion and collapse, rather we find, at large values of the scale factor, a region which cannot be interpreted in classical terms at all. Relative to an observer located in a semiclassical portion of the universe, the full story of temporal structure is then one of emergence, disappearance, re-emergence, and then at a final big crunch, disappearance again.

However, because the emergent arrow of time reverses in the model, observers on either side of the turn-around give the same description, even though it happens in opposite relative temporal directions! Thus, it is more perspicuous to take Price's "view from nowhen" again (Figure 13.3). Consider: (a) we identify the future as the direction in which entropy increases, while (b) the simple condition sets the entropy at a minimum when the universe is small.

(c) The entropy increases as the bulk degrees of freedom interact with inhomogeneities, producing increasing entanglement; (d) which occurs as α grows – that is, as the universe expands. That is, the simple condition, the identification, the Wheeler–DeWitt dynamics, and the form of the Wheeler–DeWitt potential entail

[22] Their paper contains many of the same criticisms of Hawking.

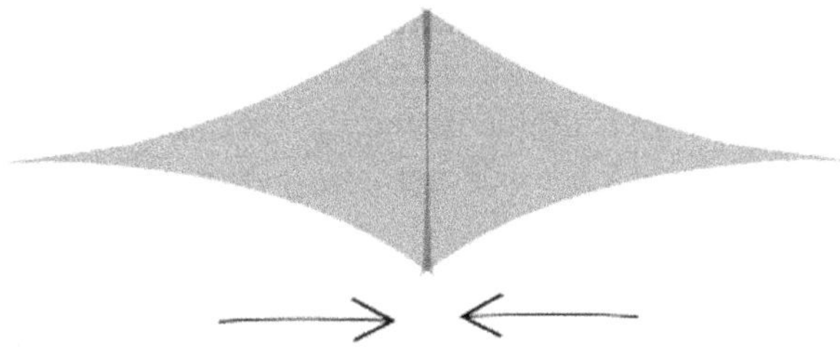

Figure 13.3 The view from "nowhen" of the double-universe. Where the temporal arrows meet, spacetime breaks down.

that the universe can only expand and never contract. We might thus, more consistently atemporally, say that the arrow of time does not *reverse*, and the universe does not *recollapse* in the model. Rather the model describes a "transition" between two semiclassical universes, each with an emergent arrow of time orientated away from the low-volume regime. The arrow in question, when it is well defined, *always* points away from the $\alpha \rightarrow -\infty$ regime and toward the $\alpha \rightarrow +\infty$ regime, and it is thus, by definition, aligned with the expansion of the universe. However, their temporal structure dissolves as they merge into a large-volume deep quantum regime (see Figure 13.3). Whilst one could hardly wish for a more symmetric treatment, the physical interpretation of the breakdown of classical physics at large scales is surely more than a little counterintuitive.

Let us finally turn to the question of physically justifying the three assumptions needed for the Kiefer–Zeh derivation of chronordinal structure in quantum cosmology – and so justify the claim that temporal structure is not merely derived, but emergent, according to the definition we gave. The first was the simple boundary condition together with the relevant form of the potential. It was assumed that the quantum state of the universe, including bulk degrees of freedom and inhomogeneities, at some boundary (initial and final) value of the internal time parameter, can be written approximately as a product state, that is, such that the bulk variables and inhomogeneities have vanishingly small entanglement. Here it is surely fair to conclude that the story provided by Kiefer and Zeh is ultimately not a satisfactory one. In particular, there is no clear *physical justification* for the simple boundary condition or the asymmetry of the potential in α. However, in this regard, the situation is not notably worse than that of the famous "Past Hypothesis" – and indeed the special boundary condition (13.14) is a form of low-entropy initial state that can be related to a quantum version of Penrose's Weyl curvature hypothesis (Kiefer 2022). In this context, one might then simply reject the demands for further justification, or indeed *explanation*, on the grounds that we do not typically expect initial conditions to be explained (Callender 2004). However, following Price (2004) once more, one might question whether universal boundary conditions are relevantly similar to local boundary conditions; if not, Callender's analogy fails. Then again, if "initial" conditions like the Kiefer–Zeh simple boundary condition should be thought of as "law-like" rather than "fact-like," once again the demand for explanation arguably vanishes. There are clearly philosophical issues to be addressed, which are outside the scope of this chapter, but the obvious conclusion is that the physical origin of such conditions is an important and incomplete explanatory project, which we would do well to pursue further.

The next crucial assumption was *relevant* coarse-graining. It was assumed that the degrees of freedom representing inhomogeneities can be divided into relevant and irrelevant subsets and that we can represent the state of the universe by a reduced density matrix where the irrelevant subset has been traced out. In this context, it is possible to draw upon parallel discussions in the context of statistical mechanics. In particular, consider the insightful analysis of Robertson (2020)[23]: according to Robertson, the crucial feature of an *objective* coarse-graining, that is, one that is not illusory or problematically anthropocentric, is that it allows us to abstract to a higher-level *autonomous* description. Autonomy here is understood in the dynamical systems sense of having independently well-posed evolution equations.

Similarly, one could argue that the coarse-grained asymmetry derived by Kiefer and Zeh provides an objective basis for the emergence of chronodirected structure. To do so would involve showing that the irrelevant inhomogeneous degrees of freedom are such that the dynamics of the rest of the universe are suitably autonomous. Such a project is attempted, with some but not complete success, in Kiefer (1987), following Halliwell and Hawking (1985).

The third and final assumption is the entropic-arrow assumption. This assumption amounts to the idea that chronodirected structure can be defined as the von Neumann entropy of the reduced density matrix. In this context, one can note that the monotonic (or secular) increase of a entropy function does not in fact require or enforce chronodirected structure but can instead be perfectly compatible with a theory of time in which only chronordinal structure obtains (Farr 2016, 2020). Moreover, if the aim in providing approximate derivation of chronodirected structure was to provide an explanation for the full gamut of temporally directed phenomena required by our functionalism, it is not clear that models such as that of Kiefer and Zeh are capable of addressing such an ambitious explanatory project, see Ryder (2022). In particular, no resources have been provided to connect the derived global entropic arrow with local temporally directed processes, such as equilibration of thermal systems or memory formation. In this sense, the model,

[23] See Wallace (2011) and Zeh (1989) for related ideas. For work on coarse-graining in classical cosmology see Te Vrugt et al. (2021). For a quantum cosmological model studied in the inflationary context, see Hollowood and McDonald (2017).

although remarkable, still falls short of explaining the arrow of time, until and unless such connections are provided.

13.5 Conclusion

In this chapter we have conducted a case study analysis of a proposal for the emergence of time based upon the approximate derivation of three grades of temporal structure within an explicit quantum cosmological model which obeys a Wheeler–DeWitt type equation without an extrinsic time parameter. Our conclusion is that the model provides a self-consistent account of the emergence of chronordinal, chronometric and chronodirected structure.

Our analysis has demonstrated that the Born–Oppenheimer and WKB approximations can be justified without presupposing the chronordinal and chronometric structures that one is aiming to derive. Further, regarding emergent chronodirected structure, we have shown that the Kiefer–Zeh model avoids Price's temporal double standards argument due to its temporal symmetry. In particular, since the model includes two symmetrically situated emergent arrows of time, it allows us to avoid asymmetry by not having an end of the universe, but rather two beginnings. There is evidently an exciting explanatory project with regard to the emergence of chronodirected structure that should be actively pursued in the context of more realistic cosmological models. Moreover, the even more difficult challenge remains of interpreting the model in the context of the cosmological measurement problem.

Acknowledgments

We are extremely appreciative to audiences in Buenos Aires, Surrey, Ovronnaz, Milan, and York for feedback, and to Henrqiue Gomes, James Ladyman, Huw Price and especially Claus Kiefer for comments and discussion. Work on this project was supported by a Bristol Benjamin Meaker Distinguished Visiting Professorship, grant IZSEZ0 214119 from the Swiss NSF, a Visiting Fellowship at Magdalen College, Oxford, and by John Templeton Foundation grant numbers 61387 and 62210. The views expressed may not represent the opinions of the Foundation.

Appendix

In this Appendix we sketch proofs of formal claims made in the discussion of the BO approximation in Section 13.3.1.

[1] Given $|\lambda_m - \lambda_n| \gg T_1$ for $m > n$, a superposition of ψ_n cannot be an eigenvector of total energy, so (5) holds. Suppose, for reductio, that (for instance) $\theta_m\psi_m + \theta_n\psi_n$ is

an eigenstate, so that it is parallel with $(\hat{T}_1 + \hat{T}_2 + \hat{W})(\theta_m\psi_m + \theta_n\psi_n) = \hat{T}_1(\theta_m\psi_m + \theta_n\psi_n) + \lambda_m\theta_m\psi_m + \lambda_n\theta_n\psi_n$. Since the ψ_n are orthonormal, such parallelism could only hold if the first term is comparable to the difference between the second two: that the kinetic energy of the heavy subsystem is comparable to the difference between the λ_n, contrary to the separation of energy levels.

[2] The adiabatic approximation, (7): $\theta_n\psi_n$ is an eigenstate of $\hat{T}_2 + \hat{W}$ from (3), and (approximately) of $\hat{T}_1 + \hat{T}_2 + \hat{W}$ from (5), hence it is also an eigenstate of $\hat{T}_1$. So, by the orthogonality of the ψ_n,$\theta_m\psi_m|\hat{T}_1|\theta_n\psi_n \propto \delta_{m,n}$. In the x_i basis $\hat{T}_1 \sim \partial^2/\partial x_1^2$, so we have (using the chain rule):

$$\int dx_1 dx_2 \theta_m^* \psi_m^* \left(\frac{\partial^2 \theta_n}{\partial x_1^2} \psi_n + 2 \frac{\partial \theta_n}{\partial x_1} \frac{\partial \psi_n}{\partial x_1^2} \right) \propto \delta_{m,n}$$

The orthogonality of the ψ_n means that (the integral of) the first term is proportional to $\delta_{m,n}$, while the second and third terms are generally not. Hence, the (approximate) proportionality requires that those terms – specifically the ψ_n derivatives that they contain – approximately vanish. Hence (7)

References

Accorinti, H. L., and J. C. M. González (2022). "Models and idealizations in quantum chemistry: The case of the Born–Oppenheimer approximation." *Philosophical Perspectives in Quantum Chemistry*, 107–124.

Arnowitt, R., S. Deser, and C. W. Misner (1960, Mar). "Canonical variables for general relativity." Physical Review, 117, 1595–1602.

Ashtekar, A., and E. Bianchi (2021). "A short review of loop quantum gravity." *Reports on Progress in Physics*, 84, 042001.

Ashtekar, A., T. Pawlowski, and P. Singh (2006). "Quantum nature of the big bang." *Physical Review Letters* 96 (14), 141301.

Ashtekar, A., and P. Singh (2011). "Loop quantum cosmology: A status report." *Classical and Quantum Gravity* 28 (21), 213001.

Banks, T. (1985). "TCP, quantum gravity, the cosmological constant and all that ... " *Nuclear Physics B* 249 (2), 332–360.

Barbero G. J. F., and E. J. Villaseñor, (2010). "Quantization of midisuperspace models." *Living Reviews in Relativity* 13 (1), 1–55.

Black, M. (1959). "The 'direction' of time." *Analysis* 19 (3): 54–63.

Blyth, W., and C. Isham (1975). "Quantization of a Friedmann universe filled with a scalar field." *Physical Review D* 11 (4), 768.

Bojowald, M. (2001). "Absence of a singularity in loop quantum cosmology." *Physical Review Letters* 86 (23), 5227.

Born, M., K. Huang, and M. Lax (1955). "Dynamical theory of crystal lattices." *American Journal of Physics* 23 (7), 474.

Born, M., and R. Oppenheimer (1927). "Zur quantentheorie der molekeln." *Annalen der Physik* 389 (20), 457–484.

Bravetti, A., C. Jackman, and D. Sloan (2022). "Scaling symmetries, contact reduction and Poincaré's dream." *arXiv preprint arXiv:2206.09911*.

Butterfield, J. (2011). "Less is different: Emergence and reduction reconciled." *Foundations of Physics* 41: 1065–1135.

Callender, C. (2004). "Measures, explanations and the past: Should 'special' initial conditions be explained?" *The British Journal for the Philosophy of Science* 55 (2): 195–217.

Callender, C. (2021). "Thermodynamic Asymmetry in Time." In E. N. Zalta (Ed.), *The Stanford Encyclopedia of Philosophy* (Summer 2021 ed.). Metaphysics Research Lab, Stanford University.

Cartwright, N. (2022). *A Philosopher Looks at Science*. Cambridge: Cambridge University Press.

Chang, H. (2015). "Reductionism and the relation between chemistry and physics." In T. Arabatzis, J. Renn, and A. Simões (eds.), *Relocating the History of Science: Essays in Honor of Kostas Gavroglu*, Boston Studies in the Philosophy and History of Science (BSPS, volume 312),193–209.

Chua, E. Y., and C. Callender (2021). "No time for time from no-time." *Philosophy of Science 88* (5): 1172–1184.

Claverie, P., and S. Diner (1980). "The concept of molecular structure in quantum theory: Interpretation problems." *Israel Journal of Chemistry* 19 (1-4): 54–81.

Cuffaro, M. E., and S. Hartmann (2021). "The open systems view." *arXiv preprint arXiv:2112.11095*.

de Blas, D. M., J. Olmedo, and T. Pawlowski (2017). "Loop quantization of the Gowdy model with local rotational symmetry." *Physical Review D* 96 (10), 106016.

DeWitt, B. (1967). "Quantum theory of gravity. I. The canonical theory." *Physical Review* 160: 1113–1148.

Dirac, P. A. M. (1958). "The theory of gravitation in Hamiltonian form." *Proceedings of the Royal Society of London. Series A, Mathematical and Physical Sciences* 246: 333–343.

Dirac, P. A. M. (1964). *Lectures on Quantum Mechanics*. New York: Dover Publications.

Dittrich, B., P. A. Höhn, T. A. Koslowski, and M. I. Nelson (2017). "Can chaos be observed in quantum gravity?" *Physics Letters B* 769: 554–560.

Farr, M. (2012). *Towards a C Theory of Time*. PhD thesis, University of Bristol.

Farr, M. (2016). "Causation and time reversal." *The British Journal for the Philosophy of Science*, 71 (1): 1–30.

Farr, M. (2020). "C-theories of time: On the adirectionality of time." *Philosophy Compass* 15 (12): e12714.

Fortin, S., and O. Lombardi (2021). "Is the problem of molecular structure just the quantum measurement problem?" *Foundations of Chemistry* 23 (3): 379–395.

Franklin, A., and V. A. Seifert (2020). "The problem of molecular structure just is the measurement problem." *The British Journal for the Philosophy of Science* 75 (1): 31–59.

Gerlach, U. H. (1969). "Derivation of the ten Einstein field equations from the semiclassical approximation to quantum geometrodynamics." Physical Review 177 (5): 1929–1941.

Gielen, S., and L. Menéndez-Pidal (2022). "Unitarity, clock dependence and quantum recollapse in quantum cosmology." *Classical and Quantum Gravity*, 39: 075011.

González, J. C. M., S. Fortin, and O. Lombardi (2019). "Why molecular structure cannot be strictly reduced to quantum mechanics." *Foundations of Chemistry* 21: 31–45.

Gotay, M. J., and J. A. Isenberg (1980). "Geometric quantization and gravitational collapse." *Physical Review D* 22 (2), 235.

Gryb, S., and K. P. Thébault (2016). "Regarding the 'hole argument' and the 'problem of time'." *Philosophy of Science* 83 (4): 563–584.

Gryb, S., and K. P. Thébault (2018). "Superpositions of the cosmological constant allow for singularity resolution and unitary evolution in quantum cosmology." *Physics Letters B* 784: 324–329.

Gryb, S., and K. P. Y. Thébault (2011). "The role of time in relational quantum theories." *Foundations of Physics* 42: 1–29.

Gryb, S., and K. P. Y. Thébault (2014). "Symmetry and evolution in quantum gravity." *Foundations of Physics* 44 (3): 305–348.

Gryb, S., and K. P. Y. Thébault (2016a). "Schrödinger evolution for the universe: Reparametrization." *Classical and Quantum Gravity* 33 (6): 065004.

Gryb, S., and K. P. Y. Thébault (2016b). "Time remains." *British Journal for the Philosophy of Science* 67 (3): 663–705.

Gryb, S., and K. P. Y. Thébault (2017). "Bouncing unitary cosmology I. Mini-superspace general solution." Preprint.

Gryb, S., and K. P. Y. Thébault (2024). *Time Regained: Symmetry and Evolution in Classical Mechanics*. Oxford: Oxford University Press.

Halliwell, J. J., and S. W. Hawking (1985). "Origin of structure in the universe." *Physical Review D* 31 (8), 1777.

Hartle, J. B. (1993). "Spacetime quantum mechanics and the quantum mechanics of spacetime." *arXiv preprint gr-qc/9304006*.

Hartle, J. B., and S. W. Hawking (1983). "Wave function of the universe." *Physical Review D* 28 (12), 2960.

Hawking, S. W. (1985). "Arrow of time in cosmology." *Physical Review D* 32 (10): 2489–2495.

Hawking, S. W. (1994). The no boundary condition and the arrow of time. In Halliwell, Perez-Mercader, and Zurek (eds.), *Physical Origins of Time Asymmetry*, pp. 346–357. Cambridge: Cambridge University Press.

Hawking, S. W., R. Laflamme, and G. W. Lyons (1993). "Origin of time asymmetry." *Physical Review D* 47 (12), 5342.

Hendry, R. F. (1998). "Models and approximations in quantum chemistry." *Poznan Studies in the Philosophy of the Sciences and the Humanities 63*: 123–142.

Hendry, R. F. (2006). *Is There Downward Causation in Chemistry?* Dordrecht: Springer.

Hendry, R. F. (2010a). "Emergence vs. reduction in chemistry." *Emergence in Mind*: 205–221.

Hendry, R. F. (2010b). "Ontological reduction and molecular structure." *Studies in History and Philosophy of Science Part B: Studies in History and Philosophy of Modern Physics* 41 (2): 183–191.

Hendry, R. F. (2017). "Prospects for strong emergence in chemistry." In *Philosophical and Scientific Perspectives on Downward Causation*. Abingdon, Oxford: Routledge, pp. 146–163.

Henneaux, M., and C. Teitelboim (1992). *Quantization of Gauge Systems*. Princeton, NJ: Princeton University Press.

Hettema, H. (2017). *The Union of Chemistry and Physics*. Cham: Springer International .

Hollowood, T. J., and J. I. McDonald (2017). "Decoherence, discord, and the quantum master equation for cosmological perturbations." *Physical Review D* 95 (10): 103521.

Huggett, N. (2021). "Spacetime 'emergence'." In *The Routledge Companion to Philosophy of Physics*, Abingdon, Oxford: Routledge, pp. 374–385.

Huggett, N., and C. Wüthrich (2025). *Out of Nowhere: The Emergence of Spacetime in Quantum Theories of Gravity.* Oxford: Oxford University Press.

Huggett, N., Ladyman, J., and Thébault K. (2024). *On the Quantum Theory of Molecules: Rigour, Idealization, and Uncertainty. arXiv preprint arXiv: 2411.01942.*

Jecko, T. (2014). "On the mathematical treatment of the Born–Oppenheimer approximation." *Journal of Mathematical Physics* 55 (5): 053504.

Kiefer, C. (1987). "Continuous measurement of mini-superspace variables by higher multipoles." *Classical and Quantum Gravity* 4 (5): 1369.

Kiefer, C. (1988, Sep). "Wave packets in minisuperspace." *Phys. Rev. D* 38 (6): 1761–1772.

Kiefer, C. (1990). "Wave packets in quantum cosmology and the cosmological constant." *Nuclear Physics B* 341 (1): 273–293.

Kiefer, C. (2005). "The semiclassical approximation to quantum gravity." In *Canonical Gravity: From Classical to Quantum: Proceedings of the 117th WE Heraeus Seminar Held at Bad Honnef, Germany, 13–17 September 1993*, pp. 170–212. Springer.

Kiefer, C. (2009). "Quantum geometrodynamics: Whence, whither?" *General Relativity and Gravitation* 41 (4): 877–901.

Kiefer, C. (2012). *Quantum Gravity. International Series of Monographs on Physics*. Oxford: Oxford University Press.

Kiefer, C. (2013). "Conceptual problems in quantum gravity and quantum cosmology." *International Scholarly Research Notices* 2013, 509316.

Kiefer, C. (2022). "On a quantum Weyl curvature hypothesis." *AVS Quantum Science* 4 (1), 015607.

Kiefer, C., and P. Peter (2022). "Time in quantum cosmology." *Universe* 8 (1), 36.

Kiefer, C., and H. Zeh (1995). "Arrow of time in a recollapsing quantum universe." *Physical Review D* 51 (8), 4145.

Komar, A. (1971). "General-relativistic observables via Hamilton–Jacobi functionals. *Physical Review D* 4 (4): 923–927.

Kuchař, K. V. (2011). "Time and interpretations of quantum gravity." *International Journal of Modern Physics D* 20 (supp01): 3–86.

Kuchař, K. V. and M. P. Ryan (1989). "Is minisuperspace quantization valid?: Taub in mixmaster." *Physical Review D* 40 (12), 3982.

Ladyman, J., and K. P. Y. Thébault (2023). "Modelling open systems in quantum physics and cosmology." Manuscript.

Maniccia, G., M. De Angelis, and G. Montani (2022). "WKB approaches to restore time in quantum cosmology: Predictions and shortcomings." *Universe* 8 (11), 556.

McTaggart, J. E. (1908). "The unreality of time." *Mind*, New Series, 17 (68)457–474.

Messiah, A. (1962). *Quantum mechanics: volume II*. Amsterdam: North-Holland.

Misner, C., K. Thorne, and J. Wheeler (1974). *Gravitation*. New York: W. H. Freeman.

Newton-Smith, W. (1982). *The Structure of Time*. Abingdon, Oxford: Routledge & Kegan Paul.

Palacios, P. (2022). *Emergence and Reduction in Physics*. Cambridge: Cambridge University Press.

Peres, A. (1962). "On Cauchy's problem in general relativity: II." *Il Nuovo Cimento* 26 (1), 53–62.

Price, H. (1996). *Time's Arrow & Archimedes' Point: New Directions for the Physics of Time*. New York: Oxford University Press.

Price, H. (2004). *"On the origins of the arrow of time: Why there is still a puzzle about the low-entropy past?"* In C. Hitchcock (ed.), *Contemporary Debates in the Philosophy of Science*, London: Blackwell, pp. 219–239.

Reichenbach, H. (1956). *The Direction of Time*. Berkeley, CA: University of California Press.

Robertson, K. (2020). "Asymmetry, abstraction, and autonomy: Justifying coarse-graining in statistical mechanics." *The British Journal for the Philosophy of Science*, 71 (2): 547–579.

Ryder, D. J. (2022). "Directed temporal asymmetry from scale invariant dynamics: Is the problem of time's arrow solved?", http://philsci-archive.pitt.edu/21530/.

Sakurai, J. J., and E. D. Commins (1995). *Modern Quantum Mechanics*, revised edition. American Association of Physics Teachers.

Salisbury, D. (2020). "Observables and Hamilton-Jacobi approaches to general relativity." Unpublished.

Scerri, E. R. (2012). "Top-down causation regarding the chemistry–physics interface: A sceptical view." *Interface Focus 2* (1): 20–25.

Seifert, V. A. (2020). "The strong emergence of molecular structure." *European Journal for Philosophy of Science* 10 (3), 45.

Seifert, V. A. (2022). "Do molecules have structure in isolation? How models can provide the answer." In *Philosophical Perspectives in Quantum Chemistry*. Springer, pp. 125–143.

Sloan, D. (2021). "New action for cosmology." *Physical Review D* 103 (4), 043524.

Sloan, D. (2023). "Herglotz action for homogeneous cosmologies." *Classical and Quantum Gravity*, 40: 115008.

Tarrío, P., M. Fernández-Méndez, and G. A. M. Marugán (2013). "Singularity avoidance in the hybrid quantization of the gowdy model." *Physical Review D* 88 (8), 084050.

Te Vrugt, M., S. Hossenfelder, and R. Wittkowski (2021). "Morizwanzig formalism for general relativity: A new approach to the averaging problem." *Physical Review Letters 127* (23), 231101.

Thébault, K. P. (2023). "Big bang singularity resolution in quantum cosmology." *Classical and Quantum Gravity 40* (5), 055007.

Vilenkin, A. (1984). Quantum creation of universes. *Physical Review D* 30 (2), 509.

Vilenkin, A. (1989). Interpretation of the wave function of the universe. *Physical Review D* 39 (4), 1116.

Wallace, D. (2011). "The logic of the past hypothesis." To appear in B. Loewer, E. Winsberg, and B. Weslake (ed.), currently untitled volume discussing David Albert's "Time and Chance."

Warrier, N. (2022). "The case of the vanishing wavefunction." *Studies in History and Philosophy of Science* 96: 135–140.

Weatherall, J. O. (2018). "Regarding the 'hole argument'." *The British Journal for the Philosophy of Science* 69 (2): 329–350.

Wheeler, J. A. (1968). "Superspace and the Nature of Quantum Geometrodynamics." In J. A. Wheeler and C. M. De Witt (eds.), *Lectures in Mathematics and Physics*. Benjamin.

Woolley, R., and B. Sutcliffe (1977). "Molecular structure and the Born–Oppenheimer approximation." *Chemical Physics Letters* 45 (2): 393–398.

Woolley, R. G. (1978). "Must a molecule have a shape?" *Journal of the American Chemical Society* 100 (4): 1073–1078.

Zeh, H. (1986). "Emergence of classical time from a universal wavefunction." *Physics Letters A* 116 (1): 9–12.

Zeh, H.-D. (1989). *The Physical Basis of the Direction of Time*. Heidelberg: Springer.

Index

www.ingramcontent.com/pod-product-compliance
Lightning Source LLC
LaVergne TN
LVHW062207200926
848484LV00006B/1345